Geophysical Hazards

International Year of Planet Earth

Series Editors:

Eduardo F.J. de Mulder
Executive Director International Secretariat
International Year of Planet Earth

Edward Derbyshire
Goodwill Ambassador
International Year of Planet Earth

The book series is dedicated to the United Nations International Year of Planet Earth. The aim of the Year is to raise worldwide public and political awareness of the vast (but often under-used) potential of Earth sciences for improving the quality of life and safeguarding the planet. Geoscientific knowledge can save lives and protect property if threatened by natural disasters. Such knowledge is also needed to sustainably satisfy the growing need for Earth's resources by more people. Earths scientists are ready to contribute to a safer, healthier and more prosperous society. IYPE aims to develop a new generation of such experts to find new resources and to develop land more sustainably.

For further volumes:
http://www.springer.com/series/8096

Tom Beer
Editor

Geophysical Hazards

Minimizing Risk, Maximizing Awareness

Springer

Editor
Dr. Tom Beer
CSIRO Marine and Atmospheric Research
and Centre for Australian Weather and Climate Research
111 Station Street
Aspendale VIC 3195
Australia
tom.beer@csiro.au

ISBN 978-90-481-3235-5 e-ISBN 978-90-481-3236-2
DOI 10.1007/978-90-481-3236-2
Springer Dordrecht Heidelberg London New York

Library of Congress Control Number: 2009938629

Printed on acid-free paper

Springer is part of Springer Science+Business Media (www.springer.com)

Foreword

The International Year of Planet Earth (IYPE) was established as a means of raising worldwide public and political awareness of the vast, though frequently under-used, potential the Earth Sciences possess for improving the quality of life of the peoples of the world and safeguarding Earth's rich and diverse environments.

The International Year project was jointly initiated in 2000 by the International Union of Geological Sciences (IUGS) and the Earth Science Division of the United Nations Educational, Scientific and Cultural Organisation (UNESCO). IUGS, which is a Non-Governmental Organisation, and UNESCO, an Inter-Governmental Organisation, already shared a long record of productive cooperation in the natural sciences and their application to societal problems, including the International Geoscience Programme (IGCP) now in its fourth decade.

With its main goals of raising public awareness of, and enhancing research in the Earth sciences on a global scale in both the developed and less-developed countries of the world, two operational programmes were demanded. In 2002 and 2003, the Series Editors together with Dr. Ted Nield and Dr. Henk Schalke (all four being core members of the Management Team at that time) drew up outlines of a Science and an Outreach Programme. In 2005, following the UN proclamation of 2008 as the United Nations International Year of Planet Earth, the "Year" grew into a triennium (2007–2009).

The Outreach Programme, targeting all levels of human society from decision-makers to the general public, achieved considerable success in the hands of member states representing over 80% of the global population. The Science Programme concentrated on bringing together like-minded scientists from around the world to advance collaborative science in a number of areas of global concern. A strong emphasis on enhancing the role of the Earth sciences in building a healthier, safer and wealthier society was adopted – as declared in the Year's logo strap-line "Earth Sciences *for* Society".

The organisational approach adopted by the Science Programme involved recognition of ten global themes that embrace a broad range of problems of widespread national and international concern, as follows.

- Human health: this theme involves improving understanding of the processes by which geological materials affect human health as a means identifying and reducing a range of pathological effects.
- Climate: particularly emphasises improved detail and understanding of the non-human factor in climate change.

- Groundwater: considers the occurrence, quantity and quality of this vital resource for all living things against a background that includes potential political tension between competing neighbour-nations.
- Ocean: aims to improve understanding of the processes and environment of the ocean floors with relevance to the history of planet Earth and the potential for improved understanding of life and resources.
- Soils: this thin "skin" on Earth's surface is the vital source of nutrients that sustain life on the world's landmasses, but this living skin is vulnerable to degradation if not used wisely. This theme emphasizes greater use of soil science information in the selection, use and ensuring sustainability of agricultural soils so as to enhance production and diminish soil loss.
- Deep Earth: in view of the fundamental importance of deep the Earth in supplying basic needs, including mitigating the impact of certain natural hazards and controlling environmental degradation, this theme concentrates on developing scientific models that assist in the reconstruction of past processes and the forecasting of future processes that take place in the solid Earth.
- Megacities: this theme is concerned with means of building safer structures and expanding urban areas, including utilization of subsurface space.
- Geohazards: aims to reduce the risks posed to human communities by both natural and human-induced hazards using current knowledge and new information derived from research.
- Resources: involves advancing our knowledge of Earth's natural resources and their sustainable extraction.
- Earth and Life: it is over two and half billion years since the first effects of life began to affect Earth's atmosphere, oceans and landmasses. Earth's biological "cloak", known as the biosphere, makes our planet unique but it needs to be better known and protected. This theme aims to advance understanding of the dynamic processes of the biosphere and to use that understanding to help keep this global life-support system in good health for the benefit of all living things.

The first task of the leading Earth scientists appointed as Theme Leaders was the production of a set of theme brochures. Some 3500 of these were published, initially in English only but later translated into Portuguese, Chinese, Hungarian, Vietnamese, Italian, Spanish, Turkish, Lithuanian, Polish, Arabic, Japanese and Greek. Most of these were published in hard copy and all are listed on the IYPE website.

It is fitting that, as the International Year's triennium terminates at the end of 2009, the more than 100 scientists who participated in the ten science themes should bring together the results of their wide ranging international deliberations in a series of state-of-the-art volumes that will stand as a legacy of the International Year of Planet Earth. The book series was a direct result of interaction between the International Year and the Springer Verlag Company, a partnership which was formalised in 2008 during the acme of the triennium.

This IYPE-Springer book series contains the latest thinking on the chosen themes by a large number of Earth science professionals from around the world. The books are written at the advanced level demanded by a potential readership consisting of Earth science professionals and students. Thus, the series is a legacy of the Science Programme, but it is also a counterweight to the Earth science information in several media formats already delivered by the numerous National Committees of the International Year in their pursuit of world-wide popularization under the Outreach Programme.

The discerning reader will recognise that the books in this series provide not only a comprehensive account of the individual themes but also share much common ground that makes the series greater than the sum of the individual volumes. It is to be hoped that the scientific perspective thus provided will enhance the reader's appreciation of the nature and scale of Earth science as well as the guidance it can offer to governments, decision-makers and others seeking solutions to national and global problems, thereby improving everyday life for present and future residents of Planet Earth.

Eduardo F.J. de Mulder
Executive Director International Secretariat
International Year of Planet Earth

Edward Derbyshire
Goodwill Ambassador
International Year of Planet Earth

Preface

This book series is one of the many important results of the International Year of Planet Earth (IYPE), a joint initiative of UNESCO and the International Union of Geological Sciences (IUGS), launched with the aim of ensuring greater and more effective use by society of the knowledge and skills provided by the Earth Sciences.

It was originally intended that the IYPE would run from the beginning of 2007 until the end of 2009, with the core year of the triennium (2008) being proclaimed as a UN Year by the United Nations General Assembly. During all three years, a series of activities included in the IYPE's science and outreach programmes had a strong mobilizing effect around the globe, not only among Earth Scientists but also within the general public and, especially, among children and young people.

The Outreach Programme has served to enhance cooperation among earth scientists, administrators, politicians and civil society and to generate public awareness of the wide ranging importance of the geosciences for human life and prosperity. It has also helped to develop a better understanding of Planet Earth and the importance of this knowledge in the building of a safer, healthier and wealthier society.

The Scientific Programme, focused upon ten themes of relevance to society, has successfully raised geoscientists' awareness of the need to develop further the international coordination of their activities. The Programme has also led to some important updating of the main challenges the geosciences are, and will be confronting within an agenda closely focused on societal benefit.

An important outcome of the work of the IYPE's scientific themes includes this thematic book as one of the volumes making up the IYPE-Springer Series, which was designed to provide an important element of the legacy of the International Year of Planet Earth. Many prestigious scientists, drawn from different disciplines and with a wide range of nationalities, are warmly thanked for their contributions to a series of books that epitomize the most advanced, up-to-date and useful information on evolution and life, water resources, soils, changing climate, deep earth, oceans, non-renewable resources, earth and health, natural hazards, megacities.

This legacy opens a bridge to the future. It is published in the hope that the core message and the concerted actions of the International Year of Planet Earth throughout the triennium will continue and, ultimately, go some way towards helping to establish an improved equilibrium between human society and its home planet. As stated by the Director General of UNESCO, Koichiro Matsuura, "Our knowledge of the Earth system is our insurance policy for the future of our planet". This book series is an important step in that direction.

R. Missotten
Chief, Global Earth Observation Section
UNESCO

Alberto C. Riccardi
President
IUGS

Contents

Contributors

Tom Beer CSIRO Marine and Atmospheric Research, Aspendale, Victoria 3195, Australia

Susan L. Cutter Department of Geography, Hazards and Vulnerability Research Institute, University of South Carolina, Columbia, SC 29208, USA

Juan Carlos Villagrán de León, Institute for Environment and Human Security – UNU-EHS, United Nations University, Bonn, Germany

Janos J. Bogardi Institute for Environment and Human Security – UNU-EHS, United Nations University, Bonn, Germany

David Liverman Department of Natural Resources, Geological Survey of Newfoundland & Labrador, St. John's, NL, Canada A1B 4J6

Gordon A. McBean ICSU Planning Group on Natural and Human-induced Environmental Hazards and Disasters, Department of Geography, Institute for Catastrophic Loss Reduction, Social Sciences Centre, University of Western Ontario, London, ON, Canada N6A 5C2

Genene Mulugeta Coordinator of ICSU-ROA Geohazards Programme, The Baltic University Programme, CSD-Uppsala, Uppsala University, Uppsala, Sweden

Harsh Gupta National Geophysical Research Institute, (Council of Scientific and Industrial Research), Hyderabad – 500 007, India

Kuniyoshi Takeuchi Public Works Research Institute (PWRI), International Center for Water Hazard and Risk Management (ICHARM) under the auspices of UNESCO, Tsukuba 305-8516, Japan

S.A.G. Leroy Institute for the Environment, Brunel University, Uxbridge (West London) UB8 3PH, UK

S. Warny Department of Geology and Geophysics and Museum of Natural Science, Louisiana State University, Baton Rouge, LA 70803, USA

H. Lahijani Iranian National Centre for Oceanography, 1411813389 Tehran, Iran

E.L. Piovano CICTERRA-CONICET-UNC, Facultad de Ciencias Exactas, Físicas y Naturales, Universidad Nacional de Córdoba, 5016 Córdoba, Argentina

D. Fanetti Dipartimento di Scienze Chimiche e Ambientali, Università degli Studi dell'Insubria, I-22100 Como, Italy

A.R. Berger Sir Wilfred Grenfell College, Memorial University of Newfoundland and Labrador, Wolfville, NS, Canada

R.K. Chadha National Geophysical Research Institute (CSIR), Hyderabad – 500007, India

Alik Ismail-Zadeh Geophysical Institute, University of Karlsruhe, Karlsruhe, Germany

Y.A. Kontar Institute of Natural Resource Sustainability, University of Illinois at Urbana-Champaign, Champaign, Illinois 61820, USA

Yu.R. Ozorovich Space Research Institute, Russian Academy of Sciences, Moscow 117997, Russian Federation

A.T. Salokhiddinnov Tashkent Institute of the Engineers of Irrigation and Agricultural Mechanization, Tashkent 700000, Uzbekistan

Viacheslav Gusiakov Tsunami Laboratory, ICMMG SD RAS, Novosibirsk 630090, Russia

Dallas H. Abbott Lamont-Doherty Earth Observatory of Columbia University, Palisades, NY 10964, USA

Edward A. Bryant School of Geoscience, University of Wollongong, Wollongong, NSW 2522, Australia

W. Bruce Masse Los Alamos National Laboratory, Los Alamos, NM 87545, USA

Dee Breger Centralized Research Facilities, Drexel University, Philadelphia, PA 19104, USA

G. Ventura Istituto Nazionale di Geofisica e Vulcanologia, Roma, Italy

S. Vinciguerra Istituto Nazionale di Geofisica e Vulcanologia, Roma, Italy

S. Moretti Department of Earth Science, Universita' di Firenze, Florence, Italy

P. Meredith Rock & Ice Physics Laboratory, Department of Earth Sciences, University College London, London, UK

M. Heap Rock & Ice Physics Laboratory, Department of Earth Sciences, University College London, London, UK

P. Baud Laboratoire de Physique des Roches, EOST, Strasbourg, France

S.A. Shapiro Freie Universität Berlin, Berlin, Germany

C. Dinske Freie Universität Berlin, Berlin, Germany

J. Kummerow Freie Universität Berlin, Berlin, Germany

Raisa Gracheva Institute of Geography of RAS, Moscow 119017, Russia

Alexandra Golyeva Institute of Geography of RAS, Moscow 119017, Russia

Reviewers

Dr. Ian Barton CSIRO Marine and Atmospheric Research, GPO Box 1538, Hobart, Tasmania, Australia

Dr. Peter Bobrowsky Senior Scientist in Natural Hazards, Geological Survey of Canada, 601 Booth Street, Ottawa, Ontario

Dr. Rajender K. Chadha National Geophysical Research Institute, Hyderabad – 500007, India

Prof. Harsh Gupta Raja Ramanna Fellow, Formerly, Secretary to Government of India, Department of Ocean Development, Vice Chancellor, Cochin University of Science & Technology, Director, National Geophysical Research Institute, National Geophysical Research Institute (NGRI), Uppal Road, Hyderabad 500 007, India

Dr. Viacheslav K. Gusiakov Head, Tsunami Laboratory, Institute of Computational Mathematics, and Mathematical Geophysics, Siberian Division, Russian Academy of Sciences, Pr.Lavrentieva, 6, Novosibirsk 630090, Russia

Dr. Roger Jones CSIRO Marine and Atmospheric Research, PB1, Aspendale, Victoria 3195, Australia

Mr. Oddvar Kjekstad Norges Geotekniske Institutt, PO Box 3930, Ullevaal Stadion, NO-0806, Oslo, Norway

Dr. Yevgeniy Kontar Institute of Natural Resource Sustainability, University of Illinois at Urbana-Champaign, 615 East Peabody Drive, Champaign, Illinois 61820, USA

Dr. David Liverman Geological Survey of Newfoundland & Labrador, Department of Natural Resources, PO Box 8700, St. John's, NL, Canada A1B 4J6

Prof. Gordon McBean Institute for Catastrophic Loss Reduction, Department of Geography, Social Sciences Centre, University of Western Ontario, London, ON, N6A 5C2 Canada

Dr. Hormoz Modaressi Head, Development Planning and Natural Risks Division, BRGM, BP 6009, 45060 Orleans Cedex 2, France

Prof. Genene Mulugeta The Baltic University Programme, CSD-Uppsala, Uppsala University, Sweden

Dr. Benjamin Preston CSIRO Marine and Atmospheric Research, PB1, Aspendale, Victoria 3195, Australia

Dr. John Schneider Group Leader, Risk & Impact Analysis Group, Geospatial & Earth Monitoring Division, Geoscience Australia, GPO Box 378, Canberra, ACT 2601, Australia

Prof. Ramesh Singh Center for Earth Observing and Space Research, College of Science, George Mason University, 4400 University Drive, Fairfax VA 22030, USA

Prof. Kuniyoshi Takeuchi Director, International Center for Water Hazard and Risk Management (ICHARM), under the auspices of UNESCO, Public Works Research Institute (PWRI), Minamihara 1-6, Tsukuba 305-8516, Japan

Prof. Dr. G. Tetzlaff Institut für Meteorologie, Universität Leipzig, Stephanstr. 3, D-04103 Leipzig

Dr. Ti Le-Huu Sustainable Development and Water Resources Section, Environment and Sustainable Development Division, UNESCAP, Bangkok, Thailand

Dr. Emma Yuen CSIRO Marine and Atmospheric Research, PB1, Aspendale, Victoria 3195, Australia

Introduction

The United Nations General Assembly declared 2008 to be the International Year of Planet Earth (IYPE), though the activities of the IYPE ran for a triennium from 2007 to 2009. There were ten science themes and this volume documents the work of the Hazards Theme of the IYPE. These activities included the preparation of a hazards brochure that specified the four key research questions; the subsequent translation of the brochure into Portuguese and Hungarian; a contribution to the Planet Earth book that was produced for the launch in Paris of the IYPE and the organisation of a Hazards Megasymposium that was held in Oslo, Norway as part of the 33rd International Geological Congress.

The Hazards Megasymposium was an invitation only two-day session devoted to an examination of the four key research questions and how they link in to the Hyogo Framework for Action and other international and local activities. It was convened by Kuni Takeuchi (Chair of the IUGG Commission on Geophysical Risk and Sustainability), R.K. Chadha (President, Natural Hazards Society) and myself. We labelled it a Megasymposium because it was four times larger than the IYPE symposium allocation originally permitted by the IGC organisers. The Megasymposium also received reports of the results of the science projects that were conducted under the auspices of the IYPE Hazards Theme.

The presentations from the Megasymposium have been collated in this volume. They fall into four parts:

Part 1 deals with the IYPE Hazards Theme and its relation to society and incorporates papers by Beer, Cutter, Villagrán & Bogardi, and Liverman. These papers collectively emphasise that implementation of the Hyogo Framework for Action (the key document of the UN Strategy for Disaster Reduction) requires a strong emphasis on human societies, human security, and that this can be achieved only through appropriate communication.

Part 2 focuses on the response of the international scientific community and incorporates papers by McBean, Mulugeta, and Gupta. The International Council for Science (ICSU) has established an international research program known as Integrated Research on Disaster Risk that will act as the de facto successor to the IYPE Hazards Program. ICSU has also established regional offices that have identified Hazards as being core science activities. Mulugeta gives the perspective from the ICSU Regional Office for Africa. Gupta gives the perspective from the ICSU Regional Office for Asia and the Pacific.

Part 3 is titled Geophysical Risk and Sustainability, and is divided into two sections. The first section, with papers by Takeuchi and by Leroy et al.

considers climate and climate change with Takeuchi concentrating on floods and Leroy et al. concentrating on case studies to examine past variations. The second section, with papers by Chadha, Ismail-Zadeh, and Kontar et al., considers the theory and the praxis involved in hazards, their prevention and mitigation. Chadha uses the Indian response to hazards and disasters as his case study; Ismail-Zadeh uses seismology as his case study; and Kontar et al. use the disturbing situation in the Aral Sea as their primary case study.

Part 4 relates to GeoHazards and Risk – Observation and Assessment and incorporates papers by Gusiakov et al., Ventura et al. and by Gracheva & Golyeva. Gusiakov et al. present a persuasive argument that past bolide impacts have had a more significant effect than is generally recognised. Ventura et al. seek better understanding of landslides by examining slow deformation of the soil. Gracheva & Golyeva also examine landslides, in particular those in mountain regions.

I would like to acknowledge the assistance of the IGC Megasymposium co-convenors without whose assistance we would not have managed to bring together such an eminent group. In this respect the financial support from the Netherlands UNESCO Commission and from Springer is greatly appreciated.

I would also like to acknowledge the assistance from the members of both the GeoUnions Hazards Team, and the IYPE Hazards Team – who acted as one IYPE Hazards Committee:

Tom Beer (Chair, Australia and IUGG)
Peter Bobrowsky (Canada and IUGS)
Piero Boccardo (ISPRS)
R.K. Chadha (India)
Susan Cutter (USA and IGU)
Rob Fitzpatrick (IUSS)
Francois Heuze (USA)
Alik Ismail-Zadeh (IUGG)
Stuart Marsh (UK)
Marcello Pagliai (Italy and IUSS)
Seree Supharatid (Thailand)
Zhongliang Wu (China)

In terms of the actual publication, I express gratitude for all of the assistance provided by Springer, the publisher. I am also very thankful to all the reviewers who spent a significant amount of time to improve the quality of the manuscripts that were submitted and thus helped guarantee the success of this publication.

Finally, I write these words shortly after Black Saturday, 7 February 2009, when 170 people lost their lives as a result of wildfires in the outskirts of Melbourne, Australia. It is sobering to realise that in spite of the marked superiority in weather forecasting, telecommunication, suppression techniques and fire-fighting technology more people died in these fires than in the fires of Friday 13 January 1939, which had previously been Australia's worst wildfire disaster. We have known since 1988 that one of the results of climate change would be increased fire danger[1]. Can the logic be inverted so that the increased frequency of hydro-meteorological disasters becomes

[1] Beer, T., A. M. Gill and P. H. R. Moore (1988) *Australian bushfire danger under changing climate regimes*, in Greenhouse, ed. G. Pearman (CSIRO), pp 421–427.

evidence for climate change? Some say yes. Some say no. We do not know for sure, but it is important to find out. I hope that it will be one of the issues considered at the XXV General Assembly of the International Union of Geodesy and Geophysics to be held in Melbourne 27 June – 8 July 2011.

CSIRO, Australia Tom Beer
Chair, IYPE Hazards Science Theme
Senior Advisor, IYPE Board.

Part I
The IYPE Hazards Theme: Minimising Risk, Maximising Awareness

The Hazards Theme of the International Year of Planet Earth

Tom Beer

Abstract "Hazards – minimising risk, maximising awareness" is one of ten broad themes that make up the science programme of the International Year. This theme focuses on four key questions:

1. How have humans altered the geosphere, the biosphere and the landscape, thereby creating long-term changes detrimental to life and the environment and triggering certain hazards, while increasing societal vulnerability to geophysical (geological, geomorphological and hydrometeorological) hazards?
2. What technologies and methodologies are required to assess the vulnerability of people and places to hazards and how might these be used at a variety of spatial scales?
3. How do geophysical hazards compare relative to each other regarding current capabilities for monitoring, prediction and mitigation and what can be done in the short term to improve these statistics?
4. What barriers exist to the utilization of risk and vulnerability information by governments (and other entities) for risk and vulnerability reduction policies and planning (including mitigation) from each of the geophysical hazards?

To answer these questions the theme aims to closely integrate with parallel programmes at various levels within other international organizations such as UNESCO-IGCP, UN-ISDR, IGBP, IGOS and the Geoscience Unions' Consortium with a primary focus being on how the four key questions of the hazards theme can be linked to the five action items of the UN-ISDR Hyogo Framework for Action.

Keywords International Year of Planet Earth · Geohazards · Georisks

T. Beer (✉)
CSIRO Marine and Atmospheric Research, Aspendale, Victoria 3195, Australia
e-mail: tom.beer@csiro.au

Introduction

Geohazards and Georisks are topics that are international in scope, but often local in their effects. The knowledge and wisdom accumulated over time and in many different parts of the world can assist the research worker and the disaster manager to provide better information and to make better decisions. This chapter briefly reviews the International Year of Planet Earth (IYPE) and its Hazards Science Theme, examines the four key research questions of the Hazards Science Theme, places them within an international research context, and notes that to mesh them with the action items of the Hyogo Framework for Action will require an ongoing international effort well beyond the IYPE.

International Year of Planet Earth

The International Year of Planet Earth is an initiative organised and promulgated by the International Union of Geological Sciences (IUGS) with the support of the Earth Science Division of the United National Educational, Scientific and Cultural

T. Beer (ed.), *Geophysical Hazards*, International Year of Planet Earth,
DOI 10.1007/978-90-481-3236-2_1, © Springer Science+Business Media B.V. 2010

Organisation (UNESCO).[1] The concept has twelve Founding Partners and 26 Associate Partners that include all major international geoscientific and other relevant organisations.

The organisers of the International Year of Planet Earth (IUGS and UNESCO) sought proclamation of the International Year of Planet Earth by the General Assembly of the UN, which was accomplished on 22 December 2005. The UN press release reads:

> *"By a draft on the International Year of Planet Earth, 2008, which the Committee approved without a vote on 11 November, the Assembly would declare 2008 the International Year of Planet Earth. It would also designate the United Nations Educational, Scientific and Cultural Organization (UNESCO) to organize activities to be undertaken during the Year, in collaboration with UNEP and other relevant United Nations bodies, the International Union of Geological Sciences and other Earth sciences societies and groups throughout the world. Also by that draft, the Assembly would encourage Member States, the United Nations system and other actors to use the Year to increase awareness of the importance of Earth sciences in achieving sustainable development and promoting local, national, regional and international action".*

[1] http://www.yearofplanetearth.org

The Year's activities span the three years 2007–2009.

The Science Programme Committee determined ten science themes, all relevant to society. Specialist Key Text Teams focused on specific questions (key research questions) within the themes to be addressed in the triennium, given that "tangible deliverables" are a binding condition for any such UN proclamation. The Science Themes were:

1. Groundwater – towards sustainable use
2. Earth & health – building a safer environment
3. Climate – the "stone tape"
4. Resources – sustainable power for sustainable development
5. Megacities – going deeper building safer
6. Deep Earth – from crust to core
7. Ocean – abyss of time
8. Hazards – minimising risk, maximising awareness
9. Soils – the living skin of the Earth
10. Earth & Life – the origins of diversity.

Brochures were published on all of the science themes. In fact there were twelve brochures available for download from this site. As at 8 November 2008 they were all available as colour PDF downloads

Fig. 1 New Zealand stamps depicting weather extremes issued as part of the IYPE outreach program

from the web site at http://www.yearofplanetearth.org/content/downloads.html

The hazards theme brochure has been translated into Portuguese (available from the same web site) and Hungarian (available from http://www.foldev.hu/geofifika.htm).

There was also a general brochure describing the overall concept of the Year of Planet Earth called *Planet Earth in our hands*, and a brochure on the outreach program.

The subtitle of the outreach brochure is "bringing Earth sciences to everyone." The outreach program complemented the science program and certain parts of the outreach program directly targeted the Hazards Theme. For example, on 5 March 2008, New Zealand issued a set of postage stamps depicting weather extremes (Fig. 1).

Hazards Theme Science Team

The Science Team of the Hazards Theme noted that the Budapest Manifesto on Risk Science and Sustainability[2] provided a generic framework suitable for environmental risk management across a variety of disciplines, including hazards studies. Details of the manifesto may be found in Beer and Ismail-Zadeh (2003). It can be summarised by the following list of items that need to be examined:

- Consultation
- Concerns
- Consequences
- Calculations
- Certainties, uncertainties, probabilities
- Comparing against pre-determined criteria
- Control, mitigate and adapt
- Communicate
- Monitor
- Review

There are 10 items listed above. Risk management praxis consists in undertaking steps 2–8 in sequential

[2] http://www.iugg.org/publications/reports/budapest.pdf

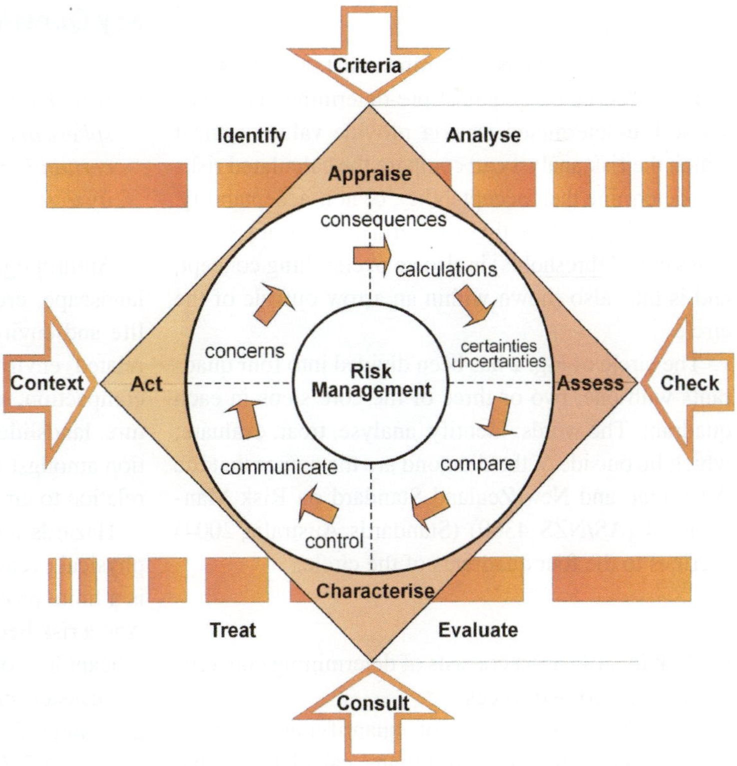

Fig. 2 Diagrammatic representation of the steps involved in the Budapest Manifesto and the Australian/New Zealand Standard on Risk Management. The steps themselves are shown by arrows within the circle. The inner circle, with the words risk management, indicates that all of the words in the diagram refer to various aspects of risk management

order, and then re-checking to make sure that any residual risk has been dealt with adequately. Thus there are seven core steps that logically follow on from each other in a linear, but circular fashion.

A diagrammatic representation of the manifesto is shown in Fig. 2, in which the two last items (Monitor and Review) have been replaced by the word "Check." The diagram divides the steps of the Budapest Manifesto into those that are overarching, such as consultation, monitoring and review, and those that are core. The seven core steps are shown within a circle, with each step marked by an arrow to denote the order in which a risk management exercise needs to be carried out. The overarching concepts are shown within arrows outside the main diagram.

When the Budapest Manifesto framework is compared to other risk management frameworks – such as the of the Australian and New Zealand Standard on Risk Management (AS/NZS 4360) (Standards Australia 2004) – then certain terms have not been incorporated into the Budapest Manifesto framework. In particular, the idea that one needs to determine the context within which a risk management activity takes place. The first step in risk management is always to establish the context in which to operate. Diagrammatically this is treated as an overarching concept and shown in an arrow outside the circle.

Item 6 of the Budapest Manifesto list of items given above is "comparing against pre-determined criteria." These pre-determined criteria provide values against which the risk analyst can compare the calculated risks and determine their acceptability, or non-acceptability. The development of these criteria (which are sometimes called thresholds) is also an overarching concept, and is thus also shown within an arrow outside of the circle.

The circle of Fig. 2 has been divided into four quadrants with one, two or three of the core steps in each quadrant. The words: identify, analyse, treat, evaluate; which lie outside of the diamond are the names that the Australian and New Zealand Standard on Risk Management (AS/NZS 4360) (Standards Australia 2004) ascribes to the four quadrants of the circle.

- Risk Identification consists of determining concerns and their consequences.
- Risk Analysis consists of quantifying the consequences through appropriate calculations that examine the certainties and the uncertainties involved.
- Risk Evaluation consists of comparison of the calculated risk against pre-determined criteria.
- Risk Treatment consists of imposing risk controls and communication of the results.

A diamond had been drawn around the circle so as to link the four arrows containing the names of the overarching concepts. The words within the vertices of the diamond refer to two quadrants of the circle. Risk appraisal consists of identification and analysis. Risk assessment is risk analysis and risk evaluation. Risk characterisation is evaluation and treatment. Action consists of identification and treatment.

Key Research Questions

The committees dealing with the Hazards Theme identified four key research questions:

Key Question 1

- *How have humans altered the geosphere, the biosphere and the landscape, thereby helping to trigger certain hazards and increasing societal vulnerability to them?*

Anthropogenic activity has strongly altered the soil landscape, creating long-term changes detrimental to life and environment. The main aspects of such soil-related environmental degradation are erosion, soil compaction, soil crusting, deterioration of soil structure, landslides, losses of organic matter, and salinisation amongst others. Similar lists could be provided in relation to air quality and in relation to water quality.

Hazards are an intrinsic property of a severe geophysical event. Hazard becomes a risk only when there is a finite probability of a manifestation of the hazard. And a risk becomes a disaster only when there is a significant loss of life or property.

Disasters can arise from multiple disturbances to the geosphere. For example, flooding (e.g.: the flooding in France in 2001 and in Central Europe in 2002) and

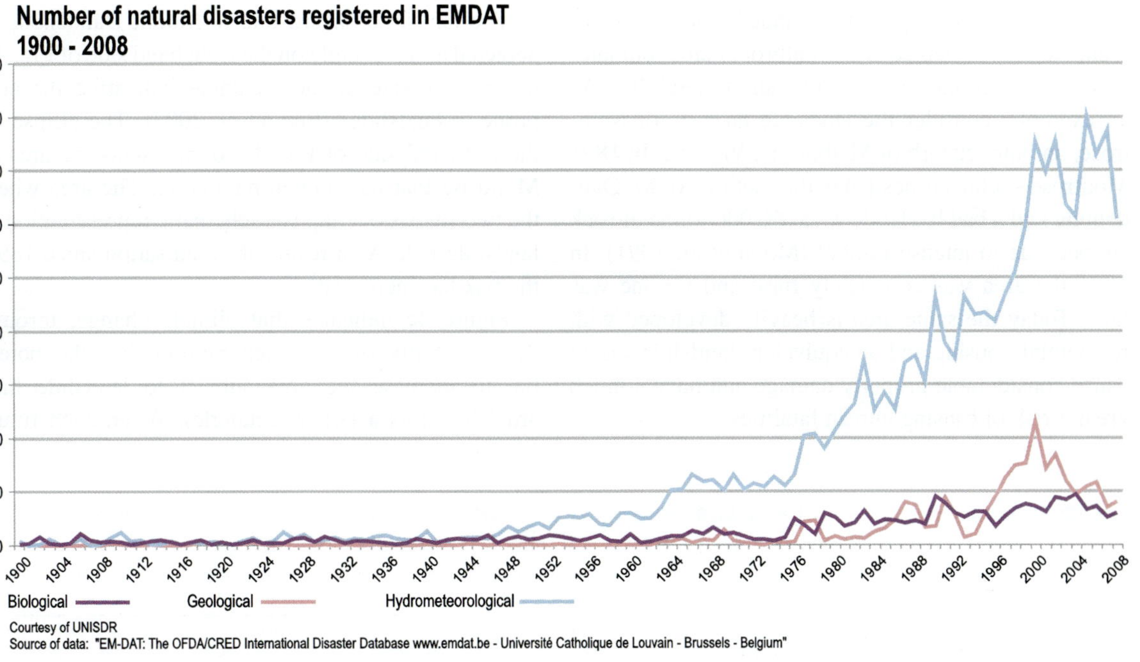

Fig. 3 The number of hydrometeorological, geological and biological disasters from 1900 to 2005 registered in the EMDAT database[3]

landslides (e.g.: South Italy, Sarno 1998 and Ischia 2006) arise from incorrect soil use and management. Reservoir-induced seismicity is an example of a direct human influence on a natural hazard. Human clearing of mangroves has led to coastal erosion.

Figure 3 depicts the number of hydrometeorological, geological and biological disasters registered in the EMDAT database[4] as processed by the UN International Strategy for Disaster Reduction (UN-ISDR). The upward increase in the hydrometeorological graph is evident. The reason for this is less evident, but will involve some combination of a number of factors:

- the increased world population, and especially the growing urbanisation of the developing world means, for example, that a flood event that would have affected very few people a century ago may now affect a megacity (Changnon 2003);
- the greater spatial spread of the population means that humans now live in areas that were recently farmlands or uninhabited. A violent storm in such a location which, in the past, would not have affected anyone, can nowadays lead to a disaster;
- increased communication, news reporting and data collection and collation means that disaster events, especially in the developed world, are now recognised, reported and categorised. In the past, central government authorities may never have known of disasters that affected outlying areas or in some cases been sent misleading reports by district governors (Davis 2001);
- human-induced changes to the climate may also be responsible for some of the upward trend in the hydrometeorological disasters. This issue is hotly debated. Analyses from global circulation models suggest an increase in heavy rainfall and a decrease in light rainfall (e.g. Gordon et al. 1992) that in Australia can lead to increased wildfires (Beer and Williams 1995). However, it is difficult from the existing climate record (Haylock and Nicholls 2000) unequivocally to determine whether the modelled effects have indeed occurred.

[3] http://www.unisdr.org/disaster-statistics/occurrence-trends-century.htm

[4] http://www.emdat.be/

There is increasing concern that human development activity, coupled with anthropogenic climate-change, may further increase human vulnerability. As an example, consider the landslide hazards of Montrose, an outer suburb of Melbourne, Victoria. In 1891 Montrose – which is nestled in the foothills of Mt Dandenong had a bad landslide when 60,000 tonne of rock slipped due to intense rainfall (Moon et al. 1991). In 1891 the area was completely rural and no-one was hurt. Today the same area is heavily developed with residential housing and an equivalent landslide would cause considerable property damage and have a much greater risk of causing human fatalities.

The landslide hazard due to Mount Dandenong is depicted as a rectangle on the right hand side of Fig. 4a, b, with the size of the rectangle indicating the area prone to landslides (Pike et al. 2003). The ellipse on the left hand side of Fig. 4a, b represents the area of Montrose that has human habitation. The area where the two shapes overlap is a schematic representation of landslide risk. As a result of urbanisation since 1891, the risk has increased.

Figure 4c indicates that climate change, through the possibility of increased rainfall, has the potential to increase the area subject to landslide hazard (shown as a larger rectangle). When such future

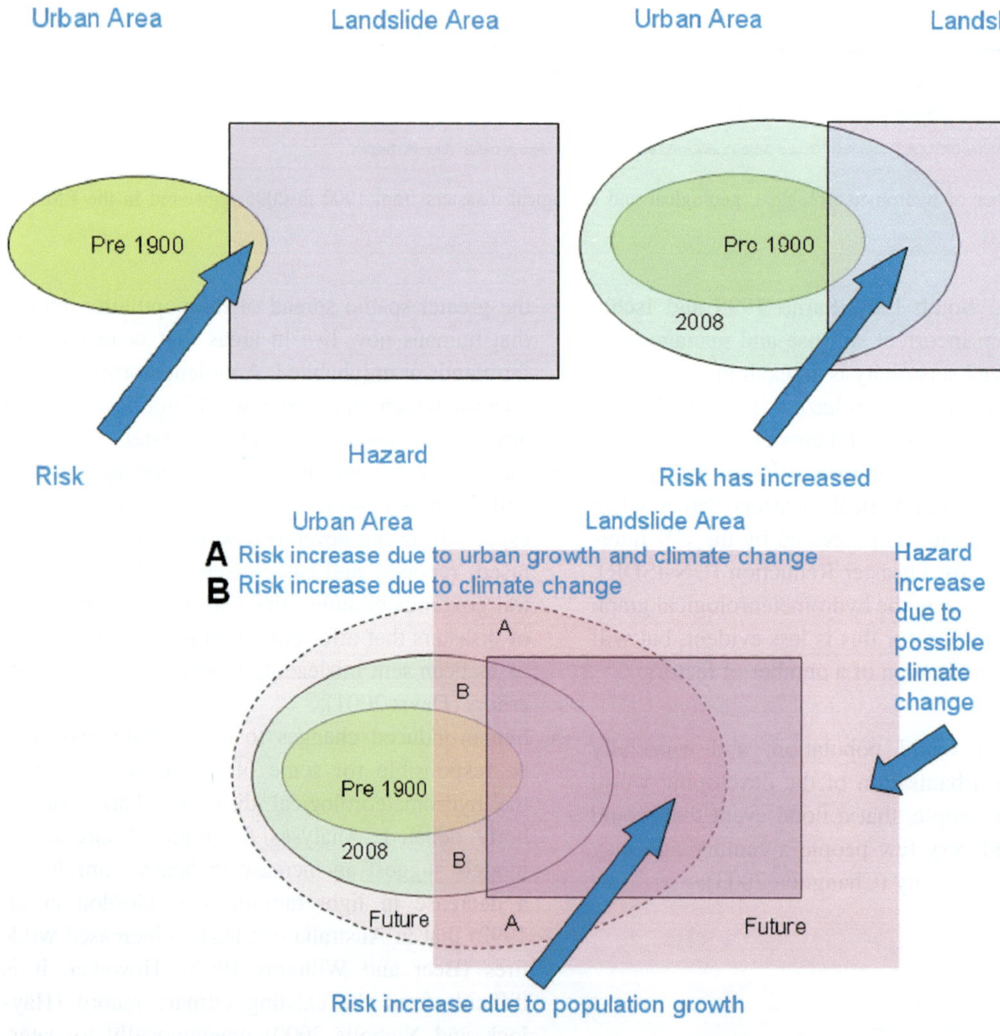

Fig. 4 (**a**) A diagrammatic representation of the risk associated with a landslide hazard in 1891. (**b**) A diagrammatic representation of the risk associated with a landslide hazard in 2008 showing that the spread of urbanisation has increased the risk. (**c**) Schematic illustration of the increase in risk due to urban growth and climate change

climate change effects are combined with expected future urban growth, then the risk increases due to two separate effects. The increased risk due to climate change alone, is shown by the letter B in the diagram. The increased risk due to both climate change and urban growth is shown by the letter A. The combination of the two effects greatly increases the area of potential risk.

Key Question 2

- *What technologies and methodologies are required to assess the vulnerability of people and places to hazards - and how might these be used at a variety of spatial scales?*

There are traditional technologies and, more recently, remote sensing methods that can be used to assess vulnerability. The combination of remote sensing, Geographical Information Systems (GIS) and Global Positioning Systems (GPS) provides a powerful method to syncretise different technologies and methods. In addition modern techniques in image analysis, computer tomography, and the use of synchrotrons provide new opportunities. Birkmann (2006) provides a comprehensive review of the field.

Specific examples of the use of these technologies in relation to specific hazards can be found in Hossain (2006), who examined the use of space-borne systems for flood warning and in Yu et al. (2006) who delimited potential debris flow areas using GIS.

The IGOS GeoHazards Initiative[5] was established to utilise these technologies to help provide the scientific and operational geospatial information needs for the prediction and monitoring of geological hazards, namely earthquakes, tsunamis, volcanoes and land instability (Marsh et al. 2004; Salichon et al. 2007)

However, the most frequent use of these remote sensing methods has been as a means to compare, at an international scale, the relative risk of different natural hazards. The reinsurance company, Munich Re, has used these methods to construct a World Map of Natural Hazards (Berz et al. 2001). More recently, this hazards information was combined with population information to examine natural disasters. The Natural Disaster Hotspots research program (Dilley et al. 2005) was funded by the World Bank and undertaken by Columbia University. Figure 5 reproduces one of the outputs – in this case depicting the regions of the world that have the highest mortality risk due to natural disasters.

Key Question 3

- *How does our current capability to monitor, predict and mitigate vary from one geohazard to another? What methodologies and new technologies can improve such capabilities and so help civil protection locally and globally?*

This key question deals with capabilities for monitoring, prediction and mitigation. These are unevenly spread over the globe which means that fatalities from natural hazards are far greater in developing countries than in developed countries.

Consider some recent examples. Tropical Cyclone Nargis struck Myanmar between 3 and 6 May 2008 leaving 78,000 people dead and 56,000 people homeless. At 6am on 2 May, just before landfall, it was a Category 4 Tropical Cyclone. Compare this with Hurricane Katrina that struck New Orleans, USA on 29 August 2005. This had been a Category 5 Hurricane but was a Category 3 Hurricane when it made landfall in Lousiana. The final death toll from Hurricane Katrina was 1,836 but the damage bill was estimated to be US$81.2 billion.

Earthquake tabulations (Yong et al. 2001) tell a similar story. On 29–30 September 1993, the Latur earthquake of magnitude 6.3 in north-western India caused substantial damage. Yong et al. (2001) state that 30,000 people died whereas Jain (1994) claims that there were about 10,000 deaths and Wikipedia[6] claims that 7,928 people died and 30,000 were injured. Regardless of which figure is correct, the death toll in this magnitude 6.3 earthquake was at least two orders of mag-

5 http://igosg.brgm.fr/

6 http://en.wikipedia.org/wiki/Killari accessed on 8 November 2008.

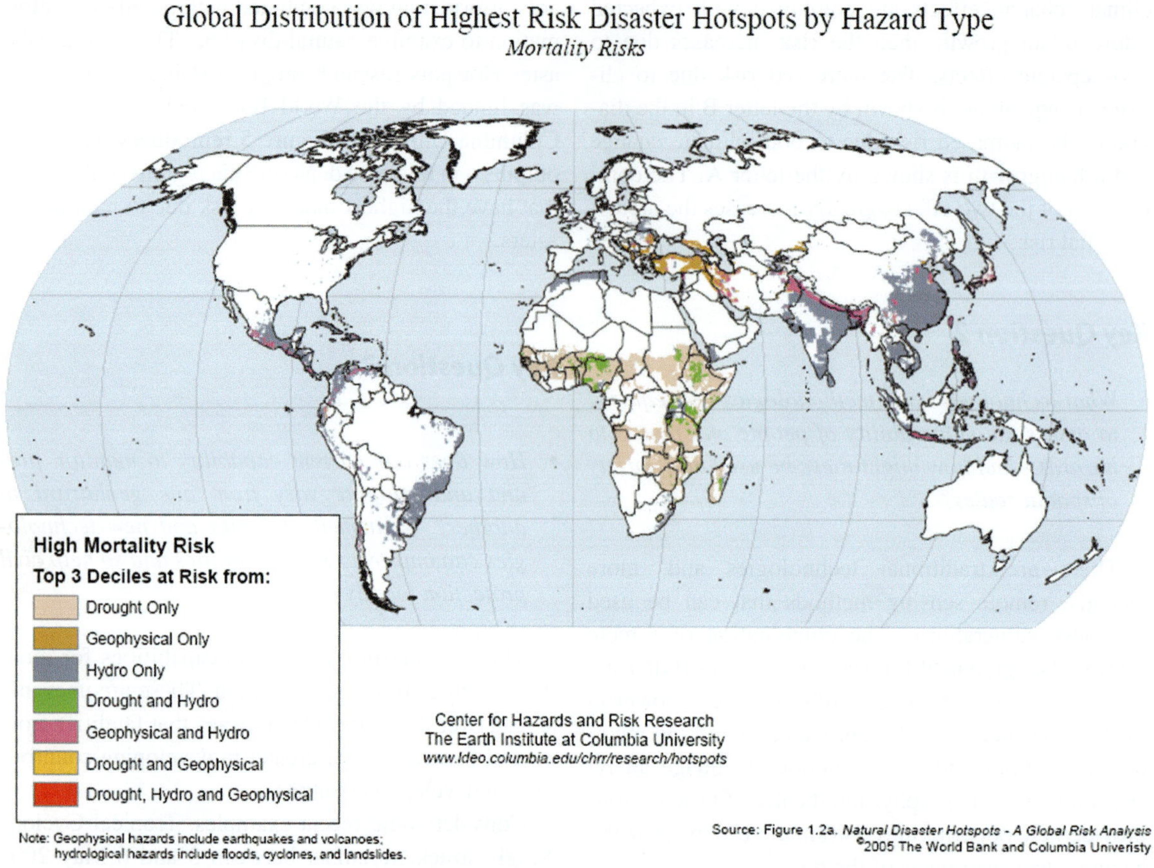

Fig. 5 Sample output from the hotspots program (Dilley et al. 2005) showing the global distribution of the regions in the world with the highest mortality risk

nitude greater than the death toll from the Northridge Earthquake that struck California, USA on 17 January 1994, which Yong et al. (2001) estimate to be 61 people. But the estimated loss for the Indian earthquake was US$80 million, whereas that of the US earthquake was over two orders of magnitude greater at US$15 billion.

In developed countries the monitoring, prediction and mitigation systems in place mean that fatalities are reduced in disasters, but total infrastructure losses are greater. In developing countries the reverse is the case. However, as pointed out by Cutter (2010) even though absolute financial impacts may be greater in developed countries, this may not necessarily be so on a relative basis (when the financial loss is expressed as a percentage of GDP). Furthermore, when one considers coping capacities of developing nations, the loss of limited but critical infrastructure can have a substantial consequence for the human population of developing countries.

Nevertheless, the costs of disasters are increasing in both developed and developing countries (International Council for Science 2008) indicating that reducing the risks from hazards is not simply a matter of economic growth and development.

But it must be acknowledged that technological innovation has greatly assisted civil protection. Meteorology offers the best case study on the way in which technology has managed to reduce the risk of hazards. In the case of Hurricane Katrina, mentioned earlier, which devastated New Orleans – by using satellites, virtually real-time, exact information on the hurricane position could be obtained. This information was fed into super-computers programmed with numerical weather prediction models that enabled meteorologists to predict more than two days before the hurricane struck the coast that New Orleans was a likely target and that evacuation measures needed to take place.

Within communications technology, one of the most important technological advances has been the widespread use of the mobile phone (also known as the cell phone). This has been accompanied by technological improvements in materials and batteries that have reduced the size and weight of the mobile phones, and increased their capability and versatility (Fig. 6); many of them now also act as small portable cameras. Additionally, the mass production of mobile phones has reduced their cost to such a level that in some societies they are becoming ubiquitous. Several countries now have more mobile phones than people. Luxembourg has the highest mobile phones per capita in the world, at 1.64 in December 2001.

The total number of mobile phone subscribers in the world was estimated as 3.3 billion by November, 2007 thus reaching an equivalent of over half the planet's population. Around 80% of the world's population enjoys mobile phone coverage as of 2006. This figure is expected to increase to 90% by the year 2010. Cell phone use in developing countries has quadrupled in the last decade.

The Finnish government decided in 2005 that the fastest way to warn citizens of disasters was the mobile phone network. In Japan, mobile phone companies provide immediate notification of earthquakes and other natural disasters to their customers free of charge. In the event of an emergency, disaster response crews can locate trapped or injured people using the signals from their mobile phones. An interactive menu accessible through the phone's Internet browser notifies the company if the user is safe or in distress. In Finland, rescue services suggest hikers carry mobile phones in case of emergency even when deep in the forests beyond cellular coverage, as the radio signal of a mobile phone attempting to connect to a base station can be detected by rescue aircraft with special detection gear.

However, most mobile telephone networks operate close to capacity during normal times and spikes in call volumes caused by widespread emergencies often overload the system just when it is needed the most. Examples reported in the media where this has occurred include the 9/11 attacks of September 2001, the 2003 Northeast blackouts, Hurricane Katrina, and the 2007 Minnesota bridge collapse. Thus mobile phones, though able to assist in small-scale emergencies, are of limited use in large-scale disasters where many of the fixed transmitters of the cellular network may have been damaged resulting in a diminished capacity of the remaining network to handle the increased volume of calls.

Key Question 4

- *What are the barriers, for each geohazard, that prevent governments (and other entities) from using risk and vulnerability information to create policies and plans to reduce both?*

Fig. 6 One of the most significant technological advances permitting widespread dissemination of information, and enabling emergency workers to maintain contact, has been the widespread use of the mobile phone (also known as the cell phone). The basic model depicted on the left retails for less than US$100. The more expensive Blackberry on the right permits portable e-mail communication

In many countries disaster prevention is completely lacking. Scientific information that could aid decision-makers to plan sustainable development and to prevent disasters is not available, not disseminated or not sufficiently considered. We have already noted previously that estimates of the fatalities from an earthquake such as the 1993 Latur Earthquake can vary from 7,928 to 30,000. Such discrepancies can arise from different scientific methodologies – or from political exigencies that can lead nations to understate or overstate the effects of a major disaster.

To plan sustainable development requires knowledge of physical variables such as the soil, the land, the environment, and to integrate this knowledge within the local social and political system. National Governments need to dedicate attention and funds to scientific research to improve and disseminate the knowledge, but often the benefits of strong support for scientific research are not immediate especially when the research relates to natural hazards and disasters that are themselves rare occurrences.

The barriers that prevent governments from using risk and vulnerability information will differ depending on the geohazard, and on the government. It is thus instructive to consider the 2004 Indian Ocean Tsunami as a case study to illustrate the situation where the barriers were speedily identified (after the event, unfortunately) and removed.

On 26 December 2004 the Aceh region of Indonesia experienced an earthquake of magnitude 9.3 that led to the countries bordering the Indian Ocean being struck by a powerful tsunami that led to 226,000 deaths (Rossetto et al. 2007). Three weeks later, the UN-ISDR organised the World Conference on Disaster Reduction (WCDR)[7] in Kobe from 18 to 22 January 2005 where the assembled nations decided to implement an Indian Ocean Tsunami Warning system analogous to that of the Pacific Ocean (Bernard 2005). Nevertheless, Alverson (2005) warned that unless the Indian Ocean Tsunami Warning System is embedded into a broader effort to observe the oceans, there is a possibility that it will not remain operational for the many decades that may be required before the next significant tsunami.

The nations assembled at the WCDR also issued the Hyogo Framework for Action (HFA), a set of five priority action items needed to build the resilience of nations and communities to disasters over the decade covering 2005–2015.

Hyogo Framework for Action

The five action items of the Hyogo Framework for Action[8] are:

1. Make Disaster Risk Reduction a Priority – ensure that disaster risk reduction is a national and a local priority with a strong institutional basis for implementation.
2. Know the Risks and Take Action – Identify, assess and monitor disaster risks – and enhance early warning.
3. Build Understanding and Awareness – Use knowledge, innovation, and education to build a culture of safety and resilience at all levels.
4. Reduce Risk – Reduce the underlying risk factors.
5. Be Prepared and Ready to Act – Strengthen disaster preparedness for effective response at all levels.

From the perspective of the scientific and engineering community, there are two key issues related to the Hyogo Framework for Action: (1) How can answers to the four key questions of the IYPE Hazards Theme, as presented above, be coupled to the five priority action items of the HFA so as to energise both those undertaking research related to disaster reduction and those implementing disaster reduction policies and plans? And (2) how can the scientific and engineering community mobilise itself so as to work together with policy-makers, politicians, government officials and emergency workers so as to produce relevant and useful science?

The answer to the first question can be found within the HFA itself. Each of the priority action items has a short, pithy message associated with it that can be used to provide links to the type of scientific research needed to implement each action item:

[7] http://www.unisdr.org/wcdr/

[8] http://www.unisdr.org/eng/hfa/hfa.htm

1. Make Disaster Risk Reduction a Priority – *Collaboration is the key*
2. Know the Risks and Take Action – *Early warning saves lives*
3. Build Understanding and Awareness – *Local knowledge is critical for disaster reduction*
4. Reduce Risk – *Building resilience protects communities*
5. Be Prepared and Ready to Act – *Disaster preparedness takes practice*

The HFA was envisaged as a ten-year plan of action, but since its inception the ever-increasing awareness that climate change and disaster risk reduction are closely linked (UN-ISDR 2008) means that even longer time scales need to be considered. Figure 7 (Sheehan et al. 2008) indicates that even slight increases in temperature will lead to disruptions in coral reefs (in blue), lead to the melting of the Greenland ice cap (in light green), and destroy ecological species (red). The black lines (falling from high values on the left of the figure to low values on the right) give the probabilities of various temperature changes if global greenhouse gas emissions peak in 2010, 2015, 2020, 2025 or – as was assumed in the Reference case – in 2030. This gives a 50% probability of a 4 degree warming, which gives a 99% probability of destruction of the area of the Great Barrier coral reef above this critical threshold, 95% probability of the Greenland ice sheet melting, and 40–60% probability of species facing extinction risk.

It is evident that a short duration scientific activity, such as the IYPE with a life of one to three years, cannot provide the scientific input that is needed to strengthen all five priority action items. Furthermore, disaster risk is generally not well assessed and monitored (Basher 2008). This was also evident to the International Council for Science (ICSU), which decided in October 2008 to initiate a long term (5–10 years) research program on hazards to be known as Integrated Research on Disaster Risk (International Council for Science 2008). This program can be considered to be the successor of the Hazards Theme of the IYPE.

Because of the ongoing work on the Hyogo Framework for Action, and the new IRDR program, this section comments on the consequences of natural hazards in developing nations, but does not actually answer the question posed: namely, what are the barriers for each hazard? These successor programs will need to undertake the systematic study necessary to answer the fourth key research question.

Integrated Research on Disaster Risk (IRDR)

The science plan of the IRDR Programme will focus on hazards related to geophysical, oceanographic and hydro-meteorological trigger events, i.e., earthquakes; volcanoes; flooding; storms (hurricanes, typhoons, etc.); heat waves; droughts and fires; tsunamis; coastal

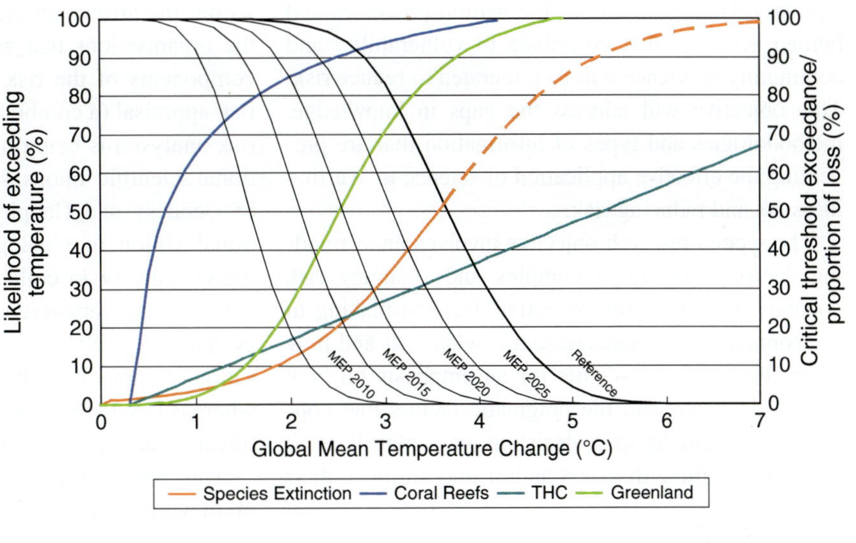

Fig. 7 The probability of new risks being introduced as a result of climate change are depicted for species extinction, coral reef bleaching, major alteration of the thermohaline circulation, and a melting of the Greenland ice sheets

erosion; landslides; aspects of climate change; space weather and impact by near-Earth objects. The effects of human activities on creating or enhancing hazards, including land-use practices, will be included. The focus on risk reduction and the understanding of risk patterns and risk-management decisions and their promotion will require consideration of scales from the local through to the international level.

There is a great shortfall in current research on how science is used to shape social and political decision-making in the context of hazards and disasters. These issues also highlight the need for more systematic and reliable information on such events. An aim of the Programme will be to both generate new information and data and to leave a legacy of coordinated and integrated global data and information sets across hazards and disciplines, with unprecedented degrees of access.

The hope is that IRDR will leave the legacy of an enhanced capacity around the world to address hazards and make informed decisions on actions to reduce their impacts, such that in ten years, when comparable events occur, there would be a reduction in loss of life, fewer people adversely impacted, and wiser investments and choices made by governments, the private sector and civil society.

The IRDR Programme will have three research objectives, the first of which deals with the characterization of hazards, vulnerability and risk. The identification and assessment of risks from natural hazards on global, regional and local scales, and the development of the capability to forecast hazardous events and their consequences will be, of necessity, interdisciplinary. Understanding of the natural processes and human activities that contribute to vulnerability and community resilience will be integrated to reduce risk. This objective will address the gaps in knowledge, methodologies and types of information that are preventing the effective application of science to averting disasters and reducing risk.

The second research objective involves understanding decision-making in complex and changing risk contexts. Understanding effective decision-making in the context of risk management – what is it and how it can be improved – calls for an emphasis on how human decisions and the pragmatic factors that constrain or facilitate such decisions can contribute to hazards becoming disasters and/or may mitigate their effects.

The third research objective, on reducing risk and curbing losses through knowledge-based actions, will require integration of outputs from the first two and could only be achieved through implementing and monitoring informed risk reduction decisions and through reductions in vulnerability or exposure. Processes of human adjustment or adaptation can be used to reduce vulnerability and increase resilience.

Three cross-cutting themes will support these objectives: capacity building, including mapping capacity for disaster reduction and building self-sustaining capacity at various levels for different hazards; the development of case studies and demonstration projects; and assessment, data management and monitoring of hazards, risks and disasters.

ICSU has also decided to open Regional Offices. During 2006, regional offices for Africa, Latin America & the Caribbean, and Asia & the Pacific were inaugurated. All three of these regions have identified Natural and Human-induced Hazards as being one of their priority research programs.

Discussion and Conclusion

This paper has reported on the procedures that are being implemented to prepare for, and to organise, large-scale programs to study the scientific and social aspects of natural hazards and disasters as a prelude to implement action to minimise their risks.

Figure 8 repeats the risk management framework of Fig. 2, but has replaced the overarching concepts within the arrows on the outside of the diagram with the organisations that are able to implement various components of the risk management diamond. Thus risk appraisal (a combination of risk identification and risk analysis) is being undertaken by various international scientific unions, such as the International Union of Geodesy and Geophysics (IUGG) or the International Union of Geological Sciences (IUGS) either individually or in collaboration. The IYPE has dealt with risk assessment (comprising risk analysis and risk evaluation). ICSU intends to deal with risk characterisation (comprising risk evaluation and risk treatment), whereas UN-ISDR deals with action through risk identification and risk treatment.

Thus the mechanisms are in place for future collaboration between the various international and multi-

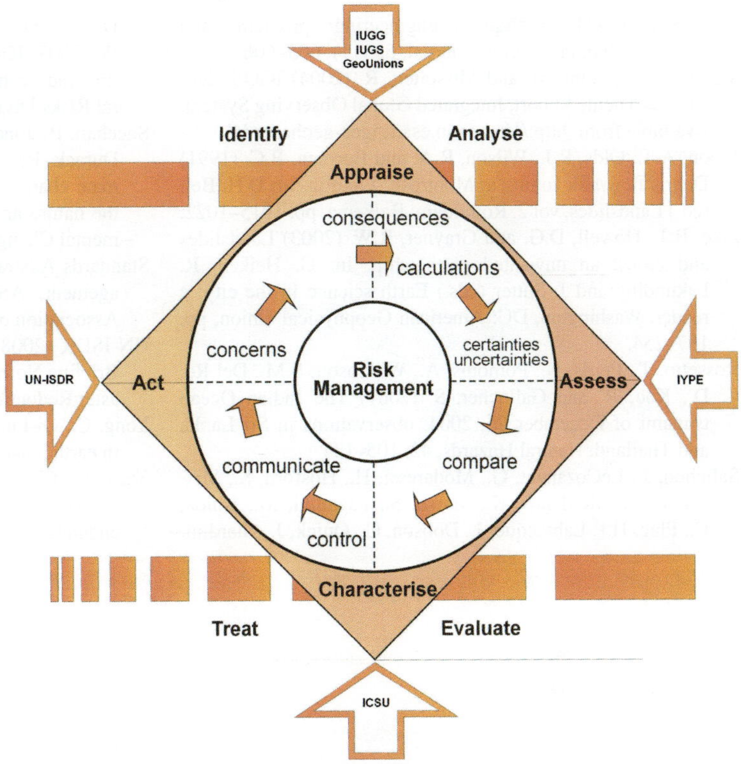

Fig. 8 This diagram repeats the risk management framework of Fig. 2, but has replaced the overarching concepts within the arrows on the outside of the diagram with the organisations that are able to implement various components of the risk management diamond

national hazards initiatives to collaborate, to share information, and to undertake joint activities. We can look forward to a future where those undertaking work on hazards and risk reduction will think global, talk regional, and act local.

References

Alverson, K. (2005) Watching over the world's oceans. Nature, 434, 19–20

Basher, R. (2008) Disasters impacts: implications and policy responses. Social Research, 75 (3), 937–954.

Beer, T. and Ismail-Zadeh, A. (eds.) (2003): Risk science and sustainability: science for reduction of risk and sustainable development for society. (NATO Science Series. Series II, Mathematics, Physics, and Chemistry; 112) Dordrecht: Kluwer Academic. xvi, p. 240.

Beer, T. and Williams, A.M. (1995) Estimating Australian forest fire danger under conditions of doubled carbon dioxide concentrations. Climatic Change, 29, 169–188.

Bernard, E. (ed.) (2005) Developing tsunami-resilient communities: the national tsunami hazard mitigation program. Dordrecht: Springer, p. 184.

Berz, G., Kron, W., Loster, T., Rauch, E., Schimetschek, J., Schmieder, J., Siebert, A., Smolka, A. and Wirtz, A. (2001) World map of natural hazards – a global view of the distribution and intensity of significant exposures. Natural Hazards, 23, 443–465.

Birkmann, J. (ed.) (2006) Measuring vunerability to natural hazards: towards disaster resilient societies. Tokyo: United Nations University Press, p. 524.

Changnon, S.A. (2003) Shifting economic impacts from weather extreme in the United States: a result of societal changes, not global warming. Natural Hazards, 29, 273–290.

Cutter, S.L. (2010) Social science perspective on hazards and vulnerability science (this volume)

Davis, M. (2001) Late Victorian holocausts: El Niño famines and the making of the third world. London: Verso, p. 464.

Dilley, M., Chen, R.S., Deichmann, U., Lerner-Lam, A.L. and Arnold, M. (2005) Natural disaster hotspots: a global risk analysis, Washington DC: World Bank, p. 145.

Gordon, H.B., Whetton, P.H., Pittock, A.B., Fowler, A.M. and Haylock, M.R. (1992) Simulated changes in daily rainfall intensity due to the enhanced greenhouse effect: implications for extreme rainfall events. Climate Dynamics, 8, 83–102.

Haylock, M. and Nicholls, N. (2000) Trends in extreme rainfall indices for an updated high quality data set for Australia, 1910–1998. International Journal of Climatology, 20: 1533–1541

Hossain, F. (2006) Towards formulation of a space-borne system for early warning on floods: can cost-effectiveness outweigh prediction uncertainty? Natural Hazards, 37, 263–276.

International Council for Science (2008) A science plan for integrated research on disaster risk: addressing the challenge of natural and human-induced environmental hazards. Paris: ICSU, p. 64.

Jain, S.K. (1994) Earthquake engineering: problems and prospects. Indian Concrete Journal, 68 (11), 605–606.
Marsh, S., Paganini, M. and Missotten, R. (2004) IGOS Geohazards Theme Report, Integrated Global Observing System. Available from: http://dup.esrin.esa.it/igos-geohazards/
Moon, A.T., Olds, R.J., Wilson, R.A. and Burman, B.C. (1991) Debris flow risk zoning at Montrose, Victoria. In: D.H. Bell (ed.) Landslides, vol 2. Rotterdam: Balkema, pp. 1015–1022.
Pike, R.J., Howell, D.G. and Grayner, R.W. (2003) Landslides and cities: an unwanted partnership. In: G. Heiken, R. Fakundiny and J. Sutter (eds.) Earth science in the city: a reader. Washington, DC: American Geophysical Union, pp. 187–254.
Rossetto, T., Peiris, N., Pomonis, A., Wilkinson, S.M., Del Re, D., Koo, R. and Gallocher, S. (2007) The Indian Ocean tsunami of December 26, 2004: observations in Sri Lanka and Thailand. Natural Hazards, 42, 105–124
Salichon, J., LeCozannet, G., Modaressi, H., Hosford, S., Missotten, R., McManus, K., Marsh, S., Paganini, M., Ishida, C., Plag, H.P. Labrecque, J., Dobson, C., Quick, J., Giardini, D., Takara, K., Fukuoka, H., Casagli, N. and Marzocchi, W. (2007) IGOS GeoHazards Theme Report, BRGM Scientific and Technical Centre Development Planning and Natural Risks Division, Orléans, France.
Sheehan, P., Jones, R.N., Jolley, A., Preston, B.L., Clarke, M., Durack, P.J., Islam, S.N.M. and Whetton, P.H. (2008) Climate change and the new world economy: implications for the nature and timing of policy responses. Global Environmental Change, 18, 380–396.
Standards Australia (2004) Standards New Zealand: risk management, AS/NZS 4360:2004. Sydney, NSW: Standards Association of Australia.
UN-ISDR (2008) Climate change and disaster risk reduction, Briefing Note 01. Geneva: UN International Strategy for Disaster Reduction, p. 11. www.unisdr.org/climate-change
Yong, C., Qi-Fu, C. and Ling, C. (2001) Vulnerability analysis in earthquake loss estimate, Natural Hazards, 23, 349–364.
Yu, F.-C., Chen, C.-Y., Chen, T.-C., Hung, F.-Y. and Lin, S.-C. (2006) A GIS process for delimitating areas potentially endangered by debris flow, Natural Hazards, 37, 169–189.

Social Science Perspectives on Hazards and Vulnerability Science

Susan L. Cutter

Abstract What makes people and places vulnerable to natural hazards? What technologies and methods are required to assess this vulnerability? These questions are used to illustrate the circumstances that place people and localities at risk, and those circumstances that enhance or reduce the ability of people and places to respond to environmental threats. Vulnerability science is an emerging interdisciplinary perspective that builds on the integrated tradition of risk, hazards, and disasters research. It incorporates qualitative and quantitative approaches, local to global geography, historic to future temporal domains, and best practices. It utilizes technological sophistication and analytical capabilities, especially in the realm of the geo-spatial and computation sciences (making extensive use of GPS, GIS, remote sensing, and spatial decision support systems), and integrates these with perspectives from the natural, social, health, and engineering sciences.

Vulnerability research focuses on the intersection of natural systems, social systems, and the built environment. These three component areas intersect with the spatial social sciences to play a critical role in advancing vulnerability science through improvements in geospatial data, basic science, and application. The environment, individuals, and societies have varying levels of vulnerability that directly influence their ability to cope, rebound, and adapt to environmental threats. At present, we lack some of the basic operational understanding of the fundamental concepts of vulnerability, as well as models and methods for analyzing them. The focus on place-based applications and the differential susceptibility of populations to hazards is a key contribution of vulnerability science. Using examples derived from recent disasters, the role of the spatial social sciences in advancing vulnerability science are reviewed.

Keywords Vulnerability science · EM-DAT · SHELDUS

Introduction

We know geohazards are distributed unevenly across the earth's surface – some regions are seismically active, while others are not; certain regions experience hydro-meteorological hazards with regularity, while others endure water deficits for most of the year. Understanding physical processes and their variability across the landscape provides the fundamental science of hazards – where, when, and why they occur, and the risks posed to society (Beer et al. 2004). While we possess some level of understanding on the distribution of natural hazards and their historical frequency, we know less about the risk (probability of an event occurring) or its likely impact on society.

Just like geohazards, the Earth's population also is unevenly distributed on the natural landscape, often clustered near coastlines, along rivers, or in seismically active zones. Population size and location coupled with the socio-economic and demographic characteristics of that population are the drivers of societal impacts, and help explain why the same geophysical event produces quite different impacts at the local level. Societal

S.L. Cutter (✉)
Department of Geography, Hazards and Vulnerability Research Institute, University of South Carolina, Columbia, SC 29208, USA
e-mail: scutter@sc.edu

T. Beer (ed.), *Geophysical Hazards*, International Year of Planet Earth,
DOI 10.1007/978-90-481-3236-2_2, © Springer Science+Business Media B.V. 2010

factors intervene between nature (and the natural processes) and the built environment to redistribute the risk prior to an event, and to amplify or attenuate the losses after an event. The interaction of society, nature, and the built environment creates dangerous places (Reisner 2004), and often elevates routine hazards into disasters.

Vulnerability Science

Simply stated, vulnerability is the potential for harm or loss. It examines those circumstances that place people and localities at risk as well as those characteristics that enhance or reduce the ability of society to respond to environmental threats. While there is considerable theoretical and conceptual development within the field of vulnerability science in many different literatures (engineering, social sciences, natural sciences), most agree on the underlying construct – vulnerability is the susceptibility to harm. Understanding vulnerability to hazards, especially geohazards, requires three separate, but intersecting knowledge domains: natural systems, social systems (and the built environment), and local places (Fig. 1).

Natural processes, independent of human agency, do not produce hazards, it is only when these geophysical processes interact with human populations that hazards arise (Burton, Kates and White 1993).[1] Humans create hazards by altering the natural landscape and affecting natural system processes, such as locating in floodplains and altering flow regimes and runoff through landscape modification. Interactions between social systems and the built environment also contribute to vulnerability. The pressing demand for shelter in many of the world's megacities can result in shoddy construction and poor placement of housing, much of it in high-risk areas such as floodplains or steep, unstable slopes. Once a landslide, earthquake, or flood occurs it destroys homes and human lives, yet this unsustainable pattern continues with increasing demand for new housing by the influx of new residents. The third knowledge component is the understanding of the local places – landscape, history, economics, culture, demographics, politics – in other words, the local geography. For example, in New Orleans the topography of the city reflects its social geography: wealthier residents live in higher elevations and have historically done so; while the poor and minority residents occupy the low-lying areas, the latter often subjected to the most intense flooding (Colten 2005; Kates et al. 2006).

The integration of all three knowledge domains (Fig. 1) provides the intellectual basis for vulnerability science, an interdisciplinary field that entails the development of methods and metrics for analyzing societal vulnerability and resilience to environmental hazards and extreme events (Cutter 2003). One of the primary goals of vulnerability science is to provide the scientific basis for disaster and hazard reduction policies.

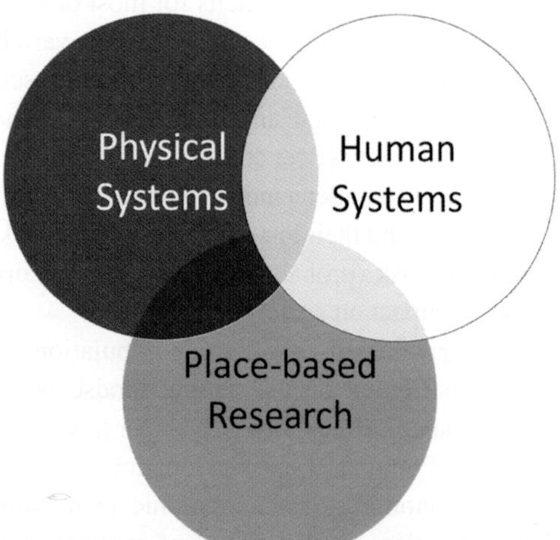

Fig. 1 Vulnerability science incorporates knowledge from the intersection of physical systems, human systems, and place-based research

The Recipe for Disaster

Prior to the implementation of any risk or vulnerability reduction policy, we must first understand the

[1] There is a difference in usage of the terms risk and hazards between the geophysical and social science community that studies disasters, hazards and risk. Rather than going into the nuances of these linguistic differences, I chose to adhere to the definitions from the social science community. From the social science viewpoint, risk is the probability of an event occurring, while hazards include the probability of an event happening as well as the impact of that event on society. In other words, natural hazards are threats to people and the things they value and arise from the interaction between human systems and the natural processes. We follow the terminology developed by Gilbert F. White and his colleagues.

geographic variability in the hazards and their impacts; the geographic variability in the populations at risk; the vulnerability of populations; and the context in which these three interact at the local or regional scale. There are rich sources of general information on the spatial distribution of hazards and disaster risks (Dilley et al. 2005; International Strategy for Disaster Reduction 2004), as well as annual overviews of disasters, embodied in the yearly World Disasters Report (International Federation of Red Cross and Red Crescent Societies 2008). The development of robust monitoring and surveillance networks have advanced our understanding of the earth sciences and provide a rich data source for hazard event parameters such as location, magnitude, intensity, duration. Global seismic and tsunami monitoring systems, and the international suite of weather-related satellites produce enormous data streams on precipitation patterns (that can lead to floods and droughts), tropical cyclones and other wind events, and severe weather (heat and cold). The US Geological Survey, the National Climatic Data Center (NCDC), the National Geophysical Data Center (NGDC), and the Smithsonian's Global Volcanism program routinely catalog data describing the physical attributes of natural hazards in the U.S. However, the societal impact of such events (lives lost or economic losses) is not consistently included in such databases (Gall et al. 2009).

Hazard Losses

At present, there is no systematic information on the societal losses to geohazards. These losses include direct monetary losses such as the value of destroyed homes, businesses, or infrastructure (e.g. roads and bridges). Yet societal losses also include human losses – loss of life, injuries, loss of livelihoods, and displacements from one area to another. If we are to reduce vulnerability to natural hazards, a good starting point is to determine the annual losses associated with natural hazards for each nation. Are such losses dispersed throughout the country or are they concentrated in one or more specific areas? Are the losses a consequence of a singular, infrequent catastrophic event, or are they the result of periodic smaller scale (and impact) events that overtime add up to significant losses for the country? In other words, what are the hazard loss profiles of nations?

Proprietary data collected by reinsurance companies such as Munich Re and Swiss Re computes insured losses from large-scale disaster events. These insurance databases provide only a limited picture of losses since low density and low capitalized countries are often excluded, especially countries in Asia, Africa, and Latin America. Within the more developed world regions, there is better coverage of urban versus rural areas, but the monetary impact of slow onset hazards such as drought are not included because of the difficulty in assessing monetary damages. At present, there is no systematic open-access inventory or accounting for individual nations or aggregated to the global scale of hazard events and losses by location or by hazard agent. How can we reduce the societal impact of natural hazards when we lack fundamental information on how large the losses are and where they occur?

There are two on-going efforts to produce such an accounting – the EM-DAT disaster database managed by CRED for nations with global coverage; and SHELDUS, a US-centric database for hazard events and losses at the county scale. These are described next.

EM-DAT

EM-DAT is a global emergency events database developed and maintained by The Centre for Research on the Epidemiology of Disasters (CRED) at the Université Catholique de Louvain (Brussels, Belgium) (http://www.emdat.be/index.html). This country-level database covers the period 1900–2007, but the majority of entries are from 1975 to present. The database includes information on people killed, people affected, direct and indirect economic damages. It can be sorted by date and by hazard category using three main groupings: natural hazards, technological hazards, and complex emergencies. Under the natural hazards category, the specific hazards include earthquakes, tsunamis, epidemics (viral and parasitic), floods (flash, general, storm surge/coastal), mass movements (landslides), storms (local, tropical cyclones), volcanic, and wildfire (forest fires and shrub/grassland fires). Beginning in 2003, each disaster event was issued a GLIDE (GLobal IDEntifer) number. GLIDE numbers provide a consistent identifier so scholars and practitioners can link information across databases on specific disasters and their impacts (http://gs.adrc.or.jp/glide/public/search/search.jsp). In EM-DAT, where the disaster affected more than one country, the same GLIDE number is used, but the losses are attributed to the country where they occurred. EM-DAT employs a threshold for inclusion into the database that each event must meet. The criteria include more than 10 people killed, more than 100 affected, a declared state of emergency, or a call for international assistance.

EM-DAT is a searchable database that can create either country profiles or hazard specific profiles. The database is user friendly, although there are limitations in accessing the raw data. Instead, country profiles contain summary statistics by event type for a specified time period (1900 to present; 1975 to present). For example, as shown in Fig. 2, there is an upward

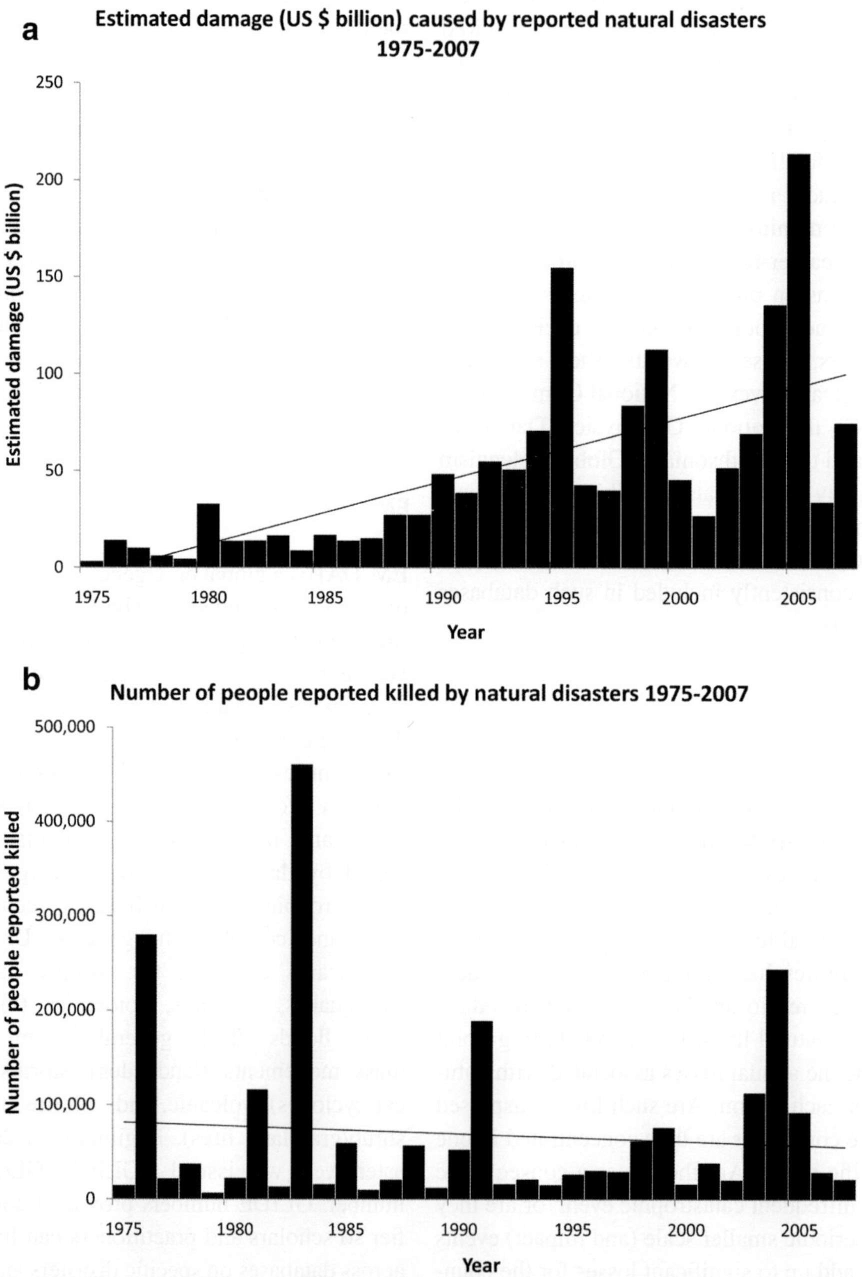

Fig. 2 International trends in disaster losses.
Source: EM-DAT: The OFDA/CRED International Disaster Database, Université Catholique de Louvain, Brussels, Belgium (www.edmdat.be)

trend in disaster damages, globally, and a downward trend in hazard-related fatalities. While the data are the best available, globally, it is often difficult to download individual year data by country, but this functionality should improve with new versions and updates on the web site.

The CRED EM-DAT database is the best available for country-level comparisons of disaster impacts, both temporally and spatially. The data are routinely used by international aid and disaster relief organizations. In fact, EM-DAT has been part of the *World Disasters Report* published annually (IFRCRC 2008), and provided the statistical background for the World Bank's *Natural Disaster Hotspots* report (Dilley et al. 2005).

SHELDUS

In the U.S. many different federal and state agencies collect hazard data such as the US Geological Survey (geophysical and hydrological data), and NOAA (atmospheric and hydrometerological data). As noted above, only some event parameters are monitored (magnitude, frequency, location in long/lat), while others (deaths, injuries, or economic losses) languish. At present, the US lacks any baseline information on such loss patterns. Despite repeated calls beginning in 1999 for a national inventory of hazard-losses (Mileti 1999; National Research Council 1999), and despite its experience with the costliest single disaster ever (Hurricane Katrina estimated at more than $100b in losses), the U.S. still does not have standardized loss inventory.

An independent research effort has tried to remedy this situation through the development of the **S**patial **H**azard **E**vents and **L**osses **D**atabase, **U.S.** (SHELDUS) (www.sheldus.org). This is a geo-referenced (to the county level) database of natural hazard events and losses for the U.S. from 1960 to present. It includes 18 different natural hazards (Table 1), location (state and county), deaths, injuries, property losses, crop losses, and beginning/end dates. The database is searchable by hazard type, location (state and county), date, major event names (e.g. Hurricane Katrina), Presidential Disaster Declaration number, and GLIDE number. The latter is an important feature, as it links SHELDUS into international databases using a globally common identifier for specific disaster events. In SHELDUS, the output from the user-initiated query includes beginning date, hazard type, state, county, injuries, fatalities, property damage, and crop damage. The economic damages reported are in period dollars, but there is an inflation adjustment, which computes the losses to current dollars (e.g. $2007). The records (which now exceed 450,000) are downloadable in multiple formats for ease of use in statistical programs. Lastly, the SHELDUS website (www.sheldus.org) provides metadata (compliant with the U.S. Federal Geographic Data Committee standards), frequently asked questions (FAQs) to aid navigation through the site, and an annual year in disasters report, called the SHELDUS Clock.

Table 1 Hazards included in SHELDUS database

Avalanche	Landslide
Coastal	Lightning
Drought	Severe Storm/Thunderstorm
Earthquake	Tornado
Flooding	Tsunami/Seiche
Fog	Volcano
Hail	Wildfire
Heat	Wind
Hurricane/Tropical Storm	Winter Weather

SHELDUS was constructed from U.S. federal government data sources, notably from USGS and National Climatic Data Center records. It includes any event that resulted in more than $50,000 in economic losses or any death during the time period (1960 to present). When a single event affected a number of different counties, the losses were attributed equally across the counties when no other specific spatial information was available on where the loss occurred. This results in fractional deaths, injuries, and dollar losses. Given changing data collection procedures over the time period by the federal agencies, SHELDUS represents a very conservative estimate of the losses (Gall et al. 2009), especially when compared to other databases. Nevertheless, SHELDUS represents the first centralized approximation of a national inventory for U.S. natural hazard losses.

Hazard losses (economic) in the U.S. are escalating (Fig. 3a) and weather-related losses account for most of them. It is equally clear from the timeline, that Hurricane Katrina was (and remains) an unprecedented event in terms of dollars lost. The large spike in 1994 represents the losses from the Northridge earthquake. The pattern for fatalities (Fig. 3b) shows the opposite trend, an overall declining pattern, but one with some

Focusing on the Environment and Human Security Nexus

Juan Carlos Villagrán de León and Janos J. Bogardi

Abstract In recent years, UNDP, UN-ISDR, Munich-Re and other institutions have been pointing out the fact that the number of reported disasters as well as the economic losses associated with such disasters have been growing steadily in recent decades. But while in developed countries risk-reduction and risk-transfer mechanisms such as insurance allow citizens to cope with such disasters, the persistence of disasters in developing countries manifests existing incapacities to cope with such events and their impacts. In this context, UNU-EHS is taking a detailed look at the long-range implications of such a trend, particularly highlighting the issue of environmental migration triggered by environmental degradation and disasters.

In the scope of the Hyogo Framework of Action (HFA), research activities carried out within UNU-EHS on topics of vulnerability and risk assessment and early warning; as well as complementary activities targeting education and capacity building, may help visualize the link between the Main Theme and the HFA.

In concordance with the mission statement of the Institute: "Advancing human security through knowledge-based approaches to reduce vulnerability and environmental risks"; this paper reviews some of the research carried out by UNU-EHS which addresses key research questions like vulnerability assessment in case of tsunamis and floods in Europe and the Indian Ocean; institutional efforts at local, national, and regional levels on early warning and preparedness; the impacts of urbanization on the modification of hazards such as floods and landslides in capital cities of Latin America; and perceptions which may promote or inhibit the establishment of policies and plans to manage existing risks. Such examples should serve to visualize the links between the Main Theme: Minimizing Risk: Maximizing Awareness, and the four key research questions.

The paper concludes with a brief outlook concerning the current dilemmas and critical issues which need to be addressed in the context of human security in a changing environment, as well as the supporting role which UNU-EHS may have in researching such critical issues.

Keywords Risk-transfer mechanisms · Risk assessment

Introduction

In recent years, the United Nations Development Programme (UNDP 2004), the International Strategy for Disaster Reduction of the United Nations (ISDR 2004); Munich-Re (2003); and the Center for Research on the Epidemiology of Disasters, CRED (Guha-Sapir et al. 2004) have all pointed out the increase in magnitude and frequency of disasters that are taking place throughout the world. The main conclusions to draw when reviewing the data presented in these reports are:

- Extreme environmental events such as droughts and hurricanes affect more people in developing nations than in developed nations.

J.C.Villagrán de León (✉)
Institute for Environment and Human Security – UNU-EHS, United Nations University, Bonn, Germany
e-mail: bogardi@ehs.unu.edu; villagran@ehs.unu.edu

T. Beer (ed.), *Geophysical Hazards*, International Year of Planet Earth,
DOI 10.1007/978-90-481-3236-2_3,

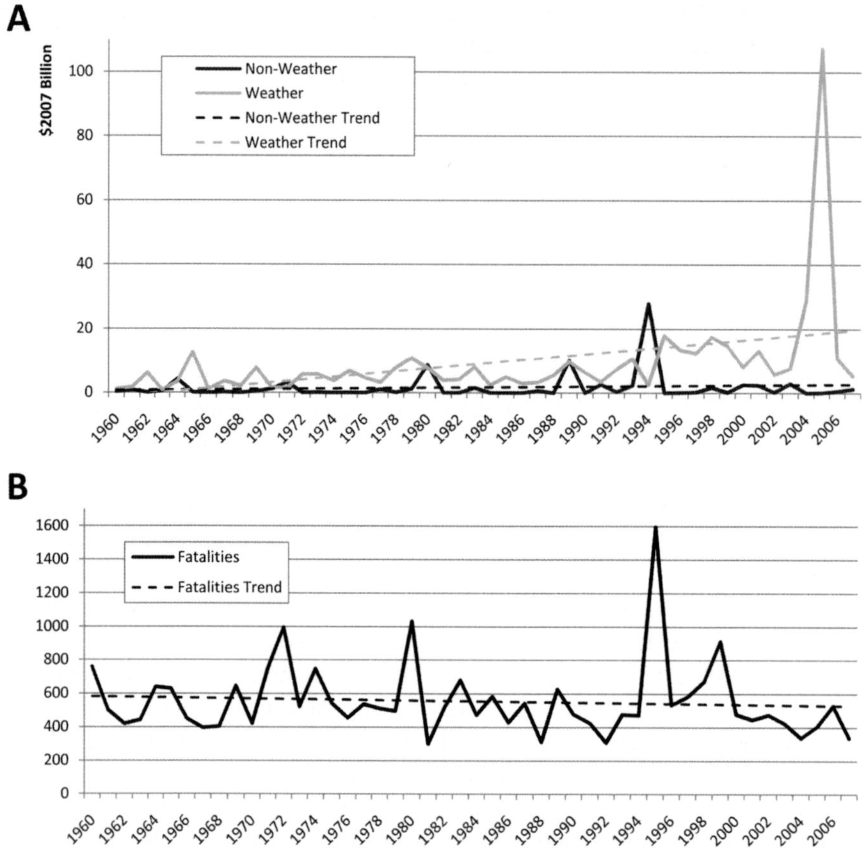

Fig. 3 Temporal patterns of losses in the United States, 1960–2007 for (**a**) property and crops damages (in billions of $2007), and (**b**) human fatalities. Source: Spatial Hazards Event Loss Database for the United States (http://sheldus.org)

periodic spikes, most notably the Chicago heat wave (1995). Other fatality peaks include the 1972 heat wave in Baltimore and flooding fatalities in the Rapid City flash flood event in South Dakota. The 1980 peak has no singular event, but rather represents a multitude of events. The leading cause of property and crop damage in the U.S. is tropical storms and hurricanes, followed by severe weather, flooding, and geophysical events (earthquakes) (Fig. 4a). For mortality, the leading hazard causes are severe summer weather (including lightning strikes), heat, and winter weather (Fig. 4b).

Spatial trends in economic losses clearly illustrate the impact of tropical storms and hurricanes along the U.S. Gulf and Atlantic Coasts (Fig. 5). The greatest losses appear in coastal counties and in selected regions in the west – notably southern California (earthquakes and wildfires); Northern California (earthquakes); Washington state (earthquakes, flooding, and volcanic eruptions), and in Idaho (wildfire). When taking the economic losses and aggregating by causal agent for each state a hazard profile can be developed. This state-level profile (Fig. 6) graphically illustrates the leading cause of losses. As can be seen, geophysical sources dominant in the western US, while severe weather dominates in the Great Plains and Midwest. Flooding is a ubiquitous hazard, and dominates in many eastern states. Finally, the US hurricane-prone states in the Southeast and along the Gulf of Mexico are clearly visible on the map. The geographic representation of such a hazard profile (based on actual losses) is a first step in the development of targeted mitigation strategies, designed to address the most costly hazard for that place, be it a county or a state.

Understanding the distribution of deaths and economic damages, globally and locally, is the first step towards building disaster resilient communities. The need for baselines against which to evaluate risk reduction options is the first step. Yet we need more than just

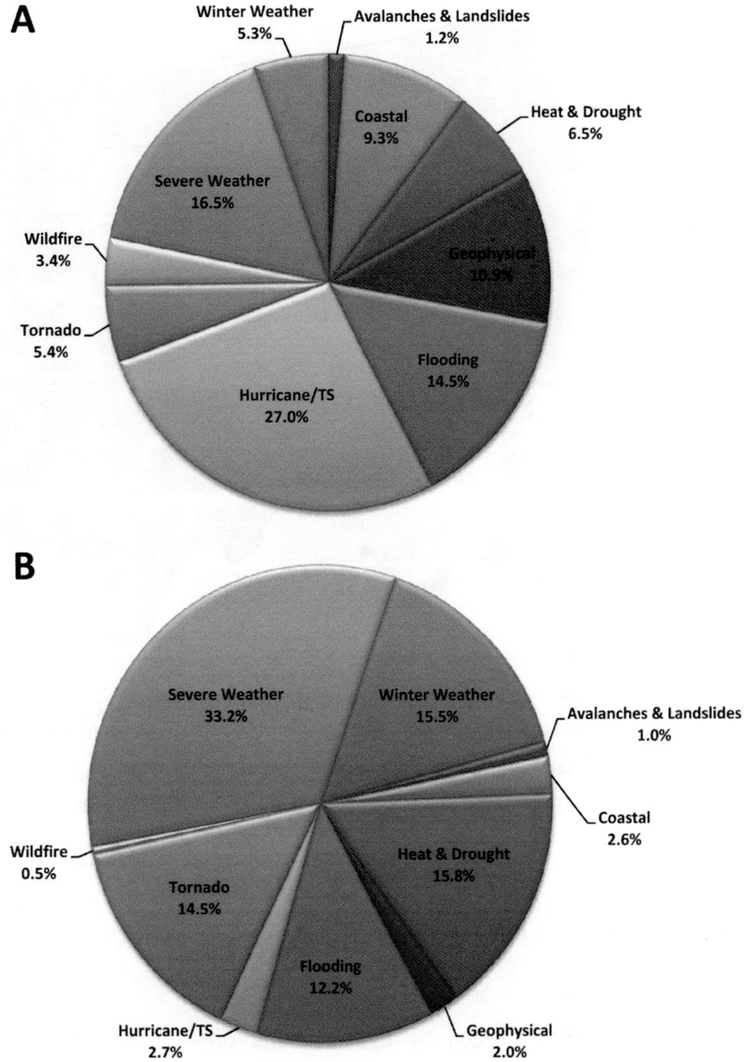

Fig. 4 Loss-causing hazards in the United States, 1960–2007 by type. The proportion of losses attributed to each hazard type for (**a**) $2007 reported damage of property and crops, and (**b**) human fatalities. Source: Spatial Hazards Event Loss Database for the United States (http://sheldus.org)

event-driven baseline data for the increasing trend and pattern of losses (economic and human life) is likely a function of two different phenomena. The first is the increasing value and density of property in harm's way (or exposure), which is driving the escalating losses. The second is the increasing vulnerability of the population that could partially explain injuries and deaths.

Populations at Risk

To assess the populations at risk from natural hazards requires not only estimates of the number of people potentially affected, but also those characteristics of the population that contribute to the social burdens of risk. The latter is often termed "social vulnerability".

For many world regions, the timely response to and delivery of disaster relief is challenging, and complicated by the lack of data about people in need of assistance. It is not uncommon for relief teams to be deployed to a disaster area without full knowledge of how many people will need aid or where they're located relative to the impact area, let alone have information on age and gender – characteristics that are vital in the delivery of food, water, shelter, and other assistance needs. As noted by a US National Research Council report, "How many people, their characteristics, and where they are constitute

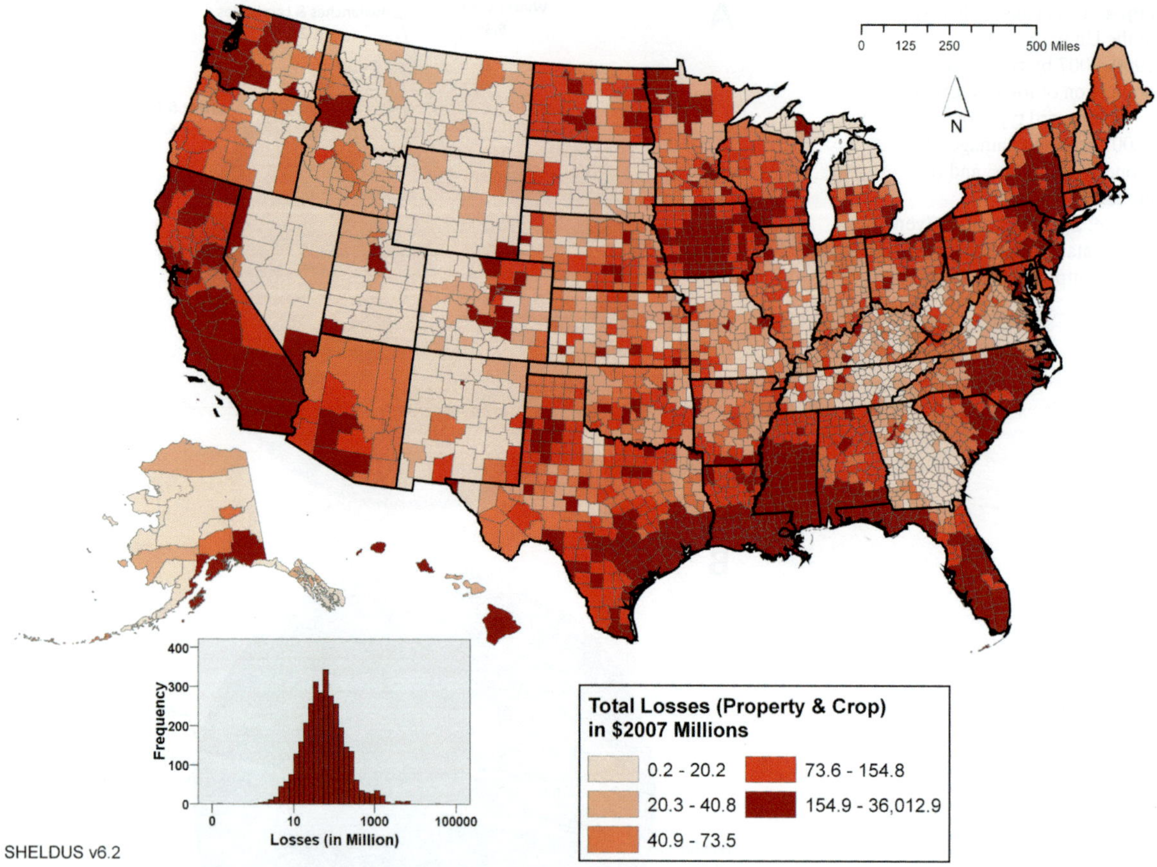

Fig. 5 Geographic variability in the pattern of hazard losses (property and crop) in the United States, 1960–2007. Source: Spatial Hazards Event Loss Database for the United States (http://sheldus.org)

critical information needed by agencies and organizations charged with disaster response. Inaccurate numbers and locations for populations can slow the relief effort and literally mean the difference between life and death (NRC 2007: 16)". National census of population data provide the backbone for estimating populations at risk and in planning for risk management and risk reduction policies. To be useful, such data must be collected at sub-national levels and include demographic characteristics such as age, gender, race or ethnicity, and economic well-being. Such geo-referenced data must be collected at periodic intervals to insure the most accurate and timely data are available. More than 85% of the world's population has been enumerated within a national census since 2000 (NRC 2007). Unfortunately, population growth and migration can render such censuses obsolete in just a few years. Furthermore, censuses are normally conducted for where people live (at night), not necessarily, where they work or go to school (daytime). If the disaster occurs during the day, the census counts may severely underestimate the likely affected population.

A range of methods is employed to improve such estimations of populations at risk. Remote sensing imagery can assist in determining settlement patterns and housing (Lo 2006), but is ineffective in determining how many people actually live in each structure. Global population databases such as the Gridded Population of the World (GPW) product from the CIESIN (http://sedac.ciesin.columbia.edu/gpw) (Balk et al. 2006), and Oak Ridge National Laboratory's Landscan (Dobson 2007) help to model the distribution and density of populations, thus providing an approximation of the sub-national at risk population.

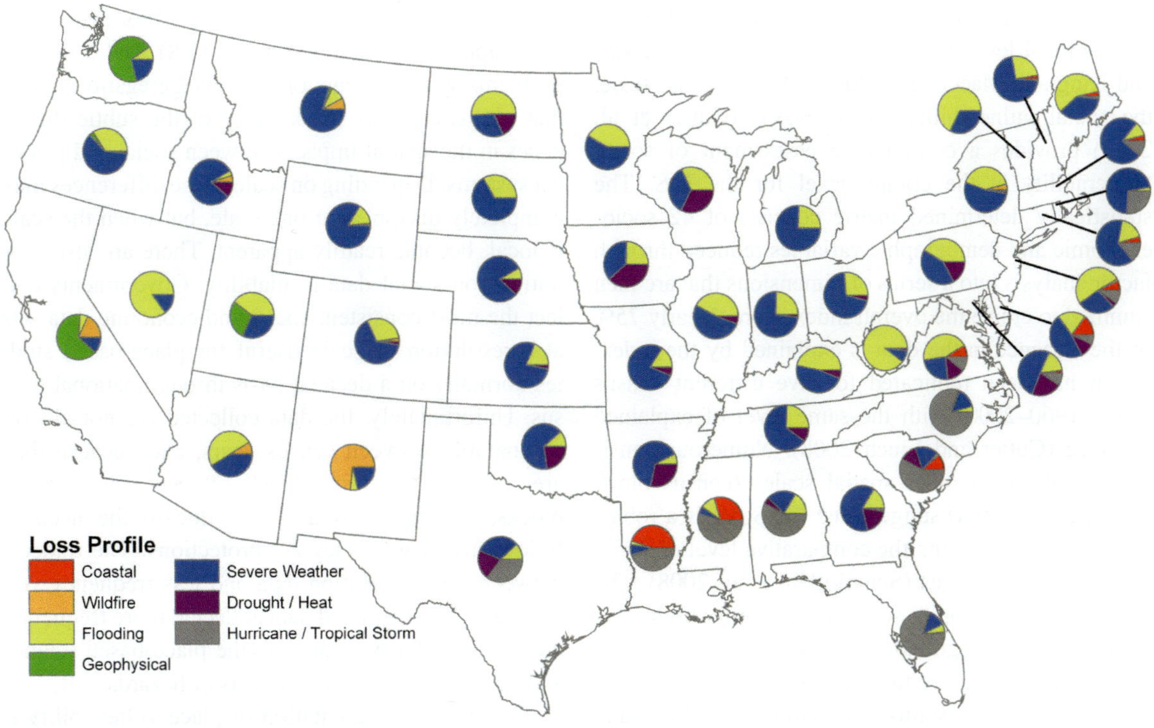

Fig. 6 Individual states hazard profile.
Source: Spatial Hazards Event Loss Database for the United States (http://sheldus.org)

Social Vulnerability

Social vulnerability describes those pre-existing characteristics of groups or conditions within communities that make them more susceptible to the impacts of hazards, in other words, those factors that shape the social burdens of risk. Social vulnerability also influences the uneven capacity of individuals, groups, or communities to prepare for, respond to, and recover from disasters. There is a wealth of empirical data (much of it based on post-event field studies of affected populations conducted over the past half century) (National Research Council 2006), which help us understand which segments of society or sections of the community are more susceptible to disaster impacts than others. For example, age is one such characteristic. Age not only affects the mobility to get out of harm's way without some additional assistance, but it also entails the need for special care for both children and elderly, especially if they are infirmed or frail. Thus, both ends of the age spectrum, the very old and the very young, tend to increase social vulnerability in communities where they constitute a significant proportion of the population (Heinz Center 2002). Another example is socioeconomic status. Wealth influences the ability to absorb losses and recover from disasters. Wealthier groups and communities have greater assets but they also have insurance and other financial reserves to absorb the loss, enabling faster recovery after the disaster. Poorer subpopulations and communities have less material goods to lose, but the impacts of any loss are disproportionally greater, and the ability to recover compromised. Thus, poorer communities are more socially vulnerable than wealthier places. The impact of Hurricane Katrina on the poor residents of New Orleans illustrates the differential social vulnerability and the role of inequality in disasters (Laska and Morrow 2006; Brunsma et al. 2007). A third example is gender, long recognized as influencing vulnerability. Gender-specific employment, lower wages, lower status, and women's roles as caregivers combine to create disproportionate impacts from hazards and disasters on women (Enarson and Morrow 1998).

There is considerable interest in the development of robust metrics to measure the multidimensional concept of social vulnerability. The development of social vulnerability indicators has progressed at the international level with the work of Birkmann (2006) and King and MacGregor (2000). At a national scale, the social vulnerability index (SoVI) (Cutter et al. 2003) provides a comparative assessment of social vulnerability at the county level for the U.S. The statistically determined metric consists of 42 socioeconomic and demographic variables reduced through factor analysis into a series of dimensions that are then summed to create the overall index score. Nearly 75% of the variance in the data is explained by the index, which has been replicated for five different census dates (1960–2000) with the same level of explained variance (Cutter and Finch 2008). Numerous sensitivity analyses at other spatial scales (census block group, census tract) suggest that the SoVI is a robust algorithm for assessing the comparative level of social vulnerability of places (Schmidtlein et al. 2008). The geographic depiction of social vulnerability highlights the clustering of high vulnerability counties where the driving factors that produce such vulnerability include low socioeconomic status, age extremes (children and elderly), and higher levels of density in the built environment. It is worth noting that Orleans parish (where New Orleans is located) was among the most socially vulnerable counties in the US and thus, it came as no surprise the enormity of the impact on that population (Cutter and Emrich 2006). The SoVI has wide applicability ranging from research and replication, to applied emergency management practice as a component of disaster mitigation plans (http://sovius.org).

Place-Based Science

People and the communities where they live and work are integral parts of the natural hazards system as are the physical systems (Haque and Etkin 2007). The mechanism for integration of physical processes and human systems is through the suite of geospatial tools to describe such places. There are many considerations in understanding the nexus of physical and social vulnerability – some are unique to the geosciences, others are limited to the social sciences. In both perspectives the role and impact of scale is critical. Physical features are generally point or line patterns, while some census enumeration unit (or polygon) collects social information. Reconciling the overlapping geometries is an important function, thus the need for and use of Geographic Information Systems (GIS). There are also problems with aggregation and disaggregation biases that ultimately may mask many of the subtle differences in the spatial impacts between social and physical systems. Depending on scale, these differences may completely disappear at one scale, but when the scale is local, become readily apparent. There are also limitations on social data availability. Governments collect the most consistent social and economic data sets at a resolution or scale useful for place-based studies normally on a decadal basis through national census. Unfortunately, the data collected are not always comparable between census years, and because they are collected at various levels of geography (census blocks, counties, postal codes) due to the need to follow stringent privacy act protections. The tradeoff between finer resolution data and less frequency versus more coarse grained data collected more frequently dictates how far we can push the place-based science and understanding of the impacts of hazards.

The spatial representation of place vulnerability is a powerful heuristic for assessing the likely impact of hazards on society. For example, a study examining coastal erosion vulnerability in the U.S. integrated physical indicators (sea level rise, slope, mean wave height, erosion/accretion rate) and social vulnerability and found that the physical parameters explained most of the variability in vulnerability (Boruff et al. 2005). However, when examining the regional trends (Atlantic Coast, Gulf Coast, and Pacific Coast) different stories emerge for each geographic area. For the Atlantic and Pacific coasts, physical parameters explained more of the geography of vulnerability, while for the Gulf Coast social vulnerability was a more significant driver, especially age (elderly), and high birth rates. The physical hazards can be delineated using composite indices as noted above, in-situ measurements (1% chance flood zone), or modeled output such as hurricane storm surge inundation zones or peak ground acceleration. When overlain with social vulnerability indicators, the geography of vulnerability becomes readily apparent (Fig. 7).

There is now a solid body of research on how the vulnerabilities based in physical systems interact with social conditions to produce hazard vulnerability. Much of this research uses a single threat source such as drought (Polsky 2004); earthquakes and tsunamis (Rashed and Weeks 2003; Wood and Good 2004;

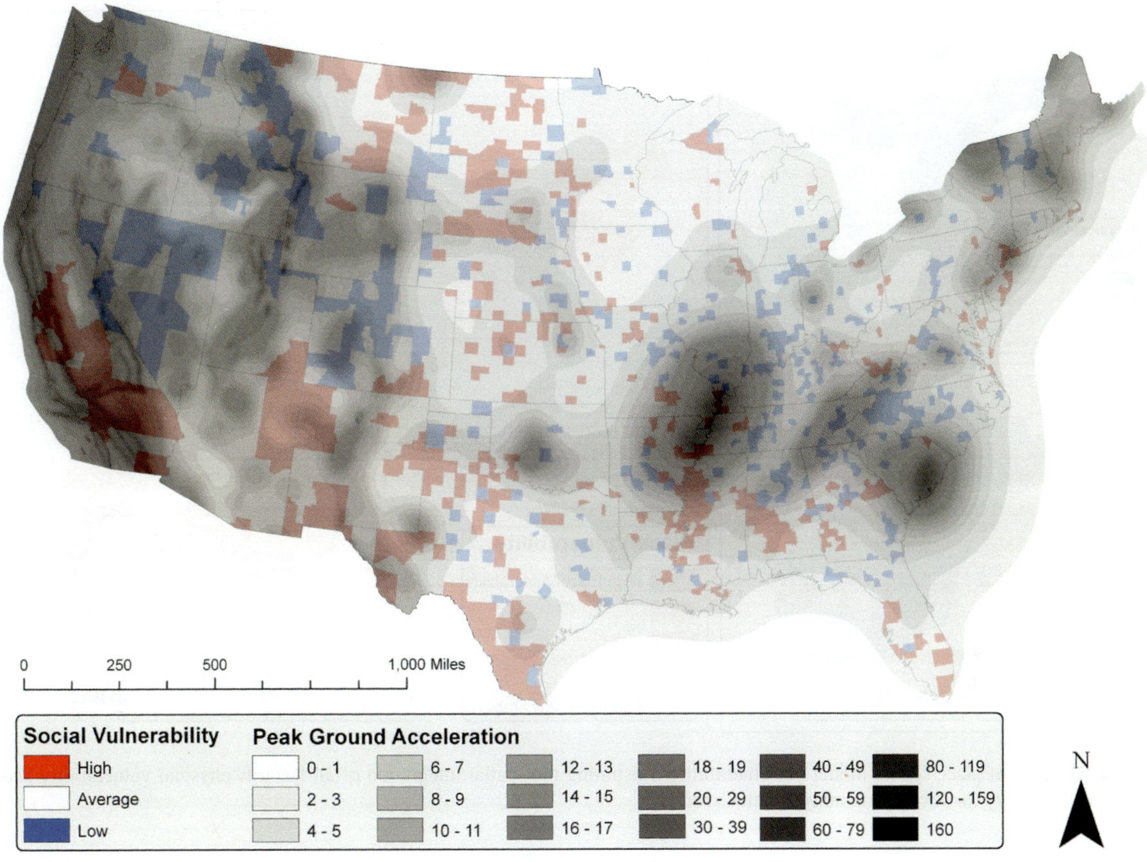

Fig. 7 Intersection of social vulnerability and modeled peak ground acceleration for the U.S.

Wood et al. 2009); sea level rise (Wu et al. 2002), hurricanes (Chakraborty et al. 2005), and levee failures (Burton and Cutter 2008).

The development of multi-hazard vulnerability assessments poses a different set of challenges in both the representation of the hazards as well as the social conditions. The different hazard zones (and geometries) must be represented as accurately as possible, which means there are often methodological and scientific questions regarding such generalizations. At the same time, there are issues in the social and demographic data in terms of quality and availability as noted earlier. For example, Fig. 8 provides a place-base assessment for Richland County in South Carolina integrating both social and physical vulnerability indicators. The circular features are protective action distances from chemical facilities, while the linear features are rail and highways that potentially carry hazardous materials. The social vulnerability is greatest in the center of the county, which has a significant transient population (a major military base), and an institutionalized population (prison). Originally developed at the county level of geography (Cutter et al. 2000), this GIS-based approach to hazard vulnerability assessment has also been implemented in small island nations (Boruff and Cutter 2007).

Vulnerability Indices: Strengths and Weaknesses

Vulnerability indices are useful constructs for developing a rough assessment of the distribution and likely impact of hazards and disasters. Their utility is exploratory and diagnostic in nature, enabling the researcher or policy maker to understand the underlying drivers of vulnerability and differences between places or among social groups (Smit and Wandel 2006). Out of necessity, indices are simplifications, in the same way that models are simplifications of

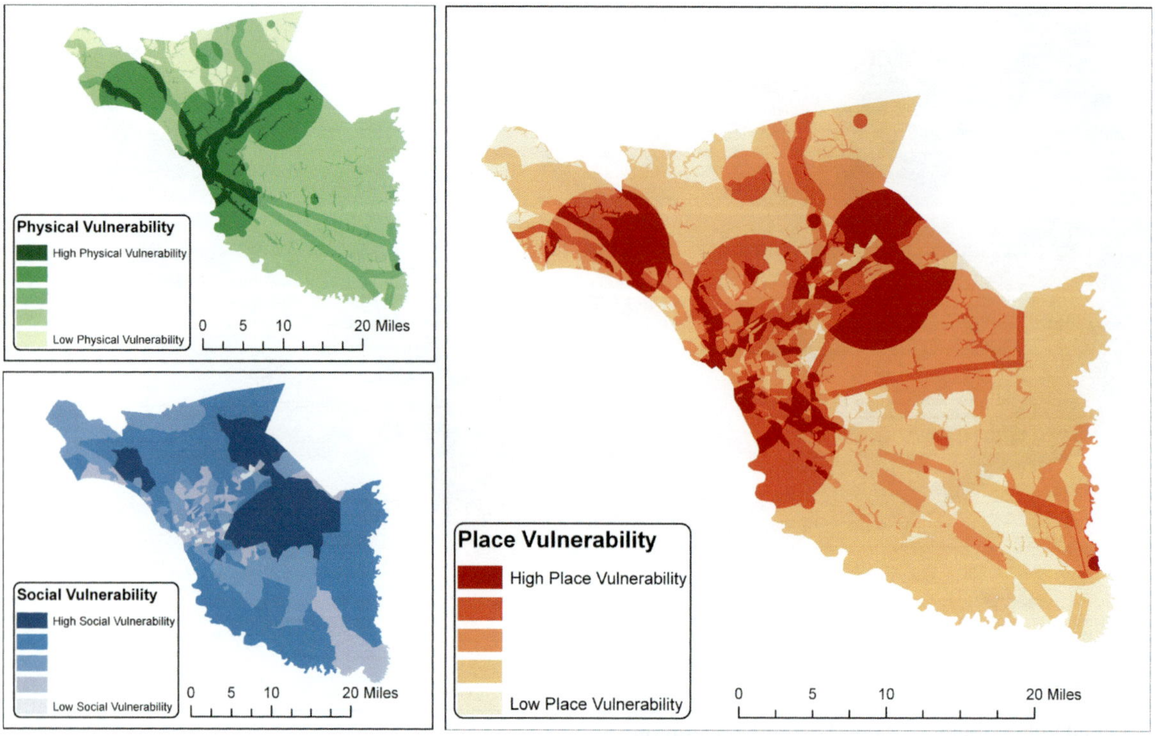

Fig. 8 Hazards of place spatial model of vulnerability illustrating the spatial integration of all hazards physical vulnerability and social vulnerability for Richland County, South Carolina

real-world processes and interactions. While there is no such thing as a perfect vulnerability index (or a sustainability index for that matter), they are useful in benchmarking or establishing baseline conditions, and tracking changes over time and across space.

At present, most of the vulnerability metrics are descriptive, not predictive indices and are mainly used as a representation of multi-dimensional phenomena, such as those pre-existing conditions in communities that make them susceptible to harm. The problem with validation still hampers the utility of vulnerability indices for there is still no good outcome measure. Dollar losses or mortality from natural hazards are inadequate, but temporal trends in population (including out and in-migration), employment, or households in need of assistance may provide more relevancy for linking pre-existing vulnerability to outcomes.

The second major issue with vulnerability indices is their application and the unit of analysis. Considerable attention focused on country-level indicators of environmental vulnerability as a companion to the UN's Human Development Index (equally fraught with many of the same of scale, adequacy of measurement, validation). This top-down approach (Barnett et al. 2008; Birkmann 2006) does not capture the intra-country variability in vulnerability or the nature of the local impacts, and as Barnett et al. correctly state. The bottom-up approach (Pelling 2003; O'Brien et al. 2004) provides the local case study (normally qualitative) but often fails to situate the local within a meso-scale analysis thus limiting comparisons across different spatial units using the same metrics. Some combination of the two is perhaps the best solution – specific enough to impart knowledge of local impacts and processes; yet measured with standardized variables that are readily transportable and comparable across different spatial units.

Applications to Policy at Home and Abroad

The differential exposure of places and people to geo-hazards results in uneven impacts and clearly influences disparities in recovery. There are many examples of such disparities in the societal impacts, not only in the U.S. but globally as well. To achieve substantial reductions in disaster losses (lives lost, social, economic and environmental assets) the World Conference on Disaster Reduction and its Hyogo Framework for Action (ISDR 2005), determined five priority actions. These include:

- Making risk reduction a national and local priority
- Identify, assess, and monitor risks
- Build a culture of safety and resilience
- Reduce the underlying risk factors
- Strengthen disaster preparedness

We cannot develop risk reduction strategies or policies in the absence of baseline geospatial and economic data on hazard events and losses. The enhancement of existing global and national databases such as EM-DAT or SHELDUS are critical to our understanding of the distribution of losses and the selective targeting of regions or places for immediate risk reduction strategies. Further, there is a critical need to consider and incorporate spatial and social inequities in risk and vulnerability and include efforts to address such disparities in risk reduction strategies. Before we can reduce the underlying risk factors and strengthen disaster preparedness, we need to have consistent information on the populations at risk and their social vulnerability. Finally, it is clear that the drivers of community vulnerability to hazards are a product of the interactions between physical systems and human systems. The development of social vulnerability metrics described in this chapter and the application of place-based hazard and social vulnerability analyses is the step in the right direction. However, more work is needed to enhance the integration between physical models and social processes – research and application that can only be achieved through multi-disciplinary research and practice where the social sciences have a key role.

References

Balk, D. L., U. Deichmann, G. Yetman, F. Pozzi, S. I. Hay, and A. Nelson, 2006. Determining the global population distribution: methods, applications and data. *Advances in Parasitology* 62: 119–156.

Barnett, J., S. Lambert, and I. Fry, 2008. The hazards of indicators: insights from the environmental vulnerability index. *Annals of the Association of American Geographers* 98 (1): 102–119.

Beer, T., P. Bobrowsky, P. Canuti, S. Cutter, and S. Marsh, 2004. *Hazards – minimising risk, maximizing awareness*. Leiden, The Netherlands: Earth Sciences for Society Foundation, International Year of Planet Earth. Online at http://yearofplanetearth.org

Birkmann, J. (ed.), 2006. *Measuring vulnerability to natural hazards: towards disaster resilient societies*. Tokyo: United Nations University Press.

Boruff, B. J. and S. L. Cutter, 2007. The environmental vulnerability of Caribbean island nations. *Geographical Review* 97 (1): 24–45.

Boruff, B. J., C. Emrich, and S. L. Cutter, 2005. Hazard vulnerability of U.S. coastal counties. *Journal of Coastal Research* 21 (5): 932–942.

Brunsma, D. L., D. Overfelt, and J. S. Picou, 2007. *The sociology of Katrina: perspectives on a modern catastrophe*. Latham, MD: Rowman & Littlefield.

Burton, C. and S. L. Cutter, 2008. Levee failures and social vulnerability in the Sacramento-San Joaquin Delta area, California. *Natural Hazards Review* 9 (3): 136–149.

Burton, I., R. W. Kates, and G. F. White, 1993. *The environment as hazard* (2nd Edition). New York: Guildford Press.

Chakraborty, J., G. A. Tobin, and B. E. Montz, 2005. Population evacuation: assessing spatial variability in geophysical risk and social vulnerability to natural hazards. *Natural Hazards Review* 6 (1): 23–33.

Colten, C. E., 2005. *An unnatural metropolis: wresting New Orleans from nature*. Baton Rouge: Louisiana State University.

Cutter, S. L., 2003. The science of vulnerability and the vulnerability of science. *Annals of the Association of American Geographers* 93 (1): 1–12.

Cutter, S. L., B. J. Boruff, and W. L. Shirley, 2003. Social vulnerability to environmental hazards. *Social Science Quarterly* 84 (1): 242–261.

Cutter, S. L. and C. T. Emrich, 2006. Moral hazard, social catastrophe: the changing face of vulnerability along the hurricane coasts. *Annals of the American Academy of Political and Social Science* 604 (1): 102–112.

Cutter, S. L. and C. Finch, 2008. Temporal and spatial changes in social vulnerability to natural hazards. *Proceedings of the National Academy of Sciences*: 105(7): 2301–2306.

Cutter, S. L., J. T. Mitchell, and M. S. Scott, 2000. Revealing the vulnerability of people and places: a case study of Georgetown County, South Carolina. *Annals of the Association of American Geographers* 90 (4): 713–737.

Dilley, M., R. S. Chen, U. Deichmann, A. L. Lerner-Lam, M. Arnold with J. Agwe, P. Buys, O. Kjekstad, B. Lyon, G. Yetman, 2005. *Natural disaster hotspots: a global risk analysis*. Washington D.C.: Hazard Management Unit, World Bank.

Dobson, J. E., 2007. In harm's way: estimating populations at risk. In National Research Council, *Tools and methods for estimating populations at risk from natural disasters and complex humanitarian crises* (pp. 183–191). Washington D.C.: National Academies Press.

Enarson, E. and B. H. Morrow, 1998. *The gendered terrain of disaster: through women's eyes*. Westport, CT: Praeger.

Gall, M., K. A. Borden, and S. L. Cutter, 2009. When do losses count? Six fallacies of natural hazard loss data. *Bulletin of the American Meteorological Society* 90 (6): 799–809.

Haque, C. and D. Etkin, 2007. People and community as constituent parts of hazards: the significance of societal dimensions in hazards analysis. *Natural Hazards* 41: 271–282.

Heinz Center, 2002. *Human links to coastal disasters*. Washington D.C.: The H. John Heinz III Center for Science, Economics, and the Environment.

International Federation of Red Cross and Red Crescent Societies, 2008. *World disasters report 2008*. Available online: http://www.ifrc.org/publicat/wdr2008/index.asp?navid=09_03, Accessed 29 October 2008.

International Strategy for Disaster Reduction, 2004. *Living with risk*. New York and Geneva: The United Nations.

International Strategy for Disaster Reduction, 2005. World Conference on Disaster Reduction, Hyogo Framework for Action 2005–2015. Available online: http://www.unisdr.org/eng/hfa/docs/Hyogo-framework-for-action-english.pdf. Accessed 30 October 2008.

Kates, R. W., C. E. Colten, S. Laska, and S. P. Leatherman, 2006. Reconstruction of New Orleans after Hurricane Katrina: a research perspective. *Proceedings of the National Academy of Sciences* 103 (40): 14653–14660.

King, D. and C. MacGregor, 2000. Using social indicators to measure community vulnerability to natural hazards. *Australian Journal of Emergency Management* 15 (3): 52–57.

Laska, S. and B. H. Morrow, 2006. Social vulnerabilities and Hurricane Katrina: an unnatural disaster in New Orleans. *Marine Technology Society Journal* 40 (4): 16–26.

Lo, C. P. 2006. Estimating population and census data. In M. K. Ridd and J. Hipple (eds.) *Remote sensing of human settlements* (pp. 337–378). Bethesda, MD: American Society for Photogrammetry and Remote Sensing.

Mileti, D. S., 1999. *Disasters by design: a reassessment of natural hazards in the United States*. Washington D.C.: Joseph Henry Press.

National Research Council, 1999. *The impacts of natural disasters: a framework for loss estimation*. Washington D.C.: National Academies Press.

National Research Council, 2006. *Facing hazards and disasters: understanding human dimensions*. Washington D.C.: National Academies Press.

National Research Council (NRC), 2007. *Tools and methods for estimating populations at risk from natural disasters and complex humanitarian crises*. Washington D.C.: National Academies Press.

O'Brien, K., R. Leichenko, U. Kelkar, H. Venema, G. Aandahl, H. Tompkins, A. Javed, S. Bhadwal, S. Barg, L. Nygaard, J. West, 2004. Mapping vulnerability to multiple stressors: climate change and globalization in India. *Global Environmental Change* 14 (4): 303–313.

Pelling, M., 2003. *The vulnerability of cities: natural disasters and social resilience*. London and Sterling, VA: Earthscan Publications.

Polsky, C., 2004. Putting space and time in Ricardian climate change impact studies: the case of agriculture in the U.S. Great Plains. *Annals of the Association of American Geographers* 94 (3): 549–564.

Rashed, T. and J. Weeks, 2003. Assessing vulnerability to earthquake hazrds through spatial multicriteria analysis of urban areas. *International Journal of Geographic Information Science* 17 (6): 547–576.

Reisner, M., 2004. *A dangerous place: California's unsettling fate*. New York: Penguin.

Schmidtlein, M. C., R. Deutsch, W. W. Piegorsch, and S. L. Cutter, 2008. A sensitivity analysis of the social vulnerability index. *Risk Analysis* 28 (4): 1099–1114.

Smit, B. and J. Wandel, 2006. Adaptation, adaptive capacity and vulnerability. *Global Environmental Change* 16 (3): 282–292.

Wood, N. J., C. G. Burton, and S. L. Cutter, 2009. Community variations in social vulnerability to Cascadia-related tsunamis in the U.S. Pacific Northwest. *Natural Hazards*, DOI 10.1007/s11069-009-9376-1. Published online: 26 March 2009.

Wood, N. J. and J. W. Good, 2004. Vulnerability of port and harbor communities to earthquake and tsunami hazards: the use of GIS in community hazard planning. *Coastal Management* 32 (3): 243–269.

Wu, S. Y., B. Yarnal, and A. Fisher, 2002. Vulnerability of coastal communities to sea level rise: a case study of Cape May, New Jersey, USA. *Climate Research* 22: 255–270.

- Economic losses associated with extreme environmental events are greater in developed nations than in developing nations.
- Losses, when expressed in relation to GDP, are greater in developing nations than in developed nations. Such losses increase the level of poverty in these nations, and inhibit efforts related to sustainable development.

In the case of developing countries, these disasters are eroding the capacity of governments and communities to recover from such hazard events, are highlighting the greater need for humanitarian assistance, and are impacting at the core of human security. In December 2003, UNU-EHS was established in Bonn, Germany, with the purpose of *Advancing human security through knowledge-based approaches to reducing vulnerability and environmental risks*. As stated in the Strategic Directions of UNU-EHS (UNU-EHS 2005), the concept of human security focuses on threats that endanger the lives and livelihoods of individuals and communities. Threats to human insecurity have been grouped in four categories (UNU-EHS 2005):

- Rapid onset hazards of natural origin.
- Rapid onset hazards of anthropogenic origin.
- Creeping changes in the environment.
- Creeping changes in socio-economic systems.

Earthquakes, tsunamis, and flash floods could be considered in the category of rapid onset hazards of natural origin; while explosions, fires, and leaks associated with toxic chemicals could be grouped in the class of rapid onset hazards of anthropogenic origin. In this context, it is important to recognize the role of globalization regarding the transfer of dangerous technologies, insecure industrial operations, and also hazardous waste disposal and recycling, to regions where governance is less effective; there is less control and even less capacity available to deal with their negative consequences. Deforestation, soil erosion, increase in soil salinity, desertification, and climate change could be associated with creeping changes in the environment; while economic decline, loss of income, shifts of markets, declining commodity prices, poverty and the negative effects of globalization are examples of creeping changes in socio-economic systems.

Safeguarding and improving human security requires a new approach that will enable better understanding of many interrelated variables – social, political, institutional, economic, cultural, technological and environmental. Deterioration of these factors amplifies the impacts of environmental change and the consequences of extreme events when they occur.

Environmental Threats to Human Security

Many environmental events – rapid onset hazards, but also creeping processes – contribute to human insecurity. Environmental calamities such as earthquakes, volcanic eruptions, floods, and droughts, have always presented a threat to human existence, and their impact on humans has increased as people have felt compelled or have been forced to move into areas exposed to different hazards. The pace of human-induced environmental degradation and resource depletion (such as deforestation, desertification, land degradation, erosion, salinisation, siltation, and climate change) are often more gradual. Nevertheless, they have significantly increased in many regions due to a combination of population pressure, increasing demand for agricultural products, and improved technological means of exploitation. The slow introduction of conservation, control, and rehabilitation measures (if any), and the lack of public awareness and participatory decision-making further contribute to human insecurity.

These processes, which undermine human security, can be observed in many parts of the world. Their systematic assessment has important spatial and temporal dimensions. Undoubtedly, one of the major challenges of policy-relevant research is to define the appropriate scale and to match methodologies and recommendations with the prevailing governance structures and time scales.

The end of the cold war around 1990 seemed to allow the world to focus on existing and newly emerging security challenges. In particular, it was conducive for the re-conceptualization of the notion of security, progressing from the old notion of *national security* based on the sovereignty of states to the notion of *human security.* In 2005, the then Secretary General of the United Nations, Kofi Annan presented to the General Assembly of the United Nations the document entitled: *In Larger Freedom: towards develop-*

ment, security and human rights for all (UNGA 2005) to emphasize the fact that the relevance of the Charter of the United Nations must be to advance the lives of individuals. His notion of larger freedom also encapsulates the idea that development, security and human rights go hand in hand. Within this document, he introduced the notion of Human Security in the following terms:

- Freedom from fear.
- Freedom from want and
- Freedom to live in dignity.

Brauch (2005) has proposed the incorporation of the environmental dimension of human security by mainstreaming three pillars to the human security concept:

- Freedom from fear.
- Freedom from want.
- Freedom from hazard impact.

The third pillar makes the explicit connection with environmental hazards. This new view of human security puts the individual, its environment and livelihood at the center of the debate. Figure 1 presents this transition from the traditional view of national security to the innovative view of human security based on the three pillars as proposed by Brauch (2005) highlighting the dimensions of this type of insecurity, and linking this notion to the context of sustainable development.

Research Agenda of UNU-EHS

Two key concepts that capture the essence of human insecurity are *vulnerability* and *risk*. To this end, UNU-EHS was established with the mission to advance human security through knowledge-based approaches to reducing vulnerability and environmental risks. The research agenda within UNU-EHS focuses on:

- Investigating the relationship between risk, vulnerability and coping capacity, and devise strategies and measures to improve the coping capacity of affected communities.
- Contributing to the development, testing and verification of vulnerability indicators.
- Exploring the links between different hazard events and "creeping" processes such as climate change and land degradation.
- Fostering a better understanding of the forces and processes of environmental degradation and their influence on the magnitude and frequency of hazards and subsequent disasters.
- Promoting a better understanding of internal displacement and trans-boundary migration due to environmental triggers (swift and creeping hazards).

While being central topics of research within UNU-EHS, vulnerability, risk, and environmental migration

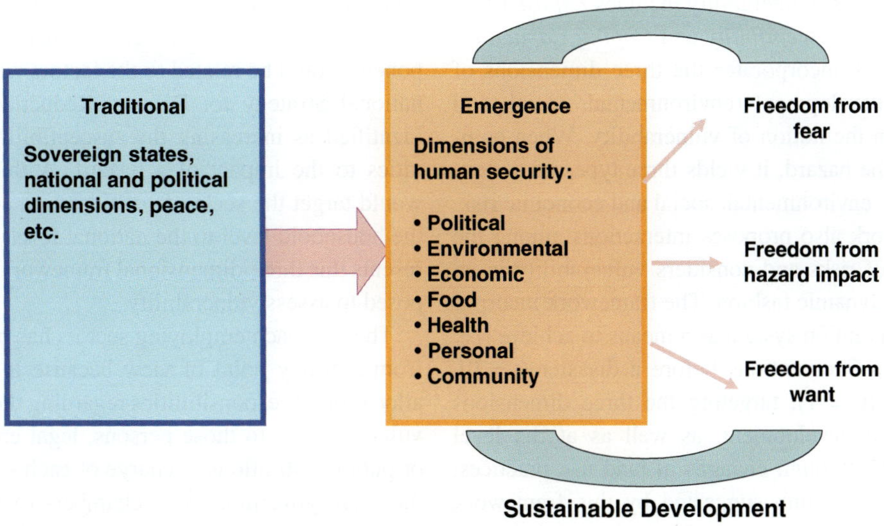

Fig. 1 Evolution of the concept of human security

remain a challenge to be addressed. First and foremost, there is no consensus on a single definition for these and other related terms nor a consensus on ways to assess them (Cardona 2001; Lavell 2003; Bogardi 2004). In an effort to systematize definitions and methods to assess vulnerability, research conducted by UNU-EHS concluded in three publications:

- A Comparative Glossary of Components of Risk in which Thywissen (2006) introduced the notion of a *Babelonian Confusion* when exploring the many definitions proposed by experts and practitioners from many institutions for a variety of terms related to disaster-risk.
- A conceptual and methodological review of notions concerning vulnerability and ways of assessing it (Villagrán de León 2006).
- A book entitled *Measuring Vulnerability to Natural Hazards* by Birkmann (2006) comprises articles concerning the basic principles and theoretical basis for vulnerability, a review of methods to assess vulnerability from the national to the local level, and elements concerning institutional vulnerability and coping capacities.

The research conducted by Birkmann (2006) and Bogardi (2004) led to the elaboration of the BBC Framework which puts vulnerability at the center of the disaster-risk cycle (Birkmann and Wisner 2006). In this framework, vulnerability includes *Exposed Elements* and their *Susceptibility* and *Coping Capacities*. The framework incorporates the three dimensions of sustainable development (environmental, social, and economic) in the notion of vulnerability. When combined with the hazard, it yields three types or dimensions of risk: environmental, social and economic risk. The framework also proposes interactions among the three types of risks and considers vulnerability in an integral and dynamic fashion. The framework incorporates an intervention system as a means to achieve risk reduction via interventions before a disaster ($t = 0$), and after it ($t = 1$); targeting the three dimensions or spheres of development, as well as at the level of the hazard, through changes in land use practices, for example. The link suggested by this framework between vulnerability and the three spheres of sustainable development emphasizes the needto understand the environment not only from the realm of hazards, but also as an essential basis of life thus specifying the link between nature and society. The BBC Framework is shown in Fig. 2. This framework has been tested in the case of tsunamis in Sri Lanka (Birkmann et al. 2007) and has been adopted by other research groups, as well. Table 1 presents a list of parameters that were employed to assess both vulnerability, as well as the intervention tools in this project in Sri Lanka. The BBC-model based on post-facts vulnerability assessment was implemented through a detailed questionnaire action.

Within UNU-EHS post-tsunami activities in Sri Lanka, a rapid assessment method was developed and tested in the coastal city of Galle. The method was developed through the systematization of observations of damages and destruction provoked by the 26 December 2004 tsunami. The outcomes from a such scenario assessment are intended to provide emergency managers and municipal authorities with information regarding potential fatalities, injuries, losses, as well as interruptions in services and life lines in case a tsunami impacts a given city. To this end, Villagran de León (2006, 2007) proposed a framework to decompose vulnerability assessment analyzing how disasters can impact the various socio-economic sectors which compose such a city; *typical sectors being: housing, communications, education, health, energy, government, industry, commerce, finance, transportation, public infrastructure, environment, tourism*, etc. In addition, the framework proposes differentiation within each sector in terms of five components: human, physical, functional, economic, and environmental. These components could be related to the factors which the International Strategy for Disaster Reduction, ISDR, has identified as increasing the susceptibility of communities to the impact of a hazard. A third dimension would target the scale of consideration spanning from the household level to the national level. Figure 3 represents this three-dimensional framework which is proposed to assess vulnerability.

The approach employing sectors has been proposed from a policy point of view because it promotes the allocation of responsibilities regarding the reduction of vulnerabilities to those persons, legal entities, private or public institutions in charge of each sector, whether these are governments or chambers of various kinds (chamber of commerce, industry, tourism, etc.) active on various political-administrative levels.

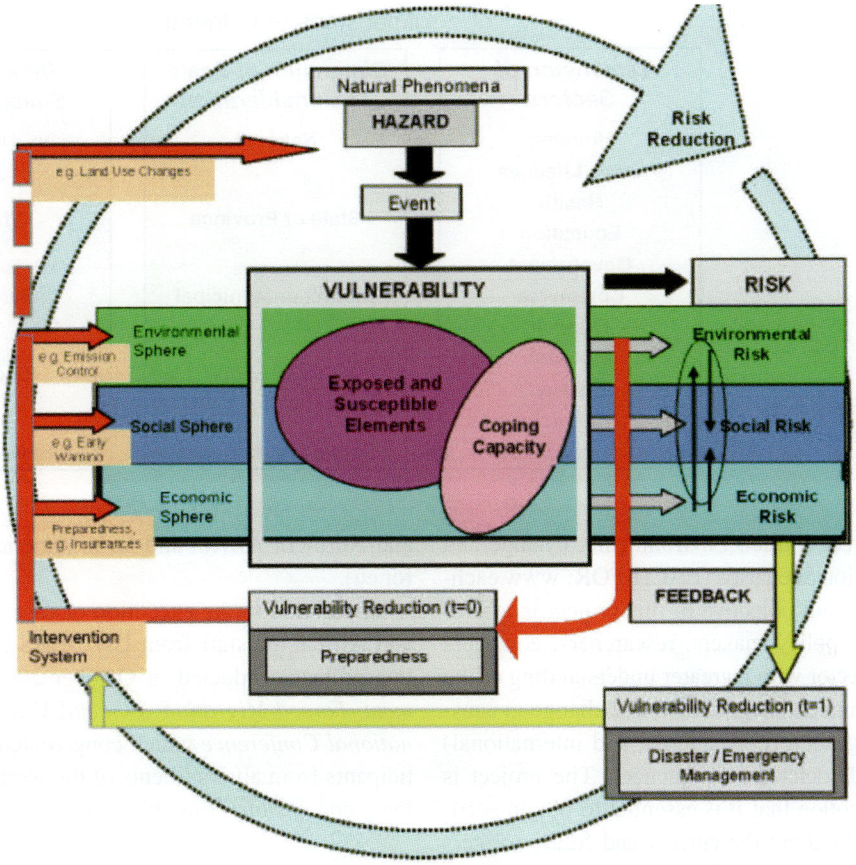

Fig. 2 The 'BBC' hazard-vulnerability-risk framework

Table 1 Parameters employed to assess vulnerability, coping capacity and intervention tools in the context of tsunamis in Galle, Sri Lanka

Vulnerability		Intervention tools
Susceptibility and degree of exposure	Coping capacity	
Impact of tsunami on household members and their assets.	Social networks.	Relocation of housing and infrastructure to inland.
Structure of the household.	Knowledge about coastal hazards and tsunami.	Early warning system.
Housing conditions and the impact of the tsunami.	Financial support from formal and informal organizations.	100 m "buffer zone" (implemented by the government).
Direct loss of possessions.	Access to information, e.g. radio.	

(Source: Birkmann et al. 2007)

Another core topic in the research agenda of UNU-EHS is the issue of environmental migration. As stated in Renaud et al. (2007), environmental degradation is a serious problem that can be exacerbated by social, economic, political and global environmental factors and could thus become one of the major "push" factors in the future leading to migration. In this context, UNU-EHS and other European institutions have been con-

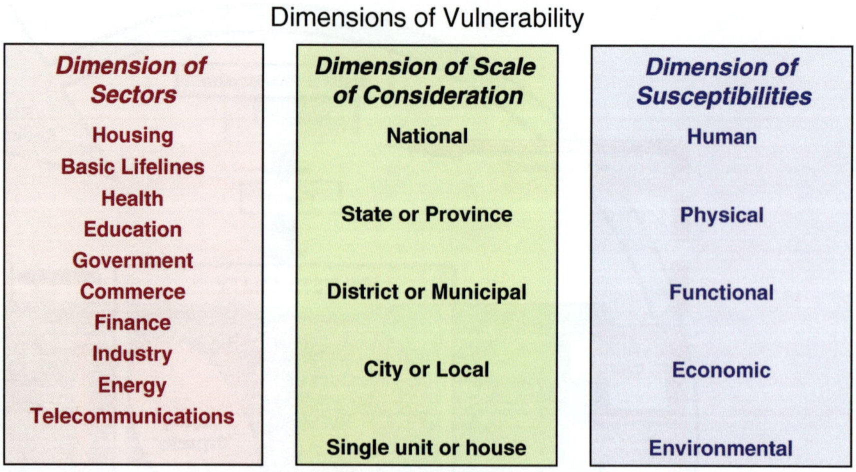

Fig. 3 Three dimensional framework for vulnerability assessment

ducting a project entitled Environmental Change and Forced Migration Scenarios (EACH-FOR; www.each-for.eu). The overall objective of this project is to provide European policy makers, researchers, educators and the civil sector with a greater understanding of the role of environmental degradation and change in causing forced migration (both internal and international) and its related societal consequences. The project is based on the notion that it is essential to obtain accurate information about the current and future triggers of forced migration in different countries of origin and, in particular, to gain an understanding of the role that environmental degradation plays in causing such migration. The EACH-FOR project aims to:

- Discover and describe in detail the causes of forced migration in relation to environmental degradation/change and their association with other social, political and economic phenomena in Europe and in the main "source" countries of migration.
- Provide plausible future scenarios of environmentally-induced forced migration.

These aims will be achieved by analysing direct (e.g. desertification) and indirect (e.g. conflicts) environmental effects on livelihoods, environmental degradation processes and migration patterns at a sub-regional or country level. Reports outlining the causes leading to forced migration, with a focus on environmental concerns and hot spots of environmental degradation and vulnerability are available for the following regions: Europe and Russia, NIS and Central Asia, Asia, Sub-Saharan Africa and Ghana, Middle East and Northern Africa, and Latin America (www.each-for.eu).

In addition to the execution of case studies in Asia and Africa, the staff from UNU-EHS associated with this project conducted in October 2008 the *Environment, Forced Migration & Social Vulnerability International Conference* which congregated over 300 participants from all continents of the world. The preliminary conclusions from this conference are:

- Migration in the future may be a part of adaption to environmental change including climate change, but more often migration is not adaptation, but rather the failure to adapt
- The poorest often cannot migrate; although these very poor groups would need to move to survive, migration may be impossible and only an act that the relatively well-off can afford
- Research needs to be carried out in a trans- and interdisciplinary way as there are many environmental, social and economic push and pull factors at play.
- It is possible to attribute single causality factors in migration, particularly when dealing with some types of rapid onset hazards. However, for slow deteriorating environments the livelihoods of the people are affected and influenced by many other factors.
- A rapid succession of events, such as recurrent flooding or drought can reach a social tipping point whereby the decision to migrate is taken. However, tipping points are more within individuals as a hidden social vulnerability dimension.

- People are attached to their original place of living and generally would prefer not to migrate. Improving livelihoods locally is a good development strategy.

In the context of deltas, UNU-EHS is conducting with other institutions the WISDOM project. The acronym stands for *Water-related Information System for the Sustainable Development of the Mekong Delta in Vietnam*. The main objective of WISDOM is to overcome gaps in the water-related information flow among research institutions of different disciplines and between these information generating agencies and decision making authorities. The specific objectives of this project are:

- The efficient collection, management, analysis and distribution of existing and newly generated data.
- The enhancement of information flow between research institutes, universities, state agencies and local authorities.
- The improvement in the understanding of the complex ecologic, hydrologic, economic and social relationships in the Delta.
- The optimisation of flood forecasting and warning systems.

Within the WISDOM project, UNU-EHS carries out research on water pollution and on vulnerability assessment in the context of floods and droughts.

Another activity focusing on environmental degradation is the Quo Vadis Aquifers (QVA) Programme, a joint activity of the United Nations University and the International Hydrological Programme of UNESCO. The programme was initiated in January 2006 and it proposes to address, through research and capacity development projects and activities, the links between groundwater resources degradation and human security. Ultimately, the various activities under the programme will:

- Advance the scientific knowledge on methods to capture the relationships between different types of pressures placed on the aquifers and the vulnerability of communities that rely on this resource, and this in a wide range of environments;
- Provide, through research results and effective communication, policy-relevant information to local and national policy-makers that would enable them to take actions to alleviate the pressures on the resources and/or reduce the vulnerability of the communities concerned;
- Through institutional and individual capacity development, provide the means for target countries to effectively address the issue at hand.

In the context of urban risks, researchers within UNU-EHS have elaborated a conceptual framework which is "distilled" from the findings, lessons learned, and conclusions expressed by researchers and practitioners around the world to identify the root causes and factors which unfortunately have led to the generation of risks, and subsequently to disasters.

With the title of GIRO Framework, it presents an integral view of risk management as a basis on which to promote actions not only to reduce risks, but also to restrain societies from increasing such risks or generating additional risks in the future. The Acronym GIRO stems from the Spanish translation of the term Integral Risk Management (Gestión Integral de Riesgo). The framework introduces the notion of a setting that influences actors in certain ways leading to the generation of such risks as unforeseen and undetected by-products of activities and processes which they carry out in relation to their livelihoods, normally executed within the traditional frameworks of development. Building on the notion of the setting and the actions carried out by actors as part of their livelihoods, the framework is able to make the connection between such actions, the risks which are generated, and subsequently the disasters which end up affecting in a negative way such livelihoods. The assessment of the links in which the chain: **setting → decisions made by actors → risks → disasters** is structured permits the detection of entry points towards the identification of potential options targeting risk management.

The framework, shown in Fig. 4, has been applied in the context of Latin American cities, in particular capital cities which are experiencing unplanned, uncontrolled growth, which is leading not only to the exposition of inhabitants to hazards, but also to the increase in hazards such as floods and landslides as a consequence of inadequate patterns of development. In particular, it is being applied to model how long-term urbanization and industrialization processes in the basin of the Quebrada Seca river in recent decades are leading to more frequent and more intense floods in the historic down-

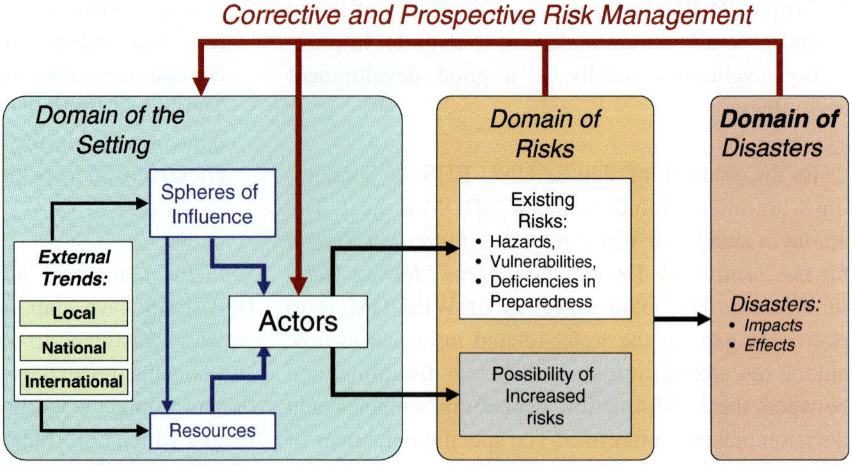

Fig. 4 The concept of 'GIRO', the integral risk management framework

town area of Belen, one of the municipal districts of the greater San Jose Metropolitan Area in Costa Rica.

Awareness and Capacity Building Activities

Results from the research conducted within the different sections of UNU-EHS find their way into policy-relevant documents such as the InterSecTions series, and into publications such as the SOURCE series, both published by UNU-EHS. In addition, researchers from the Institute publish their results in peer-reviewed publications and collaborate with UNESCO-IOC, with UNEP, and with ISDR in the elaboration of documents and guidelines which are meant to provide policy-relevant advice to government agencies, NGOs, and other institutions in the various topics currently being researched.

Capacity-building and educational activities conducted by UNU-EHS take the form of seminars, workshops, summer schools, and short term courses. These activities target Ph.D. and Master level students, as well as middle and high-ranking municipal officers from cities around the world. In particular, the Institute has made considerable efforts to support students at the Ph.D. level through research in a variety of projects (GI-TEWS, DISFLOOD, WISDOM, etc.).

Awareness is conducted through presentations in seminars, conferences, and workshops around the world organized by different agencies, and in some cases co-organized by UNU-EHS. For example, UNU-EHS provided technical advice to ISDR and organized side sessions and moderated panels in such events as the World Conference on Disaster Reduction held in Kobe, Japan, in January 2005, and in the Third Early Warning Conference held in Bonn in March, 2006.

Critical Issues and the Way Forward

While research conducted by UNU-EHS is providing answers to several burning questions, such research also highlights the need to tackle other equally important critical issues that need to be researched such as:

- Environmental degradation could become one of the major "push" factors in the future when it comes to environmental migration. Therefore, additional research is needed to dimension the full nature of this problem and its root causes.
- Disasters are becoming more frequent and their impacts more severe, eroding hard-won gains associated with sustainable development in developing countries throughout the world. While climate change can worsen such impacts; social, economic, and political trends are also contributing to such impacts. It is important to take into consideration root causes when conducting assessments of risk.
- In the future an increasing number of people worldwide will have to face more extreme weather events, sea-level rise and/or more intense weather-related hazards. In the context of climate change, it is important to identify efficient response mech-

anisms; in particular, the proper combination of preventive measures and those related to risk transfer mechanisms such as insurance and micro-insurance.

While it is important to conduct such scientific research to identify and characterize the risks faced by millions of people, it is never the whole solution. Research needs to find its way into policy-relevant advice and into common wisdom of societies through education, capacity-building and awareness. Only when all members of societies are exposed to these critical issues which surround the risks and human insecurities which they face, will such risks and insecurities be targeted and reduced for the benefit of mankind.

References

Birkmann J (2006) Measuring Vulnerability to Natural Hazards: Towards Disaster-Resilient Societies. United Nations University Press, Tokyo.

Birkmann J, Fernando N, Hettige S, Amarasingue S, Jayasingam T, Paranagama D, Nandana M D A, Nassl M, Voigt S, Grote U, Engel S, Schraven B, and Wolfertz J (2007) Rapid Vulnerability Assessment in Sri Lanka. UNU-EHS SOURCE Publication No. 7/2007, Bonn.

Birkmann J and Wisner B (2006) Measuring the Un-Measurable. The Challenge of Vulnerability. UNU-EHS SOURCE Publication No. 5/2006, Bonn.

Bogardi J J (2004) Hazards, risks and vulnerabilities in a changing environment: the unexpected onslaught on human security? Global Environment Change Part A, UN Monitor. Vol. 14, Issue 4, December 2004, S. 361–365.

Brauch H G (2005) Environment and Human Security: Towards Freedom from Hazard Impacts. UNU-EHS InterSections Publication No. 2/2005, Bonn.

Cardona O D (2001) La necesidad de repensar de manera holística los conceptos de vulnerabilidad y riesgo. Una crítica y una revisión necesaria para su gestión. http://www.desenredando.org/public/articulos/2001/repvuln/index.html. Accessed 23 February 2008.

Guha-Sapir D, Hargitt D, and Hoyois P (2004). Thirty Years of Natural Disasters. 1974–2003: The Numbers. UCL, Presses universitaires de Louvain, Belgium.

ISDR (Inter-Agency Secretariat of the International Strategy for Disaster Reduction) (2004) Living with Risk: A Global Review of Disaster Reduction Initiatives. http://www.unisdr.org/eng/about_isdr/bd-lwr-2004-eng.htm. Accessed 03 November 2008.

Lavell A (2003) Programa BID/IDEA de Indicadores para la Gestión de Riesgos, Universidad Nacional de Colombia, Manizales. http://idea.manizales.unal.edu.co/Proyectos Especiales/bid2/adminIDEA/CentroDocumentacion/DocDigi tales/documentos/AllanLavellEMBarcelonaJuly2003.pdf. Accessed 15 March 2005.

Munich-Re (2003) Topics: Annual Review: Natural Catastrophes 2002. Munich Reinsurance Company, Munich, Germany.

Renaud F, Bogardi J J, Dun O, and Warner K (2007) Control, Adapt, or Flee. How to Face Environmental Migration? UNU-EHS InterSecTions Publication No. 5/2007, Bonn.

Thywissen K (2006) Components of Risk: A Comparative Glossary. UNU-EHS SOURCE Publication No. 2/2006, Bonn.

UNDP (United Nations Development Programme) (2004) A Global Report Reducing Disaster Risk: A Challenge for Development. UNDP Bureau for Crisis Prevention and Recovery, New York.

UNGA (United Nations General Assembly) (2005) In larger freedom: towards development, security and human rights for all. Report of the Secretary-General of the United Nations as a follow-up to the outcome of the Millennium Summit. Document A/59/2005, 21 March 2005. Fifty-ninth session.

UNU-EHS (2005) Human Security in a Changing Environment: Strategic Directions 2005–2008, Bonn.

Villagran de Leon J C (2006) Vulnerability, a Conceptual and Methodological Review. UNU-EHS SOURCE Publication No. 4/2006, Bonn.

Villagran de Leon J C (2007) Rapid Assessment of Potential Impacts of a Tsunami: Lessons from the Port of Galle in Sri Lanka. UNU-EHS SOURCE Publication No. 9/2007, Bonn.

Communicating Geological Hazards: Educating, Training and Assisting Geoscientists in Communication Skills

David Liverman

Abstract Communication is important in all aspects of the geosciences but is more prominent in the area of geological hazards, as the main audience for scientific information often lacks a geoscience background; and because the implications of not communicating results effectively can be very serious. Geoscientists working in the hazards area face particular challenges in communicating the concepts of risk, probability and uncertainty. Barriers to effective communication of geoscience include the complex language used by geoscientists, restriction of dissemination of results to traditional scientific media, identification of the target audience, inability to tailor products to a variety of audiences, and lack of institutional support for communication efforts. Geoscientists who work in the area of natural hazards need training in risk communication, media relations, and communicating to non-technical audiences. Institutions need to support the efforts of geoscientists in communicating their results through providing communications training; ensuring access to communications professionals; rewarding efforts to engage the public; and devoting sufficient staff and budget to the effort of disseminating results. Geoscientists themselves have to make efforts to change attitudes towards social science, and to become involved in decision making at a community level.

Keywords Geohazards communication

D. Liverman (✉)
Department of Natural Resources, Geological Survey of Newfoundland & Labrador, St. John's, NL, Canada A1B 4J6
e-mail: dliverman@gov.nl.ca

Introduction

> "The world desperately needs to know what scientists are learning from their research endeavors. We can't stop hurricanes or tsunamis or other extremes of nature. But if we weave into the policy-making process the right mix of knowledge – integrating disciplines like environmental sciences, engineering, and health and social sciences – we can help save lives and reduce damage to property."
> Lubchenco (2005)

One of the major themes of the International Year of Planet Earth is "Hazards – minimizing risk, maximizing awareness." In the prospectus outlining the theme, four main research questions are posed (Earth Sciences for Society Foundation 2004). This paper attempts to provide a partial answer to the fourth question posed – "What are the barriers, for each geohazard, that prevent governments (and other entities) from using risk and vulnerability information to create policies and plans to reduce both?"

The need for research on geohazards and other natural disasters was highlighted by an International Council of Scientific Unions (ICSU) scoping report in 2005 that indicated, despite scientific advances in understanding causes and mitigating effects, the frequency of such events was increasing steadily. Natural disasters were reported to be at approximately 100/decade from 1900 to 1940, 650/decade in the 1960s and 2000/decade in the 1980s; it reached almost 2800/decade in the 1990s. Property damage is doubling about every seven years over the past 40 years (ICSU 2005). These data are shown in Fig. 3 of Ch. 1 of this volume.

T. Beer (ed.), *Geophysical Hazards*, International Year of Planet Earth,
DOI 10.1007/978-90-481-3236-2_4, © Springer Science+Business Media B.V. 2010

The ICSU Scoping Report recognized that a serious disconnection existed between science and action. It states:

> "We have found ample evidence to suggest that policy-makers may at times act in ignorance or disregard of the relevant scientific information and thereby significantly exacerbate damage resulting from natural hazards. Examples include: removal of mangrove swamps from vulnerable coastlines; failure to take account of foreseeable volcanic or seismic risks; land use practices that augment risks from floods, landslides or wildfires; failure to make best use of satellite data and to support networked early warning systems; failure to invest in prevention; and financial incentives that encourage short-term, localised benefits at the expense of longer-term requirements." (ICSU 2005).

It is argued here that one of the main barriers to appropriate use of geoscience information in planning and policy is that of scientific communication. Further, a major reason for problems in communication is that many geoscientists lack the skills to communicate effectively with those who need to use their knowledge and expertise.

Scientists should increase their involvement and activity so as to influence both policy and public response, yet, as it will be argued below, geoscientists working in the area of geohazards frequently lack the tools to do this effectively. The need to improve skills in this area is critical; the results of research in geosciences have direct implications for the health, safety and well-being of much of the earth's population. Scientific research can be well-funded, carried out well, and show clear direction to future policy. If it is not communicated to those who create policy, or those affected by the results, it might just as well not have been done at all.

The importance and relevance of communication in hazards research has been widely recognized in the past. The United States Geological Survey Professional Paper "Nature to be Commanded: Earth Science Maps Applied to Land and Water Management" (Robinson and Spieker 1978) was a pioneering effort in communicating the importance of earth sciences, interpreting conventional geological maps in terms that were readily understood by planners. More recently, the British Geological Survey has published a guide to geoscientists (Forster and Freeborough 2006). They emphasize targeting communication at specific audiences, identifying three main groups; professionals, informed members of the public, and potential or actual victims of a hazard. Perhaps the most important conclusion of this report relates to empowerment:-

> "It is essential that any publication that tells people that they have, or may have, a problem should include guidance on how they may, themselves, take action to: determine if they have such a problem, avoid such a problem or minimise the effect of the problem. Telling them to contact a professional for advice is not sufficient..." (Forster and Freeborough 2006).

In 2008, the Geological Society of London published a collection of papers entitled "Communicating Environmental Geoscience" with an emphasis on geological hazards (Liverman et al. 2008). This is the first volume dedicated to this topic, and reviews much previous work in the field.

Despite these attempts to assist geoscientists in communicating hazard research, it is clear that there is much to be done in this area, and with the increasing importance of geohazards on a global level, it is likely this topic will continue to be relevant in the future.

Communication of Science

> "In our society (that is, advanced western society) we have lost even the pretence of a common culture. Persons educated with the greatest intensity we know can no longer communicate with each other on the plane of their major intellectual concern. This is serious for our creative, intellectual and, above all, our normal life. It is leading us to interpret the past wrongly, to misjudge the present, and to deny our hopes of the future. It is making it difficult or impossible for us to take good action." Snow (1959)

The influential essay "The Two Cultures" quoted above discussed the growing gulf between the sciences and humanities (Snow 1959). Since Snow wrote, increasing specialization means that major communication difficulties exist within the sciences, and even within disciplines. We now have not one scientific culture, but many, each struggling to comprehend research outside of its own area of specialization.

The issues associated with communication by scientists and the effectiveness of such communication are widely recognized – perhaps most prominently in the area of health and medicine. There is an entire area of research that deals with the communication of science, and in particular the communication of risk. There are journals devoted to the subject (Science Communication, Journal of Science Communication), courses and degrees/diplomas, and many organizations employ communications professionals to deal with these issues. There is thus a considerable literature that can be applied to the specific problems of communication associated with geological hazards. This literature, however, is published mostly in social science journals, and a geoscientist is unlikely to be exposed to it through education or training.

General barriers identified in the communication of science range from the language used in science, the lack of institutional support for communication efforts, disincentives within the scientific culture, the challenges of communicating complex concepts, and the lack of media training amongst scientists. These are discussed later in this paper.

Geoscience and Geohazards Communication

> "Both empirical research and seasoned observation support the golden rule of public education for hazards: all the sophisticated materials and behavior modification techniques do not have the force of one "good" disaster to change both what people think, their behavior, and even public policy, at least in the short-term." Mileti et al. (2004)

Within the geological sciences, the issue of communication is perhaps most prominent in geological hazard research, where the concepts of risk and probability need to be explained, where uncertainty is prominent, and where the implications of research can have direct impact on the health and safety of the public. Communication, of course, is important in all aspects of geoscience but the nature of the audience for geohazard research is different to that in other areas. For example, in the field of resource exploration and development, particularly in the areas of minerals and oil and gas, the communication of geoscience faces different challenges than in the area of hazards. The challenges of communicating complex concepts certainly still exist, and the communication of risk and uncertainty is important. However, the audience has an obligation to be educated in the area of geoscience, as they are choosing to make economic decisions based on the scientific results presented. The implications of poor communication may be unwise investment, or lack of funding for worthwhile projects. Most importantly however, the audience generally chooses to undertake whatever risk might be incurred in understanding and acting on the results of research. In the area of natural hazards, however, the audience who need to understand the implications of scientific research often are forced to do so by geographical circumstance – their home, or livelihood happens to be located in a hazardous place. In some cases, those potentially affected by geological hazards may not be aware that any hazard exists. These audiences are very different from most in the geosciences, and particular methods and techniques of communication are required. A failure to adequately communicate risk in geohazards can directly result in immediate deaths, and thus intense critical scrutiny of the communication process. In addition the target audience is far less clear – should communication be directed at planners, policy makers, or the people potentially impacted by a disaster? All three are important audiences with different needs and background knowledge.

Uncertainty and risk are terms frequently used in geohazards research. They have strict definitions and means of determining their magnitude when used in a scientific context. The same words have a different meaning when used by a policy maker or politician. Faced with uncertainty, a policy maker wants more information or study in an attempt to remove uncertainty, or may discount advice.

The focus on risk and probability matches that in the field of medical research, where it is vital that informed decisions on health and well-being be made based on scientific research. Given the similar challenges it is important to note that geosciences and other historical sciences differ markedly from the experimental sciences (Cleland 2001). Researchers in the medical field define risk and probability based on controlled experiments. The nature of the earth sciences means that often such experiments are impossible to devise. When, for instance, the probability of an earthquake striking a given area is estimated, the basic method is to review knowledge of past occurrences. The record of past events is incomplete, and fragmentary. In a controlled experiment, error bounds can be reduced by increasing the number of trials or the sample size. This is not possible when looking at the variation through time of natural processes. Thus the level of uncertainty in conclusions may be greater, and hard to reduce with further research. This increases the difficulty in communicating such results outside of the scientific community.

Geoscientists tend to have different perceptions and understanding of time than those not trained in that area, and this can give rise to additional problems in communication. A process that is considered frequent by a geoscientist might well be thought to be so rare as to not merit consideration by a politician or planner. Hazards such as earthquakes or tsunamis may have recurrence intervals of centuries but the magnitude of their impacts means that it is critical that they are taken into account in planning and policy development.

The Role of the Geoscientist

> "I feel strongly that I should not go into research unless it promises results that would advance the aims of the people affected and unless I am prepared to take all practicable steps to help translate the results into action." (White 1972).

What role should geoscientists take in the communication of geological hazards? Given that there is a body of social science research dealing specifically with societal and cultural response to hazards, and a further body dealing with scientific communication, should geoscientists restrict themselves to geoscience and leave the communication of results to others?

Futerra Sustainability Communications (2005) outlined their view of the role of the scientist and professional communicators as follows:

> "Egg-head scientists are important messengers: they have authority, and reassure people that someone understands the complicated issue of climate change. But we need common-sense and likeable intermediaries as well, to translate the opaque pronouncements of scientists into practical and obvious advice."

Futerra reinforced stereotypical views of scientists as well as perhaps imagining themselves in the role of "common sense and likeable intermediaries." There is no reason, however, that scientists cannot translate their own research into practical and obvious advice, and also perhaps be viewed as likeable rather than egg-headed!

It is argued here that there is an obligation for geoscientists to "take all practical steps to translate their results into action" (White 1972) and that one of those steps is to take responsibility for communicating their results to those who need to be aware of them. There are good reasons to dispense with intermediaries and deliver the message directly.

Geoscientists spend their careers studying geological hazards. They have formal training through their education, and gain experience progressively – they have expertise in the processes, frequency, magnitude, and nature of the natural processes causing disasters. They thus have unique insight and understanding of geological hazards. Geoscientists therefore are best able to evaluate the significance of their results. They understand the limitations and uncertainty attached to their conclusions. In addition they are more likely to be considered trusted and reliable sources of information than politicians or government employees. Surveys in the United Kingdom show that scientists enjoy a high level of trust by the public, particularly if they work at a university (Corrado and Duthie 2006). Information conveyed directly by scientists carries more weight with the public than when it is interpreted by media or governing bodies.

If the results of research are communicated to those who need to know them by an intermediary – whether it is the media, a professional scientific communicator, or a policy maker, there is a potential for misunderstanding. Just as geoscientists may not have the background and training to communicate effectively, it is almost certain that any intermediary will lack the scientific background to comprehend the full implications of the research.

Geoscientists may not wish to be trained in communication, or might feel that they are not suited to that role. However it is inevitable that a geoscientist who works in the field of hazards will be put in a position where they have to communicate their work to non-scientists, and there may not be access to a communications professional to assist in an emergency situation. Not every geoscientist is suited, or able to interact with the media if a disaster occurs, but if that scientist is considered the expert on that type of hazard or is conducting research in that area, it is inevitable that they will be "discovered" by media – 30 s with an internet search engine will likely yield their name. It is thus important that even the most reluctant geoscientist has some understanding of communication issues.

If it is accepted that geoscientists should take a major role in communicating their results, then it is important that those scientists understand the barriers that are faced, and the means to overcome them. They need to have the skills and knowledge to make such communication effective.

Barriers to Communication

Language and Style

> "Vague and insignificant forms of speech, and abuse of language, have so long passed for mysteries of science; and hard and misapplied words, with little or no meaning, have, by prescription, such a right to be mistaken for deep learning and height of speculation, that it will not be easy to persuade either those who speak or those who hear them that they are but the covers of ignorance, and hindrance of true knowledge" (Locke 1690).

The standard means of communicating scientific results is through peer-reviewed articles in scientific journals. Such articles are directed at a select audience of fellow scientists who are familiar with the subject area. Explanation of scientific research demands a method of writing that allows the scientific method to be laid out clearly, and the logical steps taken presented in a standard sequence. This does not always make for easy reading, but is essential in order to demonstrate to the reader the assumptions made, the methods followed, and the logical train of thought required in the scientific method.

Scientific writing involves a number of stylistic conventions that do not improve readability, including the use of the third person, passive construction, and extensive use of acronyms. Each branch of science has developed a huge vocabulary of technical terms that are poorly understood outside of that specialized area. Scientific research involves the investigation of concepts and objects that lack common words to describe them, and so these technical terms are needed – yet to the reader from outside of the area of specialization these are seen as jargon. Some technical terms are hard to avoid but frequently scientific writing makes no attempt to explain concepts using simple language, even if this might be possible with some effort. When writing for a non-scientific audience, many scientists find it hard to understand what terms are not easily comprehended, as for them, with the terms in everyday use, they do not seem to be obscure, or difficult to understand. This problem is by no means restricted to the geosciences, and for instance can make important social science research inaccessible to the geoscientist.

The degree of specialization in scientific research, however, has made much scientific writing not just incomprehensible to non-scientists, but also to scientists themselves. Glanz (1997) discussed efforts within the physics community to develop guidelines for improving the clarity of writing, where the problem has become so serious that one physicist is quoted as suggesting that recent colloquia in his own department were so hard to understand that he was reluctant to encourage students to attend in case they would be "turned off from physics."

Hartz and Chappell (1998) presented the results of a survey of 1,400 scientists and journalists; 62% of journalists agreed with the statement that "most scientists are so intellectual and immersed in their own jargon that they can't communicate with journalists or the public." The extent to which this problem is acknowledged amongst scientists is shown by the fact 50% of scientists agreed with them.

Medium and Audience

Nearly all scientific research is published in the serial literature, presented at academic conferences, or described in technical reports. Scientific journals are rarely read by anyone other than scientists, and with the proliferation of specialized publications, it is unlikely that most journal papers are seen by anyone other than other specialists in that particular field of science. In order for geoscience to play an appropriate role in the efforts to deal with the problems posed by natural hazards, geoscientists must be able to adapt their communication skills to a variety of other means of communication – through the popular media, public awareness, public or community based consultation, and briefing of politicians and policy makers. The means and style of communication need to be adapted to the audience targeted. This audience will vary considerably depending on the type of hazard, and the nature of the research.

Take, for example, the geoscientist asked to assist in dealing with an imminent volcanic eruption – this is a crisis situation, with immediate actions required. They may have to provide advice to emergency responders in terms of evacuation, assessment of the probability and magnitude of the eruption to policy makers, be asked to brief politicians, and to meet with communities threatened. Each audience requires different information, and presented in a manner that is accessible to them (Barclay et al. 2008). A completely different approach might be warranted when attempting to communicate the impact of a low-frequency, high magnitude event such as tsunamis or earthquakes that might not be perceived as imminent. Here the emphasis may be public awareness, or communicating to planners. Thus with no imminent crisis, techniques and target audiences may be quite different. The most stressful situation is when a geoscientist is called in after a disaster, where media interest is intense. Media will be seeking to assign responsibility, emergency responders will be seeking guidance, and inevitably there will be political pressure.

Thus geoscientists must be able to tailor their communications to the situation and audience they are aimed at. The language used will vary according to audience, as will the medium of communication. Geoscientists must be prepared to provide results in a variety of formats – an internal report, a public awareness poster, a media interview/sound-bite, a brochure or pamphlet, or a verbal briefing.

Preparation of an appropriate product is only part of the task of communication. It is vital that the issue of dissemination be dealt with. An excellent product is useless unless it gets into the hands of the audience it is designed for.

Culture

A major factor in discouraging scientists from developing communication skills lies in the scientific and institutional culture within which they work. Efforts to engage the broader community can be viewed with suspicion by peers, and are not always well-regarded by granting agencies or employers.

A prime obstacle to engaging the public or media cited by those interviewed by Hartz and Chappell (1998) is a loss of status amongst their peers. There is a perception amongst scientists that scientists with a high media profile are no longer doing worthwhile research themselves, and thus turn to public engagement as being in some way less demanding. A United Kingdom survey found that 20% of scientists agreed with the statement that scientists who engaged the public were less well regarded by their peers. In qualitative interviews several scientists expressed the opinion that public engagement would be detrimental to their careers (Royal Society 2006).

Schneider (1990) recounted personal experiences of negative reactions from colleagues after seeing media coverage of his statements. Media coverage resulted in detailed explanations being reduced to brief quotes, often omitting important additional information. This undermined the accuracy of the statement, and damaged the researcher's credibility with colleagues. Schneider concluded from his early experiences that scientists have two choices – either to avoid media interaction completely or to spend enough time on it to ensure that some coverage at least was accurate, and comprehensive.

Engagement with the public or the media is rarely recognized in the academic world, where grants, tenure and promotions are linked with research published in the serial literature. In the demanding environment of modern universities, geoscientists may be reluctant to devote the significant amount of time required without incentive or reward. A major reason for not engaging the non-scientific community is the need to spend more time on research (64% of respondents, Royal Society 2006). Attempts to directly influence or advise policy are generally not encouraged, particularly in the physical sciences.

The ability to interact with media may be severely limited by the institution and political environment in which the scientist works. In some countries and organizations any communication to media or public is strictly controlled, and access to policymakers or politicians limited by bureaucratic or institutional structures. The challenges of communication are even greater in these circumstances.

The role of scientists employed within government agencies is in part to provide advice to policy-makers and politicians. Their involvement is generally limited to internal advice, and direct contact with media or the public only undertaken when directed to do so by the government agency. Few government employees are at liberty to discuss hazard issues with the media.

Uncertainty

> We conclude that advances and changes must be made in the way science is conducted and uncertainty communicated. Scientists must become more effective and compelling communicators of both what is and isn't known (Kinzig et al. 2003)

Uncertainty is an integral part of science, and forms the basis of much scientific discussion. Scientists attempt to address the assumptions made, the possible errors in experiment and make uncertainty explicit. Science advances in part by debate – when there is a consensus of agreement on most areas of a subject, research will focus on those areas that are less well understood. In this type of research, results can be amenable to more than one interpretation, and it is only by testing hypotheses that science advances. The media, policy makers or the public may find this hard to understand and in extreme cases scientific uncertainty can be portrayed as scientists in dispute over conclusions. This type of uncertainty may be used as a reason not to act on the results of research when it recommends a course of action that is contrary to political direction, or might result in the expenditure of large sums of money. Bernknopf et al. (2006) found that "the uncertainty regarding the interpretation of the science inputs can influence the development and implementation of natural hazard management policies."

Many non-scientists expect certainty when presenting the results of scientific research. A poll conducted in the United Kingdom (MORI/Science Media Centre 2002) showed that 71% of those polled "looked to scientists to give an 'agreed view' about science issues"; 61% expected science "to provide 100% guarantees about the safety of medicines". Scientists are unable to provide such certainty in the presentation of results.

Risk and Probability

> "Public reaction to risk sometimes seem bizarre, at least when compared with scientific estimates.... the suggestion that a hazard poses an annual risk of death of "one chance in x" may cause near-panic or virtual indifference." (Department of Health 1997).

It is particularly important when dealing with hazards to be able to quantify the probability of a disaster occurring. This information is a critical component of a broader risk assessment. Most geoscientists working in hazards are able to present probabilities of occurrence, but communication of those probabilities can be fraught with difficulty. Other scientists, whatever the field, usually are comfortable with probabilities, as are those involved in risk assessment. However there is often a need to communicate to those without scientific or mathematical training – the public, politicians and policy makers. Hartz and Chappell (1998) documented that 63% of journalists and 82% of scientists surveyed agreed that "most members of the news media do not understand probability and statistics well enough to explain the results of scientific research."

This difficulty is prominent in risk and hazard mapping, where geoscientists assign probability to the occurrence of hazardous events. Hazard zones are often defined on the basis of probability of recurrence. The one in one hundred year (1%) flood zone is in common usage, yet this means of communication often results in misconceptions. Rather than the correct interpretation of a 0.01 probability of a flood occurring in any given year, many people believe that this designation means that the area will flood periodically with 100 years between floods (Ogle 2004; Bella and Tobin 2007). Communication of risk is further complicated by the fact it generally incorporates two types of uncertainty; that associated with the randomness of natural phenomena (aleatory), and the other associated with lack of knowledge (epistemic). These need to be interpreted quite differently but it is not easy to communicate the difference. Mileti et al. (2004) pointed out that scientists expend much effort in defining and refining the probabilities of future hazardous events occurring, but the public interest can be expressed much more simply – will the event occur or not, and if it does will it affect me?

Interdisciplinary and Multidisciplinary Approaches

> "If we are to build up our resilience and effectively reduce the devastating effects of natural hazards, geoscientists must coordinate their efforts with engineers, emergency management professionals, policy makers, builders, investors, insurers, news media, educators, relief organizations, and the public, as well as other scientists." Geological Society of America (2005)

The field of natural hazards allows for a wide range of approaches, and many journals, conferences and workshops describe themselves as "interdisciplinary." There are however serious communication barriers between researchers working in natural hazards that compromise the ability to operate in a true interdisciplinary manner.

The following are extracts from abstracts to papers published in the last four years in a leading natural hazards journal. Each paper is a worthwhile contribution to the body of knowledge, contains results that perhaps can be used broadly by other natural hazards researchers, but are written to communicate with fellow specialists in their area of research, whether it be process sedimentology, risk analysis or political science. It should be noted that these papers were selected at random purely to illustrate the type of language used in the various sub-disciplines of natural hazards research – there is no criticism intended or implied.

> "In order to evaluate critical condition of bed sediment entrainment, a length scale which measures an effective bed shear stress is introduced. The effective bed shear stress is defined as total shear stress minus yield stress on the bed surface. The results show that critical entrainment conditions can be evaluated well in terms of Shields curve using the effective bed shear stress instead of a usual bed shear stress."
>
> "The predictive power of logistic regression, support vector machines and bootstrap-aggregated classification trees (bagging, double-bagging) is compared using misclassification error rates on independent test data sets. Based on a resampling approach that takes into account spatial auto-correlation, error rates for predicting "present" and "future" landslides are estimated within and outside the training area."
>
> "In order to capture the complexity that arises when incorporating the varieties of interests as well as impacts protection measures have on the environment, the economy and society, transparent and multidisciplinary decision support techniques are needed. This paper looks at how Cost Benefit Analysis (CBA), a tool already applied to decisions concerning protective measures, and Multi Criteria Analysis (MCA), even though new to the field as such but already successfully practiced in other environmental areas, perform according to the abovementioned criteria."

How many researchers can honestly say that they are comfortable with the language and concepts briefly expressed in all or any of these abstracts? These papers are not badly written, but are designed to communicate with a very select audience. This likely is what the authors intended, but the style chosen may mean that comparatively few people can understand the results put forward, and their implications.

In effect, this means that many research efforts in natural hazards described as interdisciplinary are in fact multidisciplinary. A multidiscliplinary approach means that a problem is addressed independently by specialists from a variety of fields. An interdisciplinary approach means that such diverse research is integrated to address a problem that a single discipline alone cannot (Schneider 1997).

Thus, an interdisciplinary approach must not mean simply that researchers from a variety of backgrounds are working on the subject. It requires co-operation and mutual understanding between the various researchers – and this in turn means that specialists must be able to communicate effectively to those outside of their area of specialization.

Geoscientists need to be able to adapt their communication methods and skills in order to operate effectively in an interdisciplinary environment. A geoscientist who is able to communicate effectively with specialists in other disciplines can ensure that their expertise is used appropriately in interdisciplinary approaches. Geoscientists also need to educate themselves in the methods and language used in other areas of natural hazards research. By doing so, they give themselves the opportunity to integrate research in a true interdisciplinary manner and to address problems that geoscience alone cannot.

Discussion

Lessons from Climate Change

Much can be learned from the interaction of science, communication, policy and politics in the area of climate change. The issue of global climate change moved from the pages of scientific journals to the front pages of newspapers in the last twenty years. Fundamentally this change has been due to the successful communication of scientific research, although the process has been anything but easy. The means of change itself has become the subject of considerable research (see Moser and Dilling 2007). Understanding how science eventually influenced public policy and perceptions in this highly visible example may provide valuable insights into methods that may apply to geohazards. The climate change debate has been highly political, because of the major economic impact of taking mitigative measures, and the perceived impacts on influential special interest groups. This has led to the politicizing of science, with intense debate between lobby groups as to the accuracy of scientific conclusions, and funding of groups that set out to undermine the authority of climate scientists by questioning motivations and research agendas. For many years the public were presented with a "balanced" view of the debate by media who gave equal weight to climate change sceptics, when in fact the scientific community was close to unanimous in ascribing global warming to emissions resulting from industrial development. Boykoff (2008) suggested that the challenges of communicating climate change science in the midst of a heated debate acted to discourage many scientists from engaging the media. This unwillingness was often based on experience of having research misinterpreted, selectively used, or quoted to advance policy or political objectives. Boykoff goes on to state "the 'battlefield' of communicating and understanding environmental geoscience is not well-served by scientists reluctant to acknowledge and act on what is an integral piece of one's contemporary responsibility: interacting with mass media."

Research in geological hazards has the potential to become highly political, as decisions are made that have profound economic and social impacts, and ultimately may result in the loss or saving of lives. Where lives have been lost, research then will be viewed in an environment where politicians and policy makers may be trying to avoid responsibility, others will be seeking scapegoats, and scientists may be caught in the middle, with their communication skills being tested in the most demanding of circumstances.

An issue that is highly relevant both in climate change and hazards research is how to portray uncertainty. This issue became prominent in communication of climate change, where uncertainty in research seized upon by some as a means of discounting research findings, and in some cases was re-framed to cast doubt on scientific competence (Williams 2000; Zehr 2000; Boykoff 2008).

Perhaps the most powerful tool used by climate scientists was the establishment of the Intergovernmental Panel on Climate Change, where an international group of credible scientists was created to inform policy and management decisions. There is little doubt that the series of IPCC reports have been highly effective in this objective.

The IPCC reports use a methodology developed and outlined by Moss and Schneider (2000) to address the communication of uncertainty. Moss and Schneider emphasize the importance of quantifying uncertainty wherever possible and provide useful guidelines as to appropriate language to describe different levels of uncertainty in a standardized manner. They point out that "there is strong experimental evidence that the same uncertainty words often have very different meanings for different people in different circumstances". Thus they suggest that "very high confidence" should be associated with a probability of 0.95–1.00, "high confidence" with 0.67–0.95 and so on. Such methods might well be applied to communication of geological hazards.

Climate change science also showcased the role of the professional or expert communicator interpreting the results of science into an accessible or popular format. Former US vice-president Al Gore's presentation series and later film "An Inconvenient Truth" did much to raise the profile of climate science. Any scientist, however, can learn from the methods used by Gore to interpret the science of climate change. His medium of choice – the computer-generated presentation – is the standard method used by nearly all scientists at academic conferences. However, the use of spectacular images, careful use of analogy, and most importantly personalizing the message by using individual experience showed how to engage a wide audience – "a paragon of clear science communication" (Minkel and Stix 2006). The approach Gore took to address the issue of climate change also offers lessons. Rather than rely on the media to interpret his message, he chose to deliver his presentation directly to as many people as he could – firstly through the presentation itself, and training of numerous other presenters, and later through the documentary film. The exclusion of the "media filter" was deliberate, and based on years of experience as a politician.

Education and Awareness

Many geoscience degrees include some training in communications. These, however, are designed to educate students in using the traditional means of scientific communication – how to write an abstract, the basics of writing a scientific paper or report, how to prepare a poster for a conference, or preparation of computer presentations. Education for geoscientists needs to address the broader aspects of communication. A brief review of M.Sc. programmes in geohazards shows courses with an emphasis on technical content, with communication issues absent from the curriculum. A similar review of the more general topic of natural hazards suggests that communication issues are rarely taught.

In training geoscientists specializing in geohazards, education should cover dealing with the media, writing for public and policy makers, and some exposure to the social science literature in natural hazards. One innovative example is the Master's International programme in "Mitigation of natural geological hazards" at Michigan Tech University in the United States. This degree includes international research through the Peace Corps programme, and courses in inter-cultural hazard communication.

There is a similar dearth of training at the professional level. In Canada, where most provinces require geoscientists to be professionally registered, the standard curriculum defining the basic education for a geoscientist does not include any component of communication with the public, and professional development opportunities are similarly sparse. This needs to be remedied.

Communicating with the Media

> "By always bearing in mind two crucial facts – that the news media are not going to change the way they work to please scientists, and that they should be approached as a branch of the entertainment industry – all subsequent decisions and behaviours on the part of scientists and their companies/institutions will be more likely to be blessed with success." Neild 2008

Communicating with the media is difficult and there is no substitute for formal training and practice. The relationship between science and media was explored by Hartz and Chappell (1998), and their report identifies numerous concerns that scientists have when dealing with journalists. A common concern of scientists dealing with the media is that findings will be portrayed in an inaccurate or misleading manner. Only 11% of scientists surveyed by Hartz and Chappell (1998) had great confidence in the press, and similar responses were interpreted as showing that scientists in general were not comfortable with media coverage of scientific research. This is perhaps due to the fact that 73% of those surveyed in a Royal Society survey in the UK had no training whatsoever in engaging with the public or media (Royal Society 2006).

The lack of media training means that scientists are unaware of the way in which journalists work, and what they are looking for. It is important that geoscientists involved with the media understand how journalists operate, and where their interest and concerns lie. They should not be viewed as a convenient means of informing the public. Expectations that they will be happy to translate and interpret research findings in an accurate and unbiased manner are frequently not met. Journalists are interested in selling stories and thus will focus on what they consider to be of interest to their readers. They will focus on drama, conflict, and human interest. Journalists may also attempt to present a balanced view of any scientific controversy – even if the consensus scientific view strongly favours one side (Boykoff and Boykoff 2004). Scientists, on the other hand, look for media coverage to educate the public about their work (over 40% cited this as an important reason for engagement with the media in the Royal Society Survey), yet misunderstand the nature of media coverage. Neild (2008) outlines common misconceptions of the way the media operates.

If a geoscientist has an understanding of how journalists operate, they can prepare themselves to better communicate their expertise. Developing a working relationship with journalists who specialize in science issues can greatly assist in public awareness of hazards but the most demanding situations are when a disaster has occurred, or when one appears imminent. Geoscientists must be well prepared to deal with the media, as there will be a wish to gain authoritative information from a trusted expert in the field. Following the advice of Mileti et al. (2004), the message conveyed must be clear, free of technical language, supported by additional more detailed information, and consistent. Conflicting interpretations, or confusing statements can give rise to frustration, or misinterpretation. Journalists may portray such conflicting messages, or expressions of uncertainty, as conflict within the scientific community. Packaging information specifically for the media can be effective – journalists often work under pressure of deadlines, so well-prepared and written background information will often appear in final coverage.

It is important that geoscientists talking to the media very clearly separate scientific findings from opinion. Schneider (2002) suggested that scientists should be able to answer the questions "What might happen?"; "How do you know?"; and "What are the odds?" based on research, but if asked "What should we do?" clearly state that the answer is an opinion, or value judgment, and needs to be considered as such. If representing a government organization, scientists are generally restricted to scientific findings only and cannot venture opinion, speculation or comment on policy.

Communicating with the Public

> "In the centre of the mainstream or standard (neoclassical) economic model of decision-making resides the anonymous rational man who performs omniscient probability calculations with unlimited cognitive resources, and maximizes expected utility in the face of scarce resources." Wang (2001)

Phrases such as public engagement, public awareness, and public education are often used yet there must be an understanding that the public is not a uniform group that can be treated as a single entity. The "public" consists of individuals who require different information according to their own personal needs and interests. The public can be a whole community or country, where the objective of communication is to raise the awareness of hazards. It can be a targeted group that needs to be warned about impending disasters, or it can be an unfortunate group of individuals who have been directly affected by a disaster. The "public" may have a completely different culture than that of the scientific researcher (see for example Pettersen et al. 2008 for case studies of adaption of communication methods based on understanding of traditional culture).

The field of science communications has undergone change in recent years. The traditional model governing public communication of science was that of "information deficit" (Burgess et al. 1998). This, in simple terms, suggested that the public would take appropriate action if only they were provided with the right information by scientists. Success of the communication process was measured by what were considered appropriate changes of behaviour by the public when provided with scientific results – there is implicitly an expectation of a rational response by the public to this new information. It has become clear, however, that simply providing information will not result in changes in behaviour (Owens 2000). This top-down model is based on the assumption that scientists are trusted, and that they, and policy makers know what is best for the "public." In its least effective form the public is only consulted in order to legitimize a predetermined course of action.

Owens (2000) states "There could hardly be a clearer demonstration of the flaws in the information deficit model than the persistent refusal of the public to have their allegedly irrational conceptions of risk 'corrected' by providing them with more information." The alternative to this model is termed by Owens the "civic" model of communication. The principle here is of dialogue, where scientists and public work together to first define the problems that exist, and to propose solutions. This requires an acceptance that expertise may be found outside of scientific research – through local or traditional knowledge for instance. It requires scientists to participate at a community level, and to not only provide information but to listen to concerns and information brought forth by others. The idea behind this type of engagement is to provide the public (in the case of hazards generally the people directly affected by a potential or actual disaster) with the information they need to take action. That information may well differ from that which the scientist believes should be taken into consideration. Owens discusses the challenges raised by this type of public engagement and questions how effective this can be in practice (notably pointing out that erecting a number of new processes for public consultation may not change the effectiveness of public engagement). If it is accepted that the civic, or deliberative model of public engagement is more effective, it requires fundamental changes in the way that scientists deal with the public. On a practical level, it is useful for the geoscientist to be aware that the information deficit model of communication has been shown to be ineffective, and that experimenting with other means of public engagement is needed.

The work of Handmer (2000) provides further insight into the challenges of communicating hazard information to the public. He analysed the particular case of flood warnings, attempting to identify reasons why warnings had proved ineffective in several major floods in the 1990s. He firstly pointed out that "failure" of flood warnings is based on inability to meet unreasonable expectations. Other areas of hazard communication might be equally ineffective, but where warnings need to be communicated effectively under intense pressure in brief time-spans, any problems are seized upon by the media. He goes on to discuss the importance of tailoring the warning to the audience at risk and highlights the importance of shared meaning. "To have any chance of 'success' warnings need to have meaning which is shared between those who draw them up and those for whom they are meant to inform. They must also appear relevant to the individual decision-maker." (Handmer 2000).

Handmer emphasizes the importance of consultation in developing warnings, suggesting the process should be more akin to negotiation, but indicates that the populations at risk are often diverse, and one warning system is unlikely to be effective in all cases.

Handmer (2000) also warned that even where effort has been made to achieve shared meaning, warnings fail for a variety of reasons – people understand the warning but do not care about the risk; people are unable to act despite the warning for economic, social or personal reasons; or people dislike being told what to do by authority, preferring to make their own decisions. He also points out that warning systems and communication need to evolve along with the society at risk – systems designed to work with a relatively homogenous group 20 years ago now may have to deal with a variety of languages and cultural backgrounds; or with people who have little experience or knowledge of previous events, and are ill-equipped to deal with them.

Communicating with Policy Makers and Politicians

Communicating with politicians or policy makers requires different techniques to communicating with the public. In order to be truly effective, the scientist needs to understand how decisions are made, how policy is developed, and where their input can be most useful. The message the geoscientist brings forward may not be welcome, as it may result in media interest, public pressure to expend funds, or be perceived as potentially embarrassing to the government. This does not make the task of communicating geoscience any easier.

The communication issues differ depending on where the geoscientist is employed. If external to the government or policy making body, then understanding the institutional and decision making structure allows scientific advice to be tendered where it is most likely to be used (Simpson 2008). Such input may be via existing institutional structures such as standing committees, public hearings, or public enquiries. In some cases the governing body may not be seeking scientific input, and then, if the geoscientist believes that their research has implications for public policy or safety, they may wish to work with the media, non-government organizations, or groups of concerned citizens to build up public interest and political pressure.

Geoscientists working within government should understand the decision-making structure that leads to policy development. The critical factor in communicating geohazards research here is understanding the audience, and their prime concerns in reaching decisions. Politicians in democratic governments may be governed by relatively short-term considerations, as the electoral process is usually on a four to five year cycle. Thus issues that might arise in that time frame may gain more attention than hazards with a recurrence interval of decades or centuries. Decisions are sometimes made from a political or economic perspective rather than a scientific one, and the geoscientist must to be able to communicate their results with an understanding of these implications. The tendency to focus on short-term issues and therefore to dismiss long-term risks (or rare events) can be reduced where government officials have a legal requirement for "duty of care" of their constituents. In this case, the nondisclosure of available hazard and risk information carries the threat of prosecution, which tends to result in more willingness to promote an open dialogue about risks.

Some Simple Rules for Communication

This is by no means intended as a comprehensive guide to communication of geohazards but provides a starting point for those aware of a need to improve their communication skills when dealing with hazards. These draw on the discussion above but also heavily on Mileti et al. (2004), essential reading for anyone involved in communicating geohazards, as well as other sources.

- Understand the culture, background and decision making structure of your audience (whether indigenous people, or government bureaucracy).
- Work with communications professionals but be prepared to lead efforts yourself.
- Differentiate between scientific fact and opinion or value judgements.
- Prepare a variety of products in a variety of formats, tailored to different audiences.
- Dissemination of information is as important as the information itself; use a variety of media – TV, radio, newspapers, distribution of brochures.
- Ensure a consistent message.
- Avoid technical terms, use analogy, commonplace examples; strive to be as clear as possible.
- Communication of hazards is an ongoing process, not a one-off effort; be prepared to repeat information numerous times.
- Don't expect provision of information to change behaviour; become involved in the discussion at all levels and be prepared to listen and learn as well as instruct.
- The geoscientists' role is to provide information to assist people in making decisions, not to make decisions for them. If the geoscientist possesses information on what people should do before, during and after an event this needs to be conveyed clearly and effectively.

Conclusions

Despite many advances in the field of communication of science, and some admirable efforts in the communication of geohazards, much needs to be done. It is clear that geoscience research with direct implications to health, safety and economic well-being is frequently not being used by the public, policy makers and politicians. Any efforts to improve the ability of geoscientists to communicate their knowledge may help in the mitigation of geological disasters.

Several recommendations are made.

1. When educating geoscientists to work in the area of hazards, communication skills must be taught and media training should be provided. Geoscientists must be exposed to the social science approach to hazards and disasters so they understand their role in the broader context of natural hazards, and be able to take part in interdisciplinary research.
2. Institutions – whether they be government, academia or industry – need to support geoscientists in developing communication skills. This can be through in-house training, employment of media and communication specialists, or encouragement to geoscientists who wish to engage the public.
3. Research findings need to be communicated in a variety of ways – certainly through the traditional scientific paper or report, but also in formats more accessible to those potentially affected by disasters, or in a position to mitigate them through policy.
4. Geohazard project development must include both time and budget for communication and dissemination of results.
5. Alternatives must be sought to top-down communication models. Geoscientists need to be able to engage in genuine dialogue at the community level.
6. Journals that wish to be viewed as interdisciplinary need to ensure that contributions are written so as to be useful to the widest range of readers, or at least include a "plain language" summary of the findings, and their implications.

The IUGS Commission "Geosciences for Environmental Management" established a working group "Communicating Environmental Geoscience" in 2006. The working group attempts to develop and improve the tools and skills environmental geoscientists need to communicate effectively with non-specialists – politicians, policy makers, regulators, educators, and the public at large. They direct a programme of workshops, training courses, meetings, and publications. The group is building on existing efforts, but to effectively reach a world-wide community of environmental geoscientists remains challenging.

Acknowledgments I thank the members of the IUGS GeoIndicators management committee and the IUGS GEM Commission for their encouragement to develop a working group to deal with this issue, and for much interesting discussion and feedback. Martin Batterson is thanked for his review of an earlier version of this paper. The paper was significantly improved by the comments of an anonymous reviewer, and through discussion following the presentation of some of this work at the 2008 International Geological Congress in Oslo. The Geological Survey of Newfoundland and Labrador is thanked for its support in my involvement with the activities of the CEG-GEM working group.

References

Barclay J, Haynes K, Mitchell T et al. (2008) Framing volcanic risk communication within disaster risk reduction: finding ways for the social and physical sciences to work together. In: Liverman DGE, Pereira CP, Marker B (eds), Communicating Environmental Geoscience, Geological Society of London Special Publication 305.

Bella HM, Tobin GA (2007) Efficient and effective? The 100-year flood in the communication and perception of flood risk. Environmental Hazards 7: 302–311.

Bernknopf RL, Rabinovici SJM, Wood NJ Dinitz, BL (2006) The influence of hazard models on GIS-based regional risk assessments and mitigation policies. International Journal of Risk Assessment and Management 6: 369–387.

Boykoff M (2008) Media and scientific communication: a case of climate change. In: Liverman DGE, Pereira CP, Marker B (eds), Communicating Environmental Geoscience, Geological Society of London Special Publication 305.

Boykoff M, Boykoff J (2004) Bias as balance: global warming and the U.S. prestige press. Global Environmental Change 14: 125–136.

Burgess J, Harrison C, Filius P (1998) Environmental communication and the cultural politics of environmental citizenship. Environment and Planning A 30: 1445–1460.

Cleland CE (2001) Historical science, experimental science, and the scientific method. Geology 29: 987–990.

Corrado M, Duthie T (2006) Opinion of Professions – Trend Data. Report for the Royal College of Physicians. Ipsos MORI, United Kingdom.

Department of Health (1997) Communicating About Risks to Public Health: Pointers to Good Practice. Department of Health, London, p. 27.

Earth Sciences for Society Foundation (2004) Hazards – Minimizing Risk, Maximizing Awareness. Earth Sciences for Society Foundation, Leiden, The Netherlands.

Forster A, Freeborough K (2006) A guide to the communication of geohazards information to the public. British Geological Survey Internal Report, IR06–009.

Futerra Sustainability Communications (2005). New Rules/New Game: Communications Tactics for Climate Change. http://www.futerra.co.uk/downloads/NewRules:NewGame.pdf, accessed 25 August 2008.

Geological Society of America (2005) Geoscience and Natural Hazards Policy. Position paper on Geological Society of America website, http://www.geosociety.org/positions/position6.htm accessed 20 July 2008.

Glanz J (1997) Cut the communications fog say physicists and editors. Science Magazine 277: 895–896.

Handmer J (2000) Are flood warnings futile? Risk communication in emergencies. Australasian Journal of Disaster and Trauma Studies 2000-2.

Hartz J, Chappell R (1998) Worlds apart: how the distance between science and journalism threatens America's future. First Amendment Center Publication 98-FO2.

ICSU Scoping Group on Natural and Human-induced Environmental Hazards Report to the ICSU 28th General Assembly, Suzhou, China, October 2005.

Kinzig AD, Starrett K, Arrow S et al. (2003) Coping with uncertainty: a call for a new science-policy forum. Ambio 32: 330–335.

Liverman DGE, Pereira CP, Marker B (2008) Communicating Environmental Geoscience, Geological Society of London Special Publication 305.

Locke J (1690) An Essay Concerning Human Understanding, 1975 edition, (ed) PH Nidditch, Clarendon Press, Oxford.

Lubchenco J (2005) Science's communication gap. International Herald Tribune, Published: Friday 11 November 2005.

Mileti D, Nathe S, Gori P et al. (2004) Public Hazards Communication and Education: The State of the Art. Natural Hazards Informer, Issue 2 (update), Natural Hazards Center, University of Colorado.

Minkel JR, Stix G (2006) Scientific American 50: Policy Leader of the Year. Scientific American website, http://www.sciam.com/article.cfm?id=scientific-american-50-po-2006-12 , accessed 25 August 2008.

MORI/Science Media Centre (2002) Science and the Media. Survey conducted for the Science Media Centre, reported at http://www.ipsos-mori.com/polls/2002/science.shtml, accessed 25 August 2008.

Moser S, Dilling L (eds) (2007) Creating a Climate for Change: Communicating Climate Change and Facilitating Social Change. Cambridge University Press, Cambridge.

Moss RH, Schneider SH (2000) Uncertainties in the IPCC TAR: Recommendations to lead authors for more consistent assessment and reporting. In: Pachauri R, Taniguchi T, Tanaka K (eds) Guidance Papers on the Cross Cutting Issues of the Third Assessment Report of the IPCC, World Meteorological Organization, Geneva, pp. 33–51.

Neild T (2008) Altered priorities ahead – or, how to develop fruitful relationships with the media. In: Liverman DGE, Pereira CP, Marker B (eds) Communicating Environmental Geoscience, Geological Society of London Special Publication 305.

Ogle R (2004) Communicating what the 1% chance flood means. In "Reducing Flood Losses: Is the 1% Chance (100-year) Flood Standard Sufficient?" Association of State Floodplain Managers; 2004 Assembly of the Gilbert F. White National Flood Policy Forum; background papers, p. 136.

Owens S (2000) 'Engaging the public': information and deliberation in environmental policy. Environment and Planning A 32: 1141–1148.

Petterson MG, Tolia D, Cronin SJ, Addison R (2008) Communicating geoscience to indigenous people: examples from the Solomon islands. In: Liverman DGE, Pereira CP, Marker B (eds) Communicating Environmental Geoscience, Geological Society of London Special Publication 305.

Robinson GD Spieker AM (1978) Nature to be Commanded: Earth Science Maps Applied to Land and Water Manage-

ment. United States Geological Survey Professional Paper 950. Government Printing Office, Washington, DC.
Royal Society (2006) Survey of Factors Affecting Science Communication by Scientists and Engineers. Royal Society, London.
Schneider SH (1990) Global Warming: Are We Entering the Greenhouse Century? Sierra Club/the Lutterworth Press, Cambridge, p. 343.
Schneider SH (1997) Defining and teaching environmental literacy. Tree 12: 457.
Schneider SH (2002) Keeping out of the box. American Scientist 90: 496–498.
Simpson C (2008) Communicating environmental geoscience – Australian communication pathways. In: Liverman DGE, Pereira CP, Marker B (eds) Communicating Environmental Geoscience, Geological Society of London Special Publication 305.
Snow CP (1959) The Two Cultures and the Scientific Revolution, 1993 edition. Cambridge University Press, New York, p. 181.
Wang XT (2001) Bounded rationality of economic man: new frontiers in evolutionary psychology and bioeconomics. Journal of Bioeconomics 3: 83–89.
White G (1972) Geography and public policy. The Professional Geographer 24: 101–104.
Williams J (2000) The phenomenology of global warming: the role of proposed solutions as competitive factors in the public arenas of discourse. Human Ecology Review 72: 63–72.
Zehr SC (2000) Public representations of scientific uncertainty about global climate change. Public Understanding of Science 9: 85–103.

Part II
The Response of the International Scientific Community

Introduction of a New International Research Program: Integrated Research on Disaster Risk – The Challenge of Natural and Human-Induced Environmental Hazards

Gordon A. McBean

Abstract Weather-climate and geophysical hazards create many disasters around the world and the impacts have been devastating on many communities and countries. Over the decades there has been significant international scientific response, much of it organized by the International Council for Science (ICSU) and its partners in the United Nations system, especially the World Meteorological Organization and UNESCO. There is also an international policy response. For example, the UN Framework Convention on Climate Change, the 2002 World Summit on Sustainable Development and the related Millennium Development Goals and the World Conference on Disaster Reduction, held in Kobe, Hyogo, Japan in 2005, which agreed on the Hyogo Framework for Action. Through the deliberations of an ISCU-sponsored process, a new international research program Integrated Research on Disaster Risk (IRDR) – the challenge of natural and human-induced environmental hazards – has now been initiated, with the support of ICSU, the International Social Sciences Council and the UN International Strategy for Disaster Reduction. The focus of the research programme is on disaster risk and disaster risk reduction and takes an integrated approach to natural and human-induced environmental hazards through a combination of natural, socio-economic, health and engineering sciences, including socio-economic analysis, understanding the role of communications, and public and political response to reduce the risk. The legacy of IRDR will be an enhanced capacity around the world to address hazards and make informed decisions on actions to reduce their impacts. The IRDR Scientific Objectives are: 1: Characterization of hazards, vulnerability and risk; 2: Understanding decision-making in complex and changing risk contexts; 3: Reducing risk and curbing losses through knowledge-based actions. There are cross-cutting themes and approaches on: Capacity building; Case studies and demonstration projects; and Assessment, data management and monitoring.

Keywords Environmental hazards · Integrated research on disaster risk

Introduction

Almost every day, some community around the globe is affected by a weather-climate hazard, such as a flood, storm or drought, resulting in that community needing outside assistance – this is called a disaster. Together with geophysical (earthquakes and possibly resulting tsunamis, volcanoes and landslides) and

G.A. McBean (✉)
ICSU Planning Group on Natural and Human-induced Environmental Hazards and Disasters, Department of Geography, Institute for Catastrophic Loss Reduction, Social Sciences Centre, University of Western Ontario, London, ON, Canada N6A 5C2
e-mail: gmcbean@eng.uwo.ca

Presentation to 33rd International Geological Congress, Oslo, August 2008

Presented on behalf of the Members of the ICSU Planning Group on Natural and Human-induced Environmental Hazards and Disasters: G. McBean (Canada, Chair); T. Beer (Australia); I. Burton (Canada); C.-J. Chen (Taiwan); O.P. Dube (Botswana); J. R. Eiser (UK); F. Lúcio (Mozambique); H. Gupta (India); W. Hooke (USA); R. Keller (USA); A. Lavell (Costa Rica); D. Murdiyarso (Indonesia); V. Osipov (Russia); S. Sparks (UK); H. Moore (ICSU)

T. Beer (ed.), *Geophysical Hazards*, International Year of Planet Earth,
DOI 10.1007/978-90-481-3236-2_5, © Springer Science+Business Media B.V. 2010

biological events, there are now about 470 disasters per year [1] with widespread impacts across the planet. These events result in great loss of human lives, livelihoods and economic assets in both developed and developing countries. The frequency of recorded disasters has been rising rapidly. From about 65 per year in the 1960s and 200 per year in the 1980s, it reached almost 280 per year in the 1990s.[2] The economic costs of these disasters are also increasing.[3]

In many parts of the world, especially hazard-prone areas, poverty and population growth mean that more people and communities are at risk. Interconnected infrastructure systems, the concentration and centralization of economic and political functions, and complex spatial and functional interrelationships all contribute to the vulnerability of populations to disruptions caused by natural hazards and these contexts are changing rapidly. Natural disasters are capable of cancelling out development gains (Handmer 2003; Mutter 2005), and the risk to development stemming from disasters was clearly recognized by UN Member States in the Millennium Declaration in 2000.[4] The world-wide growth in disaster losses poses a threat to sustainable development.

Globalization results in a world more closely interconnected, with changing senses of responsibility towards countries and localities. Hazard events can have repercussions at a great distance. Globalization also extends in new ways to the geophysical environment. The most salient, but not the only, example is climate change. Changes in the global climate will continue to alter the risks associated with natural hazards. According to the Intergovernmental Panel on Climate Change (IPCC 2007), climate change is accelerating. While the linear warming trend over the last 50 years (0.13°C per decade) is nearly twice that for the last 100 years, a warming of about 0.2°C per decade is projected for the next two decades. With that will come, over the twenty-first century, more frequent hot extremes, heat waves and heavy precipitation events (very likely), and more areas affected by drought (likely). As the tropical sea-surface temperatures increase, it is likely that future tropical cyclones (typhoons and hurricanes) will become more intense, with higher peak wind speeds and intensities of precipitation. Glacier- and permafrost-related hazards such as glacier lake outburst, ice and rock avalanches and impacts on foundations are strongly connected to climate change and increasingly threaten human settlements and infrastructure.

International Scientific Response

Part of the response of human societies to these kinds of events has been to motivate and activate the scientific community to undertake studies to better understand these events so that their characteristics and occurrences can be understood and potentially predicted in order that societies can prepare for them and reduce their impacts. The International Council for Science (ICSU)[5] and its partners in the United Nations system, mainly the World Meteorological Organization (WMO)[6] and the UN Educational, Scientific and Cultural Organization (UNESCO)[7] and its Intergovernmental Oceanographic Commission (IOC)[8] have undertaken many scientific programs on the natural environment and its hazards. The Global Atmospheric Research Programme of the period 1967–1980 was focused on weather and its prediction (Perry and O'Neill 1979). The World Climate Research Programme,[9] since 1980, has had as its objectives the study of predictability of the climate system and the role of human activities in modifying it.

[1] Centre for Research on the Epidemiology of Disasters – www.emdat.be (EM-DAT)

[2] Part of the increase in numbers of disasters reported in disaster statistics may be explained by the increasing numbers of smaller and medium-level events that are registered as being related to natural and human-induced or socio-natural phenomena (ISDR 2007) and by better reporting mechanisms.

[3] MunichRe 2006 Topics Geo – Natural catastrophes 2006 Analyses, assessments, positions. Copyright 2007 Münchener Rückversicherungs-Gesellschaft, Königinstrasse 107, 80802 München, Germany, Order number 302-05217 (available at www.Munichre.com)

[4] The Millennium Declaration of 2000 – see www.un.org XXX

[5] www.icsu.org/10_icsu75/75ANNIV_Achiev_CC.html#tit2

[6] www.wmo.int

[7] www.unesco.org

[8] ioc-unesco-org

[9] wcrp.wmo.int/wcrp-index.html

In the late 1980s, the growing increase in hazardous events led to the 1990s being declared by the UN[10] as the International Decade for Natural Disaster Reduction (IDNDR). The International Framework of Action for the IDNDR had as its Objective "*to reduce through concerted international action, especially in developing countries, the loss of life, property damage and social and economic disruption caused by natural disasters such as earthquakes, windstorms, tsunamis, floods, landslides, volcanic eruptions, wildfires, grasshopper and locust infestations, drought and desertification and other calamities of natural origin.*" The goals of the Decade included: "*(c) To foster scientific and engineering endeavours aimed at closing critical gaps in knowledge in order to reduce loss of life and property.*" It was also recognized that there was a need to devise guidelines, disseminate information and have programs for technical assistance and technology transfer.[11]

In 2004, the ICSU Panel on the Priority Area Assessment on Environment and its relation to Sustainable Development[12] recommended that an ICSU-led Natural and Human-induced Hazards Research Programme be created. A Scoping Group was established in early 2005 and it reported to the ICSU 28th General Assembly,[13] noting that research was needed on how to translate research findings about natural hazards and human behaviour into policies that are effective in minimizing the human and economic costs of hazards. Such research required a multidisciplinary approach focused on the needs of identified customers. The 28th ICSU General Assembly endorsed the recommendation that a new programme be developed and subsequently, the 29th ICSU General Assembly[14] in 2008 endorsed the scientific plan for the new research initiative, now called Integrated Research on Disaster Risk (IRDR) addressing the challenge of natural and human-induced environmental hazards. This paper further describes this Program.

The International Context and the Hyogo Framework for Action

In 1992, the UN Framework Convention on Climate Change (UN FCCC)[15] has as its objective the "*stabilization of greenhouse gas concentrations in the atmosphere at a level that would prevent dangerous anthropogenic interference with the climate system.*" The objective is stated in terms of avoiding "dangerous" anthropogenic interference. In the minds of most people, dangerous corresponds to hazardous and extreme climate-related events – such as floods, droughts, severe storms and heat-waves. The dangerous nature of these events depends in good part on the exposure and vulnerability of communities and these can be controlled and reduced by human actions. The 2002 World Summit on Sustainable Development[16] and the related Millennium Development Goals led to a Johannesburg Plan of Implementation which includes among its commitments by governments, the following selected text from Section IV, paragraph 37 and subparagraph (h):

> "*IV. Protecting and managing the natural resource base of economic and social development*
>
> *37. An integrated, multi-hazard, inclusive approach to address vulnerability, risk assessment and disaster management, including prevention, mitigation, preparedness, response and recovery, is an essential element of a safer world in the twenty-first century. Actions are required at all levels to:*
>
> *(h) Develop and strengthen early warning systems and information networks in disaster management, consistent with the International Strategy for Disaster Reduction;*

[10] www.un.org/documents/ga/res/44/a44r236.htm

[11] UN IDNDR Goals: (b) To devise appropriate guidelines and strategies for applying existing scientific and technical knowledge, taking into account the cultural and economic diversity among nations; (d) To disseminate existing and new technical information related to measures for the assessment, prediction and mitigation of natural disasters; (e) To develop measures for the assessment, prediction, prevention and mitigation of natural disasters through programmes of technical assistance and technology transfer, demonstration projects, and education and training, tailored to specific disasters and locations, and to evaluate the effectiveness of those programmes.

[12] Report of the CSPR Assessment Panel on Environment and its Relation to Sustainable Development

[13] www.icsu.org/Gestion/img/ICSU_DOC_DOWNLOAD/80_DD_FILE_27thGAResolutions.pdf

[14] www.icsu.org/3_mediacentre/PRESS_1.html

[15] unfccc.int/essential_background/items/2877.php

[16] *World Summit on Sustainable Development* www.un.org/events/wssd leading to a plan of action on the Millennium Development Goals www.un.org/millenniumgoals/

In 2005, governments attending the World Conference on Disaster Reduction[17] (Kobe, Hyogo, Japan) agreed on five priorities for action stated as part of the Hyogo Framework for Action,[18] of which the first two were focused on better understanding of disaster risk and enhancing early warning systems and the roles of knowledge, innovation and education to build a culture of safety and resilience. They agreed that: "*The starting point... lies in the knowledge of the hazards and the physical, social, economic and environmental vulnerabilities to disasters that most societies face, and of the ways in which hazards and vulnerabilities are changing in the short and long term.*" It was recommended that research: "*develop improved methods for predictive multi-risk assessments and socioeconomic cost–benefit analysis of risk reduction actions at all levels; incorporate these methods into decision-making processes at regional, national and local levels; and strengthen the technical and scientific capacity to develop and apply methodologies, studies and models to assess vulnerabilities to and the impact of geological, weather, water and climate-related hazards, including the improvement of regional monitoring capacities and assessments.*" These recommendations provided a focus for the design and development of an international research program.

The Scientific Programme of Integrated Research on Disaster Risk

The title *Integrated Research on Disaster Risk – the challenge of natural and human-induced environmental hazards* (acronym: IRDR) was chosen on basis of the rationale of the preceding sections – integration, risk and disasters. IRDR (ICSU 2008) is to be a Research Programme of ten or more years' duration. ICSU and the International Social Sciences Council (ISSC)[19] are the IRDR co-sponsors. The United Nations International Strategy for Disaster Reduction (ISDR) has endorsed the Programme. This paper is a description of that Research Programme which was developed by the ICSU Planning Group on Natural and Human-induced Environmental Hazards and Disasters, as noted earlier. An IRDR Scientific Committee has been established by the co-sponsors to further define the inter-disciplinary scientific strategy and determine its specific objectives and priorities, through an extensive consultation process. The Science Committee will develop an implementation plan and facilitate the exchange of information among the scientists participating in the Programme and the broader community in general.

Scope of the IRDR Programme

The Science Plan of IRDR focuses on natural and human-induced environmental hazards where a hazard is defined as a potentially damaging physical event, phenomenon or human activity that may cause the loss of life or injury, property damage, social and economic disruption or environmental degradation. Note that a hazard "may" cause impacts. The other factor is the vulnerability of the system which is determined by physical, social, economic, and environmental factors or processes, which increase the susceptibility of a community to the impact of hazards. Disasters result when there is the intersection of a hazard and vulnerabilities. Hence to understand the factors involved in disasters it is important to understand both the hazards with their occurrence and characteristics and the factors that make communities vulnerable.

In the last decade, more than 75% of all natural disasters[20] were triggered by weather-related events with floods being the largest single trigger event accounting for 33% of all disasters, in term of global statistics. Although the number of earthquake-triggered disasters was about 7% of the total, they have great impacts in terms of loss of life and destruction of infrastructure and property. Some events, such

[17] World Conference on Disaster Reduction www.unisdr.org/wcdr

[18] Hyogo Framework for Action http://www.unisdr.org/wcdr/intergover/official-doc/L-docs/Hyogo-framework-for-action-english.pdf

[19] International Social Sciences Council – www.unesco.org/ngo/issc/

[20] Centre for Research on the Epidemiology of Disasters – www.emdat.be (EM-DAT) and MunichRe Group information – www.munichre.com/en/publications/default.aspx?publicationLanguage=2&category=17

as impacts by near Earth objects (asteroids for example) are even more uncommon but they can have huge impacts if they occur. Hence, a research program must include examination of all hazards and vulnerabilities, including hazards related to hydrometeorological and geophysical trigger events, i.e., earthquakes; volcanoes; flooding; storms (hurricanes, typhoons, etc.); heat waves; droughts and fires; tsunamis; coastal erosion; landslides; aspects of climate change (for example, increases of extreme events); and space weather and impact by near-Earth objects. The effects of human activities on creating or enhancing hazards, including land-use practices, are included. IRDR will only deal with epidemics and other health-related contexts where they were consequences of the forenamed events. Technical and industrial hazards and warfare and associated activities are not be included per se; however, it is recognized that there is much to be learned from research in such areas and IRDR will take advantage of that knowledge and insight.

The focus of the research programme should be on disaster risk and disaster risk reduction. In order to reduce risk, there needs to be integrated risk analysis, including consideration of relevant human behaviour, its motivations, constraints and consequences, and decision-making processes in face of risks. The physical events themselves can sometimes be directly attributable to human agency (i.e. are "human-induced"), as with many cases of small- and medium-scale flooding, landslides, land subsidence and drought in rural and urban settings related to environmental degradation and human intervention in ecosystems, as well as global climate change. These human-induced or socio-natural hazards are created at the interface of natural and human processes. In addition, human actions determine whether or not an event beyond human control (e.g. heavy rain or an earthquake) will lead to disastrous flooding (e.g. through construction in a flood plain) or building collapse (through inadequate building specifications and techniques). The severity of any disaster also depends on how many people choose (or feel they have no choice but) to live and work in areas at higher risk, as well as on organizational factors relating to protection and emergency planning, and on fundamental aspects of social equity.

Disaster risk management consists of a range of policies and practices developed to prevent, manage and reduce the impacts of disasters, and includes four elements: (1) *Mitigation–prevention* – actions taken before or after a hazard event to reduce impacts on people and property; (2) *Preparedness* – policies and procedures designed to facilitate an effective response to a hazard event; (3) *Response* – actions taken immediately before, during and after a hazard event to protect people and property and to enhance recovery; and (4) *Recovery* – actions taken after a hazard event to restore critical systems and livelihoods and return a community to pre-disaster conditions. The IRDR has, as its first priority, research activities related to mitigation and prevention of disasters and, as a second priority, research on preparedness. Hence, the primary focus of the Plan is on research activities leading to the reduction and control of disaster risk factors and the impacts of natural and human-induced environmental hazards.

The scoping exercise identified the most significant research gaps to be interdisciplinary cohesion, i.e. the intersections of the natural, socio-economic, health and engineering sciences; and the issue of how knowledge about hazards is, or can be, put to use. Public perception–decision making in the context of natural hazards, risks and uncertainty is an important research area, as is the study of human behaviour and cultural contexts for vulnerability analysis.

Vision and Legacy

The IRDR Science Programme is an integrated approach to natural and human-induced environmental hazards through a combination of natural, socio-economic, health and engineering sciences, including socio-economic analysis, understanding the role of communications, and public and political response to reduce the risk.

The legacy of IRDR will be an enhanced capacity around the world to address hazards and make informed decisions on actions to reduce their impacts. This would include a shift in focus from response–recovery towards prevention–mitigation strategies and the building of resilience and reduction of risk and learning from experience and avoidance of past mistakes. Through this enhanced capacity and a shift in strategic approaches, societies, in future, would benefit from a reduction in related loss of life, with fewer people adversely impacted, and wiser investments and choices made by civil society, when comparable events occur.

This legacy clearly implies a strong commitment of IRDR to development – development of science and development of broadly-based capacity. Its partners in this development must include the national and international development assistance agencies as well as the national and international science institutions and funding councils. Capacity to address these issues requires the involvement of all countries in a meaningful way.

An important part of the legacy will be the repository of information and data that have been acquired and that will be of continuing availability and value to the global community.

Research Objectives

IRDR will undertake coordinated, international, multidisciplinary research leading to more effective global societal responses to the risks associated with natural and human-induced environmental hazards. The IRDR is guided by three broad research objectives that will, when projects make successful contributions to them, lead to understanding of hazards, risk and vulnerability and enhanced capacity to model and project risk into the future; to the understanding of the decision-making choices that lead to risk and how they may be influenced; and how this knowledge can better lead to disaster risk reduction.

Objective 1: Characterization of Hazards, Vulnerability and Risk

This objective concerns the identification and assessment of risks from natural hazards on global, regional and local scales, and the development of the capability to forecast hazardous events and their consequences. Recognizing that risk depends on hazards, exposure and vulnerability, the research will be of necessity interdisciplinary. Understanding of the natural processes and human activities that contribute to vulnerability and community resilience will be integrated to reduce risk. The objective addresses the gaps in knowledge, methodologies and types of information that are preventing the effective application of science to averting disasters and reducing risk. A challenge is the broad range of time- and space-scales for hazards and disasters. The response to these varying types of hazards leads to many challenges, and an objective of this Research programme is to understand these connections in ways that will lead to responses contributing towards a reduction in losses.

There are three sub-objectives:

- 1.1: identifying hazards and vulnerabilities leading to risks;
- 1.2: forecasting hazards and assessing risks; and
- 1.3: dynamic modelling of risk.

The natural sciences have a central role in the forecasting of natural hazards and characterizing their attendant risks, and mitigating the adverse effects. Natural and engineering sciences are the basis of technological solutions to early warning, provision of advice to authorities in areas at risk and during emergencies, and the design of effective mitigation strategies to increase community resilience and protection. However, the natural sciences cannot be effective in isolation, with no consideration given to the critical human and environmental factors that lead to disaster. Thus the social sciences have a major role in the assessment of vulnerabilities and risk, as well as developing more effective methodologies.

Objective 2: Understanding Decision-Making in Complex and Changing Risk Contexts

Over the past several decades, human knowledge and understanding of natural hazards has grown dramatically. Today, far more is known about the spatial and temporal distribution of natural hazards and the location of high-exposure areas. Forecasting capacity has also improved dramatically, especially for weather-related events. Far more is now known about the social dimensions of disaster. Yet despite this growth in knowledge, losses associated with environmental hazards have risen during past decades. As noted by the ICSU Scoping Group (2005), there is a great shortfall in current research activities on how science is used to shape social and political decision-making in the context of hazards and disasters.

Objective 2 is focused on understanding effective decision making in the context of risk management – what is it and how it can be improved. In linking with the other objectives, the emphasis is on how human decisions and the pragmatic factors that constrain or facilitate such decisions can contribute to hazards becoming disasters and/or may mitigate their effects.

The political, institutional, cultural and economic aspects of decision-making and behaviour are important and need to be explored. Many of the problems in decision-making are also political and social problems in that they involve divergent interpretations of what the problems and response options really are. There are often conflicting values and interests at work, and strikingly different opportunities to influence developments. The salience of strategic societal choices, and of competing rationalities, which cannot be subsumed within the language of risk and risk management, is recognized, so this broader context will be addressed in the Programme as the research moves beyond the management framework to lay out the complexity of the political and social challenges encountered.

There are three sub-objectives:

- 2.1: identifying relevant decision-making systems and their interactions;
- 2.2: understanding decision-making in the context of environmental hazards; and
- 2.3: improving the quality of decision-making practice.

Risk depends critically on human actions and decisions. Although many forms of human activity may increase, rather than decrease, the damage and danger from natural hazards, from the perspective of the actors themselves such decisions may often appear "rational," and even the only practicable option under the circumstances. Projecting risk into the future will depend, in part, on the choices people make, individually and collectively (through their governments at all levels), and how they implement these choices. Projects designed to meet Objective 2 would identify the decision-making systems, by whom and where the decisions are made, and how these decision-making processes can be understood to provide the basis for intervention when required. From the background and rationale sections of this science plan, it is clear that there are barriers to good decision-making that would lead to effective risk-reduction approaches. Through this process, it is expected that improvements could be made to the quality of the decision-making process. Decision making also depends on the availability of good information. For example, telecommunications and remote sensing are domains in which gaps between operational and scientific activities are easy to identify and have consequences on the decision making. Engineering sciences have a specific role to play in the adaptation of the tools to the need of the decision-makers.

Objective 3: Reducing Risk and Curbing Losses Through Knowledge-Based Actions

"Reduction of risk" refers to all the factors that are contributing to the growing hazard and disaster losses and is an overall objective of the Research Programme. Objective 3 integrates outputs from Objectives 1 and 2. The task of characterizing risk involves identification of hazards and exposure and vulnerabilities of places and people, and hence, assessing the level of risk and understanding how the risk can, may or will change with time. Risk identification requires a multi-hazards approach, since communities are commonly threatened by several different hazards that may be linked to one another.

Since risk results from the interaction of hazards with vulnerable communities, property and facilities and ecosystems which are exposed, all these variables fall within the span of the programme. Reductions in risk can be achieved through implementing and monitoring informed risk reduction decisions (this includes modification of the hazards themselves) and through reductions in vulnerability or exposure. The latter can be achieved by the prevention or discouragement of the occupation of high-hazard-risk zones and sometimes by the relocation or protection of those at risk. Also, the processes of human adjustment or adaptation can be used to reduce vulnerability and increase resilience. Reduction of loss is in the end the central objective, including attention to risk and risk management, and also to the reduction of impacts and the management of uncertainties.

The combination of factors can vary considerably from place to place, and the wide range of disasters experienced in the recent past demonstrates that there is no simple causal explanation. The central thrust of research towards Objective 3 would therefore be to use the combined understanding from many different fields of expertise into an integrated approach to the understanding of the causes of disaster in order to provide practical guidance on the reduction of risk and the curbing of losses. The approaches suggested may be described as diagnostic or forensic. At a superficial or anecdotal level many of the reasons for past failures to reduce risks and curb losses are known. What is not well understood is how these factors work together in different ways and in different places to produce the adverse consequences with which we are more and more familiar? Research towards achieving Objective 3 would develop a new approach to understanding rising risks by bringing to bear and integrating to the extent practicable all existing knowledge of risk factor in order to provide better diagnoses and to lay the scientific basis for more effective policies and actions.

Cross-Cutting Themes and Approaches

The overall global benefits of the Programme are dependent on global capacity building and recognition of the value of risk reduction activities, which are likely to come through successful case studies and demonstration projects. In addition, there would be three cross-cutting themes:

Capacity Building

Capacity or capability can be defined as a combination of all the strengths and resources available within a community, nation or region that can reduce the level of risk, or the effects of a disaster. It includes physical, institutional, social or economic means such as financial, political and technological resources, as well as skilled personal or collective attributes such as leadership and management at different levels and sectors of the society. Capacity building aims to develop human skills and societal infrastructures within a community, nation or region in order to reduce the level of risk. The objectives of the capacity building theme are to:

- Map capacity for disaster reduction;
- Build self-sustaining capacity at various levels for different hazards;
- Establish continuity in capacity building.

Case Studies and Demonstration Projects

The IRDR Scientific Committee will, over the first few years, through working with funding agencies, facilitate case studies to identify major research needs and gaps at the interface of natural and social sciences. The case studies, involving a wide range of hazards, scales, geographical regions, cultural and economic context and implemented through appropriate international and interdisciplinary teams of scientists, would aim at analysis of crises or disasters caused by natural phenomena from which lessons can be learnt. The focus of the analysis would be to establish what was done well and what caused failure. The case studies would elucidate how well methods and approaches applied at the time worked, where there were shortcomings in the science and procedures, learn from examples of good practice, and identify what integrated research is needed within the framework of IRDR. The proposed case studies provide important entry points for social science research and the projects are important as having value in their own right as well as for inputs into integrated models. Disasters and crises can rarely be characterized as complete failures or successes; real situations are always complex and simple categorization is not helpful.

Assessment, Data Management and Monitoring

Data management is an important component of any science project, and in particular, for a global and complex environmental hazards research programme of the scope and complexity of the one proposed. To ensure that the diversity of data from the Programme is collected in a consistent fashion, is preserved, properly archived and made accessible to the science community requires special efforts from the onset.

In order to be able to determine the consequences of environmental hazards and disasters in terms of their impacts and effects, one needs baseline monitoring so as to provide the characteristics of the undisturbed environment and its populations, and episodic monitoring to provide the magnitude of the environmental hazard, and the severity of the impacts and effects that led to the hazard becoming a disaster. For the disaster prevention and recovery community to use such data it is important that a mechanism be in place to permit timely production and dissemination of easy-to-use, accurate and credible information to the appropriate authorities. These assessments, data and monitoring capacities will be an important legacy of the IRDR Programme.

To be able to achieve such a goal requires both long-term ground-based and remotely sensed monitoring, pre-determined methodologies for data presentation, and identification of the gaps in our ability to rapidly provide this information to the disaster managers. This cross-cutting theme would have two objectives:

- Guidelines for consistent data management and assessments of hazards, risk and disasters;
- Applying local assessments globally and global assessments locally

Structure of and Mechanisms for Guidance and Oversight of the Programme

The IRDR Programme will be guided by the IRDR Scientific Committee. There are many international initiatives and activities already existing in the field of natural hazards and disasters including: the international and national scientific programmes either already ongoing or potentially to be initiated, on hazards research and their sponsors; international and national organizations who are involved in development, humanitarian assistance and similar issues; and, in general, governments, private sector and civil society. An ongoing consultative process is essential and the Science Committee will use other forums, where appropriate, such as the ISDR Global Platforms[21] and the International Disaster and Risk Conferences (IDRC),[22] both of which involve governments and a broad range of civil society and business. A session on Hazards and Disasters at the May 2009 World Social Science Forum[23] has be held. A guiding principle should be that creation of new stand-alone forums should be avoided, unless necessary.

IRDR will draw upon the expertise and scientific outputs of many *Partners in research.* For example, the five Geo-Unions of ICSU – IUGG,[24] IUGS,[25] IUSS,[26] IGU[27] and ISPRS[28] – collaborate on a number of issues, including natural hazards and have established the International Year of the Planet Earth (IYPE),[29] which has identified four broad, overlapping research questions which the IRDR will clearly contribute to:

- How have humans altered the geosphere, the biosphere and the landscape, thereby promoting and/or triggering certain hazards and increasing societal vulnerability to geohazards?
- What technologies and methodologies are required to assess the vulnerability of people and places to hazards and how might these be used at a variety of spatial scales?
- How does our current ability to monitor, predict and mitigate vary from one geohazard to another? What methodologies and new technologies can improve such capabilities, and so help civil protection locally and globally?
- What are the barriers, for each geohazard, that prevent governments (and other entities) from using risk and vulnerability information to create policies and plans to reduce both?

21 ISDR GP reference

22 IDRC International Disaster and Risk Conference – www.idrc.info

23 WSSF reference

24 International Union of Geodesy and Geophysics – www.iugg.org

25 International Union of Geological Sciences – www.iugs.org

26 International Union of Soil Sciences – www.iuss.org

27 International Geographical Union – www.igu-net.org

28 International Society for Photogrammetry and Remote Sensing – www.isprs.org

29 International Year of Planet Earth – Yearofplanetearth.org

There is the partnership of IUGS and UNESCO in the International Consortium on Landslides,[30] the International Geoscience Programme (IGCP)[31] and the Scientific Committee on the Lithosphere. The World Organization of Volcano Observatories (WOVO)[32] is the foremost international body dealing with volcanic eruptions. The World Climate Research Programme has identified a focus on climate extremes including floods and droughts. The Intergovernmental Oceanographic Commission of UNESCO promotes the concept of "end-to-end" tsunami warning systems, in cooperation with ISDR and WMO. The WMO Natural Disaster Prevention and Mitigation Programme[33] contributes to different stages of disaster risk reduction. The WMO World Weather Research Programme[34] is a ten-year international study aiming to reduce and mitigate natural disasters by transforming timely and accurate weather forecasts into specific and definite information in support of decisions that produce the desired societal and economic outcomes.

In the social sciences realms of science, there are programs such as the International Human Dimensions Programme on Global Environmental Change's (IHDP)[35] new Integrated Risk Governance Project. The British Psychological Society[36] has recently set up a working party to improve its own response to disasters, crises and traumas. The European Federation of Psychological Associations (EFPA)[37] is working on planning responses to disasters and terrorism at a European level. The goal of Global Risk Identification Programme (GRIP)[38] is reduced natural hazard-related losses in high-risk areas to promote sustainable development.

The Global Change System for Analysis Research and Training (START),[39] presently co-sponsored by the WCRP, IGBP and IHDP, has ongoing projects to build capacity and regional networks in Africa, Asia and Oceania. The Inter-American Institute for Global Change Research[40] has capacity building and research activities in the western hemisphere. The World Bank Global Facility for Disaster Reduction and Recovery (GFDRR)[41] is a partnership that recognizes disaster reduction as a critical dimension of the global poverty reduction agenda.

The Committee on Data for Science and Technology (CODATA)[42] has the expertise in data systems that the Programme can draw upon and one of the ten themes in the Integrated Global Observing Strategy (IGOS)[43] is Geohazards. A theme of the Group on Earth Observations (GEO)[44] is "Reducing loss of life and property from natural and human-induced disasters."

Added Value of an Internationally Integrated, Multidisciplinary, All-Hazards Research Programme

The Hyogo Framework for Action calls for all-hazards approaches, people-centred systems and overall risk assessment. The assessment of the Planning Group was that, despite all the present activities ongoing on natural hazards, there is an imperative for a research programme, sustained for a decade or more, that is integrated across the hazards, the disciplines and the geographical regions, wherein would lie

30 International Consortium on Landslides – www.iclhq.org

31 International Geoscience Programme – www.unesco.org/science/earth/igcp.shtml

32 World Organization of Volcano Observatories – www.wovo.org

33 WMO NDPMP

34 WWRP

35 International Human Dimensions Programme on Global Environmental Change – www.ihdp.unu.edu/ – new project Integrated Risk Governance

36 British Psychological Society – www.bps.org.uk

37 European Federation of Psychologists' Associations – www.efpa.eu

38 GRIP Global Risk Identification Programme – www.undp.org/cpr/we_do/disaster_global_risk_id.shtml

39 Global Change System for Analysis Research and Training (START) – www.start.org

40 IAI

41 Global Facility for Disaster Reduction and Recovery – gfdrr.org

42 International Council for Science: Committee on Data for Science and Technology – www.codata.org

43 IGOS Geohazards Initiative – igosg.brgm.fr/

44 Global Earth Observation System of Systems – www.earthobservations.org

its value-added nature. The coupling of the natural sciences' examination of hazards with the socio-economic analysis of vulnerability and mechanisms for engaging policy decision-making processes will be a major value added. Few research endeavours exist as regards decision making and policy formulation which seek to integrate, from the beginning, social and physical science aspects. Scientific information needs to be combined to more adequately understand how information and knowledge is considered, incorporated and acted on, or not. Hyogo Priority 4 is to "reduce the underlying risk factors." Significantly, the "risk factors" so identified are all socio-political and economic (basic, root causes of disaster) and the research proposed would enhance understanding of these by considering the role of decision-making at all levels, from intergovernmental and multinational organizations down to the individual citizen.

The legacy of IRDR will be an enhanced capacity around the world to address hazards and make informed decisions on actions to reduce their impacts. The legacy will also be the development of science and development of broadly-based capacity and the repository of information and data that have been acquired and that will be of continuing availability and value to the global community.

Acknowledgements The author would like to thank all the members of the ICSU Scoping and Planning groups on Natural and Human-induced Environmental Hazards and Disasters and the ICSU staff (Peter Collins, Howard Moore, Maureen Brennan and others) for their contributions and to ICSU and its supporters for the financial support to enable the Groups to meet and carry out their deliberations.

References

Handmer, J.A., 2003: Adaptive capacity: what does it mean in the context of natural hazards? In: Climate Change, Adaptive Capacity and Development, J.B. Smith, R.J.T. Klein and S. Huq (eds). Imperial College Press, London.

ICSU, 2005: Natural and human-induced environmental hazards. Report from the ICSU Scoping Group. 36 pp. Available at: www.icsu.org/2_resourcecentre/RESOURCE_list_base.php4.

ICSU, 2008: A science plan for integrated research on disaster risk addressing the challenge of natural and human-induced environmental hazards. A report of the ICSU Planning Group on Natural and Human-induced Environmental Hazards and Disasters. 66 pp. www.icsu.org/2_resourcecentre/RESOURCE_list_base.php4.

IPCC, 2007: Summary for policymakers. In: Climate Change 2007: The Physical Science Basis. Contribution of Working Group I to the Fourth Assessment Report of the Intergovernmental Panel on Climate Change, S. Solomon, D. Qin, M. Manning, Z. Chen, M. Marquis, K.B. Averyt, M.Tignor and H.L. Miller (eds). Cambridge University Press, Cambridge and New York.

ISDR (2007): Disaster risk reduction: 2007 global review. UN International Strategy for Disaster Reduction, 98pp, available from:http://www.preventionweb.net/english/professional/publications/v.php?id=1130&pid:34&pih:2

Mutter, J.C., 2005: The earth sciences, human well-being, and the reduction of global poverty. EOS, 86, 16, 157, 164–165.

Perry, J.S. and O'Neill T.H.R., 1979: The global atmospheric research program. Reviews of Geophysics, 17, 7, 1753–1762.

Building a University Network for Disaster Risk Reduction in sub-Saharan Africa

Genene Mulugeta

Abstract Africa is impacted by a multitude of natural and human-induced hazards and disasters; such as drought, flooding, landslides, volcanoes, and earthquakes. These claim thousands of lives, devastate homes and destroy livelihoods. With more than 40% of the population living below the poverty line, sub-Saharan Africa is also the least-equipped and prepared continent to cope with the impacts of these events. In addition lack of detailed information on the economic impact of natural disasters makes it even more difficult to get an accurate picture of the damage caused by natural disasters. Whereas the impacts of different hazards and disasters impacting Africa are frequently interlinked, previous mitigation efforts have usually taken a fragmented approach. Moreover, available science programs at universities do not offer systematic theoretical foundations on vulnerability/sustainability science, nor do they teach practical methods in the fields of disaster prevention and management. These efforts have also been hampered by lack of databases and difficulties with effective dissemination of information. To address and find sustainable solutions to these challenges a scoping team of African scientists, organized by the International Council for Science, Regional Office for Africa (ICSU ROA) has assessed available knowledge and produced a science plan on natural and human-induced hazards and disasters in sub-Saharan Africa. The science plan: (i) outlined the multitude of hazards which impact Africa, (ii) identified the gaps that exist in understanding the nature of this vulnerability, (iii) suggested measures that need to be taken for managing hazard risks, and (iv) proposed strategies for adapting to hazards. The science plan emphasizes the urgent need to build human and institutional capacity to fill in the knowledge gaps through a multidisciplinary involvement of academics in African universities and research institutions. ICSU ROA is now implementing the science plan while linking the planned activities with the objectives of the International Year of Planet Earth (IYPE) and other ongoing initiatives. This paper discusses current activities that are being undertaken by the hazard and disasters task teams of ICSU ROA to assess the risks posed by natural hazards in sub-Saharan Africa. The proposed initiative also stresses the importance of fostering outreach activities to strengthen the link between universities and society through carrying out joint initiatives on issues of hazards and disasters in Africa.

Keywords Natural & Human-induced hazards · ICSU-ROA

Introduction

Africa is in many ways the continent most in need of scientific knowledge to provide solutions and assist its socio-economic development. However, investment in Science, Technology, and Innovation (STI) is frequently a low priority for decision- and policy-makers, and scientific institutions have relatively weak infrastructures. The proposed programme, therefore, is intended to link a number of African universities with

G. Mulugeta (✉)
Coordinator of ICSU-ROA Geohazards Programme, The Baltic University Programme, CSD-Uppsala, Uppsala University, Uppsala, Sweden; Visiting Research Fellow, Vrije Universiteit, deBoelelan 1085, 1081 HV Amsterdam
e-mail: genene.mulugeta@csduppsala.uu.se

T. Beer (ed.), *Geophysical Hazards*, International Year of Planet Earth,
DOI 10.1007/978-90-481-3236-2_6, © Springer Science+Business Media B.V. 2010

international partners in the field of natural hazards mitigation and risks assessment. Because risk assessment is the basis of disaster reduction strategies, this situation calls for a multidisciplinary approach of different scientific institutions, as this can provide the tools and contribute to critical thinking to mitigate disasters. Given the magnitude of disasters impacting Africa, it is essential for the continent to employ better planning, and preventive and mitigation measures to reduce the toll in the first place.

The ICSU ROA hazards and disasters program (Mulugeta et al 2007) is intended to promote research, knowledge sharing and dissemination, training and capacity building to advance disaster risk reduction in sub-Saharan Africa in a multidisciplinary framework. In the long term, ICSU ROA aims to mainstream disaster risk reduction practices into knowledge management so as to reduce vulnerability to future hazards and disasters impacting Africa. In this respect, ICSU ROA will act as a regional focal point to enhance the scientific potential of African universities and research institutions. Furthermore, the organization will work towards advocacy for incorporating research findings into policies, and propose to facilitate planning guides and training activities at all levels in society.

This paper presents ICSU ROA's efforts to establish a university network so as to respond more effectively to the increasingly frequent natural disasters that impact the continent. Some activities envisioned by the ICSU-ROA Hazards and Disasters Programme programme include: (i) Building strong research and training institutions in Africa at national and regional levels (ii) Facilitating the exchange of scientific information and sharing of ideas across borders (iii) Strengthening the link between scientific research and policy- and decision-making (iv) Promoting outreach activities to build resilience to disaster risks (v) Tapping the knowledge base of rural and urban communities (vi) Facilitating international cooperation in monitoring and mitigating hazards and disasters.

Natural and Human-Induced Hazards and Disasters in Africa

Natural and human-induced hazards and disasters impacting Africa are on the rise both in frequency and intensity causing countless deaths and formidable damage to infrastructure and the environment (Fig. 1, Table 1, from Scheuren et al. 2007). The type of natural hazards that often cause major disasters in Africa concern: *Climatological:* drought, extreme temperatures, wild fires. *Hydrological:* floods, wet mass movements. *Meteorological:* storms, and *Geophysical:* earthquakes, volcanoes, dry mass movements - all based on the new classifications of the Database of the Centre for Research on the Epidemiology of Disasters (EM DAT Scheuren et al. 2007). An analysis of reported disasters from EM-DAT shows that during the past five decades the majority of those affected by hazards in sub-Saharan Africa were impacted by hydro-climatological hazards, and almost totally dominated by drought (Fig. 1). Figure 2 compares the number of victims, involving both affected (Fig. 2a) and killed (Fig. 2b), due to natural disasters including epedimics. The data show that the number of people affected by disasters show an increasing trend over the last five decades. By comparison, the total number of people killed for the same time period does not show a continuously increasing trend. This raises the question as to the causes behind the statistics? Is it a result of a better coordination effort by governments and humanitarian assistance agencies? Or is it due to a better preparedness? Such and related questions demand research and knowledge based answers, and emphasize the need for a coordinated approach to devise various disaster mitigation and adaptation strategies.

Compared to other natural disasters like cyclones and floods, droughts impact more people, are the most costly, affect a larger area, and are more frequent. In Africa, droughts contribute to loss of life, poor crop yields, famine, mass migration of people from rural areas to towns, decrease in water availability and allow for a wider spread of diseases, such as Malaria. Since the 1960s, sub Saharan Africa has suffered a series of extensive droughts with serious episodes in 1965–1966, 1972–1974, 1981–1984, 1986–1987, 1991–1992 and 1994–1995. Among the countries frequently impacted by drought are in east Africa (Ethiopia, Kenya, Mozambique). In west Africa (Burkina Faso, Niger, Mauritania, Senegal). In southern Africa (Botswana, S. Africa, Lesotho, Swaziland, Namibia); and in middle Africa, Chad and Angola.

Next to drought, hydrological hazards, such as floods and flash floods are the continents most frequent and injurious natural disaster (Fig. 1). In sub-Saharan Africa, the areas most often affected by flooding are East, West and Southern Africa. The

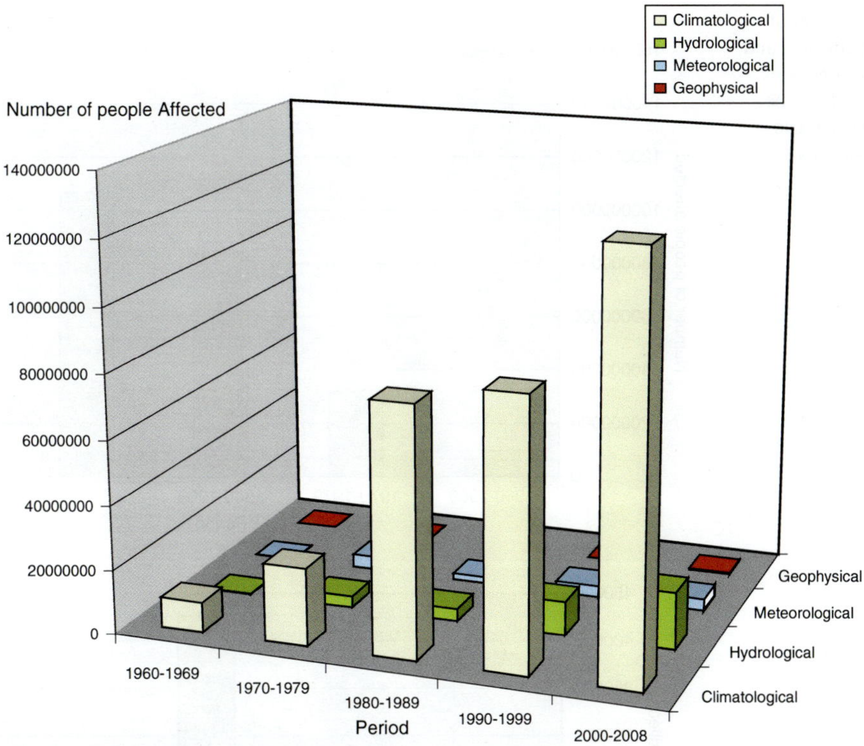

Fig. 1 Number of people affected by disasters, differentiated by hazard type (e.g. climatological, hydrological, meteorological, and geophysical) using EM DAT statistics. The number of people affected is defined to mean displaced or evacuated people requiring immediate assistance, during a period of emergency. *Note*: the last decade is one year less than the others. The Graph is drawn using the numbers shown in Table 1

Table 1 Number of people affected by hazard type

Period	Climatol.	Hydrol.	Meteorol.	Geophys.
1960–1969	9368483	173600	125000	9500
1970–1979	24187600	3633000	4109199	21000
1980–1989	77407567	3943104	1985541	32285
1990–1999	83600531	10547595	3564190	53800
2000–2008	128003578	17677428	3633404	424435

Indian Ocean islands such as Madagascar, Seychelles, Reunion, Maldives; and the coastal areas of Eastern and Southern Africa; e.g. Mozambique are frequently flooded due to the impact of cyclones. The Mozambique floods of February 2000 are a recent example of a flood disaster. Rainfall accompanying tropical cyclone Eline caused excessive flows in rivers, such as the Limpopo, with catchments in other countries. These floods affected a total of about 4.5 million people and caused 700 deaths; losses were estimated at US$500 million, and the GDP growth rate decreased from 10 to 2%.

By comparison, disasters due to geological events have a far smaller impact on Africa, accounting for less than 3% of hazards occurring on the continent (Fig. 1). The area mostly impacted by seismic and volcanic activity is the East African Rift System (EARS). Earthquakes with magnitudes greater than 6 occur almost annually in the East African Rift. Recent events include the February 2006 Mozambican M 7.5 earthquake, which was one of the largest ever recorded in southern Africa. The continent has about 140 volcanoes that have erupted during the last 10,000 years, of which 25 are still active (that is, they have erupted during recent historic times (*ca.* 500 years)). The 2002 volcanic eruption in the Democratic Republic of Congo destroyed 25% of the city of Goma and forced the evacuation of 500,000 people.

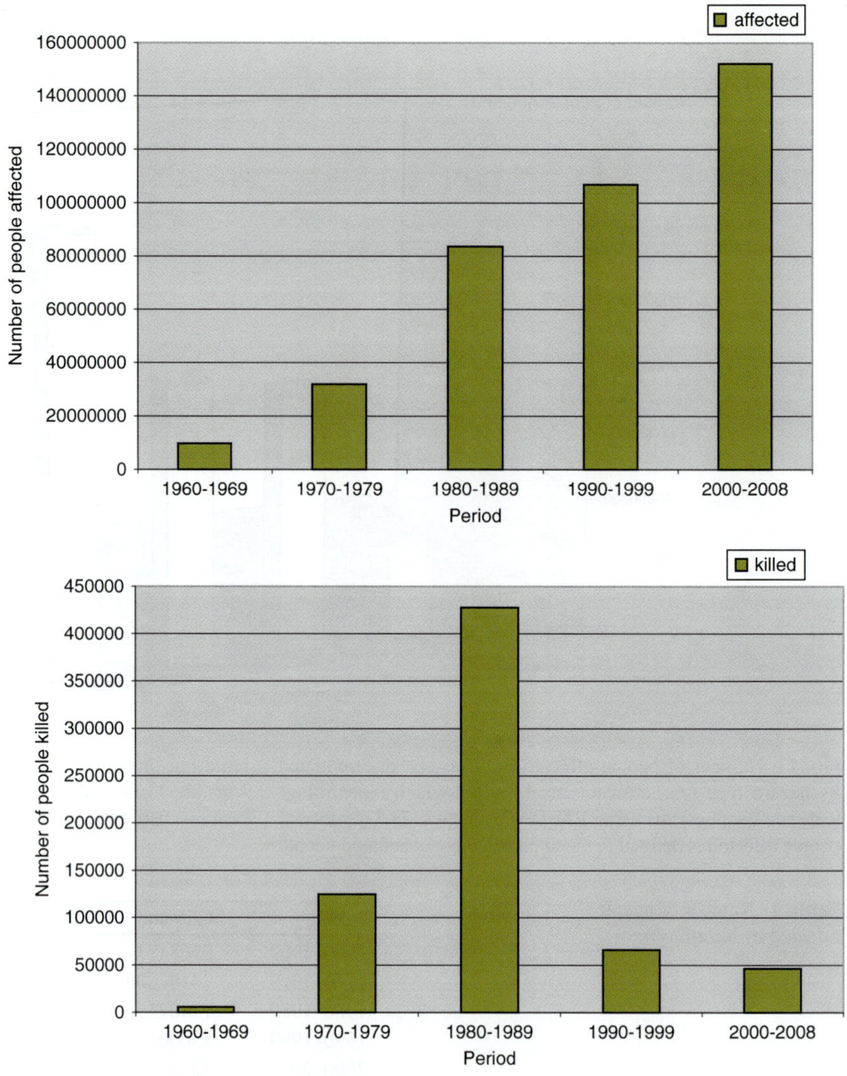

Fig. 2 (**a**) Total number of persons reported affected, and (**b**) Total number of persons reported killed for the same time period. Here, the data also includes epidemics

In addition, human induced hazards; such as climate change, environmental degradation, the HIV Pandemic, high rate of population growth (3.3–3.7%), regional conflicts, (e.g. Sudan, DRC, Somalia), rapid urbanization with consequent air and water pollution, aggravate these natural hazard impacts by increasing the vulnerability of societies and ecosystems already under stress. In order to reduce poverty and hence attain the Millennium Development Goals, sub-Saharan Africa needs to reduce its vulnerability to disasters. All these factors combined reinforce the need for building a university network for disaster risk reduction in sub-Saharan Africa.

Assessing Africa's Vulnerability to Natural Hazards and Disasters

As discussed above, and because of lack of coordinated mitigation and adaptation strategies, Africa is most vulnerable to the impact of natural and human-induced hazards. The fact that most of the sub-Saharan African countries are poor makes this continent the one that is least equipped and least prepared to cope with the impacts of disasters. The majority of the countries in sub-Saharan Africa, at least 30, show Low Human Development Index (LHDI<0.5), thus, the mortality per disaster is much higher than in countries with High Human Development Index (HHDI). For example, Analysis of 2,557 natural disasters reported since 1991, shows that two-thirds of those killed came from countries with Low Human Development, while just 2% came from highly developed nations (World Disaster Report, 2001). Thus, for the majority of the population in sub-Saharan Africa, when a disaster strikes, it destroys not only existing savings, but also income opportunities and livelihoods, thereby aggravating the vicious cycle of poverty and vulnerability. Thus, reducing disaster risk through preventive measures is a central concern for Africa's sustainable development, as well as for meeting the Millennium Development Goals.

While much of Africa is highly prone to hazards and disasters, the different regions have been too little studied, as a result gaps in knowledge are impeding effective adaptation and mitigation strategies. For most countries in Africa, there are gaps in the availability and quality of scientific data, information and skills, and where these are available they are often not sufficiently shared. Moreover, effective strategies to prevent hazards from becoming disasters require not only knowledge generation and sharing but also preventive policies to reduce contemporary and medium to long term risks. Moreover, to better address the risks associated with natural hazards and society's vulnerability to them, it is essential to shift from post disaster response to risk-based hazard mitigation planning. The vulnerabilities outlined above can only be counteracted by enhancing people's capacities to cope up, avoid, and adapt to natural and human-induced hazards and disasters.

Gaps in Knowledge and Research Needs

Multidisciplinary research is required for understanding the causes and consequences of the various hazards and disasters in sub-Saharan Africa, including the impact of climate change. For example, initial studies blamed the causes for the persistence of the drought, starting from the late 1960s on poor land use and the resulting desertification, but recent work indicates that the three decade long drought might have been due to complex interactions among the atmosphere, land, and ocean (Nicholson, 2001; Foley et al., 2003). In this respect, it is essential to carry out a multidimensional assessment of the spatial and temporal trends of droughts in the continent, as well as the relation between drought and land degradation, and the relation between fire and drought. In southern Africa, severe droughts (such as those of 1982–1983 and 1997–1998) have been linked to the El Niño–Southern Oscillation (ENSO) phenomenon. Nearly all climate change scenarios show greater chances of severe droughts over southern Africa, particularly in the central and western areas (IPCC, 2001; Scholes & Biggs, 2004). Increased climatic variability resulting from climate change is likely to exacerbate the problems in the region's recurrent exposure to drought and floods. As a result, both floods and droughts may become more frequent and more severe. Furthermore, research on the effect of climate change on the cyclicity between drought and flooding is required. Research is also required concerning the relations between drought and dust storms and fire hazards. The Sahel region is one of the largest sources of dust storms in the world. Summer storms are due to gusts associated with convective rain-bearing storm systems; whereas winter storms are associated with the Harmattan winds. The dust alters air quality, affecting animals, plants, and the weather. Various centres in Africa, such as African Centre for Meteorological Applications for Development (ACMAD), The Nairobi Drought Monitoring Centre (DMCN), the Drought Monitoring Centre-Harare (DMCH) and AGRHYMET, provide early warnings of drought, capacity building and applications for the mitigation of adverse impacts of drought. For example, the AGRHYMET Regional Centre provides meteorological, hydrological, crops and pasture conditions for West African countries. However, these centres work in isolation and research outcomes are not used for training and capacity building, for example in African universities.

At present, the accuracy and lead times of flood forecasts in sub-Saharan Africa are limited or questionable. Thus, training and research should place emphasis on the prevention of floods. New research and collaborative efforts are needed to advance flood management in the future. This also requires addressing the link between floods and landslides, which is a major hazard in the continent. In this respect, a pilot study of a river catchment system that would involve several countries and include several settings (for example, both rural and densely populated areas) is required.

Tropical cyclones cause huge economic losses, especially in Indian Ocean island states and coastal countries, by damaging dwellings, infrastructure and power plants. Countries such as Mauritius are well prepared for cyclones, while others, such as Madagascar, Comoros, and Mozambique, are more vulnerable. Therefore, further research is needed to assess the risks posed by cyclones. The World Meteorological Organization (WMO) Regional Specialised Meteorological Centre in Reunion serves the sub-region with information concerning cyclone disasters, especially the members of the South West Indian Ocean Cyclone Committee (SWIO). Cyclone warnings are broadcast on radio and television and published in the press. Warnings are also disseminated locally through, for example, schools, religious networks, and government and traditional establishments. In this way it has been possible for countries such as Mauritius to reduce the number of people killed by cyclones. By comparison, countries like Mozambique are not well prepared, as exemplified by the 2000 flood disaster.

Concerning geophysical hazards, an integrated seismic hazards research is needed to assess seismic hazards. This requires the compilation of base maps of known faults, as well as efforts to detect possible unknown faults. It is also necessary to build interactive databases of high-risk areas and integrate them with population distribution, seismic history, and vulnerability to hazards and disasters. To advance seismic research, cooperation needs to be developed amongst existing institutions and networks such as Africa Array at the University of Witwatersrand (South Africa). Research is also required for assessing the stability of structures such as dams, and of the collapse of structures due to mining activities.

The Tsunami generated by the 2004 Indian Ocean earthquake, known as the Sumatra–Andaman earthquake, also impacted coastal areas in east Africa. The hazard created an awareness of the need for a tsunami warning system for the Indian Ocean. Before the event of 26 December 2004, little research had been done to address the risk of Tsunamis in the region. A survey conducted by UNESCO's Intergovernmental Oceanographic Commission (IOC), the World Meteorological Organization (WMO), and the International Strategy for Disaster Reduction (ISDR) showed that African countries have limited capacity to implement mitigation measures for Tsunami hazards effectively. This prompted the UN to start working on an Indian Ocean Tsunami Warning System and had the first monitoring facility in place by 2005.

Africa's preparedness for monitoring proximal volcanic hazards and for responding to future disasters is inadequate. Systems have been installed to monitor seismic, thermal, and gas emissions. These need to be complemented with satellite-based monitoring systems such as global navigation satellite systems and radar imagery for better mitigation strategies. Remote sensing data (temperature, gases, geodetic, infrared) as well as telemeter monitoring of magnetic and electric fields, gases, and temperature are essential to monitor volcanoes. An ongoing project at the Royal Museum of Central Africa in Tervuren (Belgium) is studying and monitoring African active volcanoes (SAMAAV), using radar interferometry to examine the recent evolution and assess the risks associated with four active volcanoes (Mount Nyiragongo, Mount Cameroon, Mount Fogo, and Mount Oldoinyo Lengai). This work is being done in collaboration with African volcanologists, including those in Cameroon, the DRC, and Tanzania.

Thus, reducing vulnerabilities to hazards and disasters in sub-Saharan Africa requires: (i) development of early warning systems and devising efficient communication strategies for timely information on the occurrence of disasters; (ii) improving the technological preparedness in areas prone to specific types of hazards such as floods and cyclones; (iii) developing methodologies for risk assessment and management, and (iv) producing hazards and disaster vulnerability maps. Moreover, the resulting risk and probability analysis for each hazard type will need to be linked to social vulnerabilities. Such an approach is essential because early warning systems and detailed hotspot maps of risks at the regional and sub regional level are lacking in sub-Saharan Africa. Some of these measures can be addressed through establishing multidisciplinary research programmes. In this respect, universities can play leading roles to mitigate hazards and stop the trend towards increasing vulnerability from natural disasters.

ICSU ROA Planned Activities

ICSU-ROA is in the process of initiating and implementing the following activities.

Hazards and Disasters Research Programme

Underlying ICSU-ROA's hazards and disasters research programme is a concern that African research on hazards and disasters is fragmented and insufficiently integrated and developed, at all levels. As discussed above, there is a good deal of research activity in Africa, but this is not integrated and efficiently used for building mitigation and adaptive capacities. These measures can be brought about by increasing capacity of universities and institutions to train academically qualified staff that are able to apply new methodology and technology for hazard and risk assessment and management in risk reduction measures. The ICSU-ROA hazards scoping teams, through their network of scientists will also design university based programs that foster interdisciplinary education programs rather than traditional education based on single disciplines. This will require not only new tertiary programs, but also changes in the way institutions of higher education work through communicating and advancing interdisciplinary hazard education. The proposed interdisciplinary approach would also stimulate regional cooperation through the free movement of persons and ideas.

In order to get researchers from different disciplines to collaborate, ICSU-ROA has undertaken three steps; namely: (1) Arranging regional workshops to establish contacts among African researchers. (2) Establishing task teams to assess the risks posed by natural hazards, and (3) Initiating joint projects on different themes. This enables the various institutions to pool together their training and research resources.

ICSU-ROA has identified two major multidisciplinary research programmes to implement its science plan. Each programme will involve a number of research themes, and will be implemented regionally by the participating universities; thus enabling the sharing of results and information across the thematic initiatives via workshops and scientific meetings. The research programmes involve: (1) Geohazards in Africa linked to the International Year of Planet Earth (IYPE), and (2) Hydro-meteorological hazards in Africa. Each research programme will be driven by African scientists organized in the different themes; such as earthquakes, drought, fire, landslides and volcanic hazards.

The *Geohazard* research programme will focus on 5 sub-projects, all spanning over a period of five years: with an emphasis on interdisciplinary collaboration between the different themes through consultation with the participating institutions and appropriate partners. Each theme will encompass perhaps four to five African universities, and will bring together selected scientists to develop new thematic research lines as well as develop capacity building initiatives, mainly through networking and exchanging experiences. The proposed geohazard research themes concern:

- Assessing and mitigating the earthquake hazard in Africa.
- Monitoring hazards from Volcanoes and explosive crater lakes
- Mitigating the impact of Landslides in sub-Saharan Africa
- Sea-level changes and pollution in coastal areas of Africa
- Deformation of the African Lithosphere: Integrated modelling and implications for Geo-hazards

The *Hydrometeorological* Hazard Programme will focus on four themes aimed at assessing the vulnerability of socio-ecological systems in sub-Saharan Africa to hydro-meteorological hazards and disasters, and the resilience of communities to these events. Particular attention will be given to drought and flooding which are the two major hydro-meteorological hazards in the continent, but the programme also covers fires (wild and urban), and dust storms. The themes are:

- Drought hazards and disasters in Africa
- Flood and flash flood hazards and disasters in Africa
- Dust events and related hazards in Africa
- Climatic and socio-economic determinants for anthropogenic fires in African savannahs.

A summary of the contents of these research themes are as follows:

Assessing and Mitigating the Earthquake Hazard in Africa

The earthquake hazard project is aimed at expanding the human and physical capacity for seismic monitoring and hazard assessment in Africa; compiling a comprehensive catalogue of African earthquakes and a seismotectonic map of Africa; developing models of the Earth's structure and its evolution to address fundamental problems in global tectonics and assist mineral exploration; and increasing the awareness of geoscience challenges and solutions amongst decision makers and the public, through outreach activities, especially in the framework of the International Year of Planet Earth.

In this sub-project it is proposed to carry out four linked tasks in parallel; namely:

Task 1: Compile African Catalogue of Earthquakes
Task 2: Prepare Seismotectonic Map of Africa
Task 3: Seismic Hazard Assessment of the African continent
Task 4: Integrated Seismic Hazard Assessment of vulnerable areas (e.g. major African cities, and critical lifeline infrastructure).

Monitoring Hazards from Volcanoes and Explosive Crater Lakes

This subproject aims to identify and locate the most active volcanoes, review their past and current status, and characterise their behaviours; design methods of monitoring the volcanoes; assess and document the locations and periodicities and impacts of CO_2 out-gassing from crater lakes and establish structural and morphological links between such crater lakes and nearby active volcanoes; monitor the physico-chemical characteristics of water in identified explosive crater lakes as well as the types of gaseous emissions from them, develop volcanic hazard maps and conceive appropriate approaches of education and public awareness raising. Although systems have been installed to monitor seismic, thermal and gas emissions in volcanic regions of some countries, these need to be complemented with satellite based monitoring systems. Remote sensing techniques are essential in so far as monitoring eruptions of volcanoes and CO_2 gas explosions from the crater lakes are concerned.

Mitigating the Impact of Landslides in sub-Saharan Africa

The major issues of concern for this sub-project include developing policies and guidelines for landslide management; comprehensive landslide mapping and loss assessment; and regional landslide forecasting and monitoring. So far little research has been done on the causes, mechanisms and mitigations of landslides hazards in Africa. The causes for the initiations of landslides remain poorly understood. The mitigation measures which should be applied should relate to the economic, social and environmental realities of the continent to facilitate effective implementations. It is therefore very necessary to undertake and promote further research on the initiations, controlling parameters and mitigation measures of landslides in the continent.

Sea-Level Changes and Pollution in Coastal Areas of sub-Saharan Africa

The project will cover studies on vulnerability to climate change-enhanced eustatic sea level rise and viable adaptation options along sub-Saharan Africa, as well as coastal area pollution. The project aims at developing proper diagnosis of coastal erosion along critical and oceanographic distinct regions of the coastal states of sub-Saharan Africa using first-order principles of surf zone flow and beach profile interaction on the short term, and historical chart analyses of coastline changes, on the long term.

Deformation of the African Lithosphere: Integrated modelling and Implications for Geo-Hazards

The specific objective of this sub-project is to advance the understanding of deformation of the African lithosphere and its societal implications, and consequences with respect to hazards. Recent research has revealed insights into extensional and continental break-up processes of the African lithosphere. Rifting in Africa is mainly localized along the broad topographic uplift, the surface expression of which is the East African Rift System (EARS). The proposed research will involve geodynamic modelling of continental rifting along the EARS to study the link between the surface brittle deformation to processes occurring in the lower lithosphere and mantle. The dynamics of the African lithosphere controls the regional stress field, mode of deformation and type of magmatism.

Drought Hazards and Disasters in Africa

In the Sahel, precipitations have been decreasing since the beginning of the twentieth century. The year 1968 has been recorded as the driest; it has thus been used as the reference point. After 1970, drought events have increased in frequency. The project proposes to address the environmental and social impacts of drought and desertification in the Sahel. The project aims to set up a long term warning and monitoring system of climatic change, drought, and desertification. The output will be effective tools for decision makers to enable them to devise better mitigation and adaptation strategies. The project will incorporate a regional outreach dimension since desertification is not a local phenomenon. Rural and urban populations as well national research centres will also be involved in the project.

Dust Events and Related Hazards in Africa

Dust storms frequency is found to be increasing due to change in land use and land cover that is also related with the increase of population. The latest estimates put global dust emissions at about 2,000–3,000 million tons each year with half emitted from Africa, where in some parts, annual dust production has increased ten-fold in the last 50 years. It is proposed to carry out

- detailed analysis of multi satellite sensors data at spatial and temporal scales of past African dust storm events to study the aerosol parameters, meteorological parameters and air quality that will provide inter-annual variability of dust and dispersion of dust.
- analysis of aeronet data to study the characteristics of the dust.
- back trajectory modelling to study the origin of dust storms and track of these dust.

Climatic and Socio-Economic Determinants for Anthropogenic Fires in African Savannahs

Biomass burning (BB) is a typical feature of the African continent, and as a result of its spatio-temporal extent and intensity, has a wide and far reaching impacts in a number of domains at local, regional and global scales. In particular, BB is used as a farm management practice to prepare farms for the growing season. This sub-project proposes to investigate:

- the impact of Bushfire-induced atmospheric composition changes on the local-to-regional climate and hydrological cycle in Africa.
- the long-term impacts of BB on the socio-economic development of Africa based on a series of climate change scenarios.
- the physical and socio-economic determinants of the causes and effects of BB in West Africa.

Flood and Flash Flood Hazards in Africa

This project aims at enhancing the human resource base for flood and flash flood monitoring and management in Africa. The objective is to improve the scientific basis of flood forecasting, while developing tools for effective early warning systems. Floods are the most devastating natural hazards and can be predicted, except in the case of flash floods. Flash floods generate sudden and great volumes of water that flow rapidly and cause inundation. The impact of flooding can include loss of life, destruction of property, crops, and cattle. At present, the accuracy and lead times of flood forecasts in sub-Sahran Africa are limited or questionable; thus research is required on the prevetion of flood hazards.

Outreach Activities

One of the challenges of addressing risks associated with hazards and disasters in sub-Saharan Africa is raising the awareness of the impact of these to the public, decision makers, and policy makers. Thus, ICSU-ROA intends to foster outreach activities in the region, through university/community partnerships, by providing a knowledge base for adaptation to environmental and human induced hazards at all levels, national, regional, municipality and community level. This is essential as not enough is known about the interactions among the natural and social forces that impact the region.

A physical understanding of how natural systems work is not sufficient for hazard mitigation and preparedness. Thus, an integrated approach involving both natural and social sciences is also suggested to better understand the linkages between the economic, social, and political dimensions of extreme natural events. This information is currently inadequate and will also require developing research tools on risk analysis as well as require multi-hazard, risk assessment maps. Societal factors, such as how people view both hazards and mitigation effort play a critical role in determining which steps are actually taken, which are overlooked and, thus the extent of future disaster losses. This requires also involving the impacted people in the research and hazard impact assessment process. ICSU-ROA advocates that not only is it essential to work with those directly affected by disaster but that there needs to be an interdisciplinary, university/society combined approach to disaster risk reduction in which the benefits of both scientific and indigenous knowledge can be utilised to the best advantages in a culturally compatible way. This also provides a framework for interacting with and educating those affected by hazards and disasters. A preliminary effort in this direction has already being taken by ICSU-ROA for promoting public awareness of the impact of disasters. ICSU/ROA has arranged a first field workshop on disaster risk reduction in Mozambique in connection with the 29th ICSU general assembly in Mozambique.

Multidisciplinary Hazard Database

The numbers behind the disasters discussed above using EMDAT, though very useful for comparisons of the magnitude and type of disasters are estimates of recorded incidents only and the real numbers are most probably higher, due to various reasons; such as lack of communication in rural areas where the majority of the population live, and lack of equipment and capacities for effective assessment. In addition, there may exist other inconsistencies to the core data that pertain to: (i) lack of a consistent methodology for tracking and recording the data, (ii) overlap of data (as the same hazard, e.g. drought in one year can continue to the next year) and (iii) lack of a multiyear hazard and locations analysis; as discussed, for example, by Below et al (2007). This prompts the need for establishing a multidisciplinary database for hazard mitigation and preparedness.

Sub-Saharan Africa has not yet established a systematic database or a single repository for hazards and disasters data. As a result, the losses caused by African disasters are often underestimated. ICSU-ROA proposes to build a regional hazards and disasters database for collecting, analysing, and storing data on losses from past and current disasters. Disaster databases are useful, as information stored in the database can be used to analyse disaster loses, and for comparison with future losses. Databases are also useful as early warning tools and for informed decision making. However, available databases such as the Emergency Disasters Data Base (EM-DAT), managed by the Brussels-based Centre for Research on the Epidemiology of Disasters (CRED), NatCat and Sigma lack systematic, sub-national patterns of disaster loss, as well as standardized local and national disaster data collection. The proposed ICSU-ROA Database should include information on the types of losses, their locations, their specific causes, and the economic losses, and the distinction between regional, national and sub-regional impacts.

International Cooperation

International cooperation is vital in dealing with regional hazards and disasters mitigation so as to save lives, eradicate poverty, and promote sustainable development. At present, ICSU-ROA is promoting increased cooperation with key partners such as UN/ISDR (e.g UNESCO, IUGG, UNDP), the World Meteorological Organization (WMO), the International Federation of Red Cross, the United Nations Office for the Coordination of Humanitarian Affairs (OCHA) as well as with ongoing scientific programmes (e.g IYPE) to share knowledge and technology related to sustainable hazard mitigation as well as to address the root causes of vulnerability to hazards, In partnership with these and other organizations, ICSU-ROAs scientists will address among others some of the pertinent questions, outlined in the Geohazards theme of IYPE, with focus on sub-Saharan Africa; such as: (1) What technologies and methodologies are required to assess Africa's vulnerability of people and places to hazards and how might these be used at a variety of spatial scales? And (2) How do hazards compare relative to each other regarding current capabilities for monitoring, prediction and mitigation and what can be done in the short and long term? ICSU-ROA intends to engage African scientists and institutions to find solutions to these pertinent questions.

Conclusion

This paper assessed the impacts of natural hazards in sub-Saharan Africa. It also reviewed current activities that are being undertaken by the hazards team of ICSU-ROA to build the scientific knowledge base necessary for disaster risk reduction. It also outlined the necessary measures that need to be taken to build the resiliency of communities in sub-Sahara Africa to manage hazards so that these do not develop into disasters. Such an integrated and multidisciplinary approach in a university framework has many advantages by:

- Allowing joint use of resources such as data, information systems, software tools and access to stakeholders (authorities, disaster management agencies, media etc.)
- Facilitating capacity building, education, communication, training for hazard mitigation and reduction.
- Building a database for assessing the impacts of human actions on hazard risk
- Facilitating international and regional cooperation and coordination for sharing information, knowledge and experiences.
- Supporting outreach activities for hazard reduction.

In this effort the ICSU-ROA hazards and disasters programme will play a critical role in promoting the urgently needed regional and global conversations for disaster risk reduction.

Acknowledgements The author would like to acknowledge the funding support provided by the organizers of the 33rd International Geological Congress that enabled participation in the congress.

References

Below, R., Grover-Kopec, E. and Dilley, M. (2007). Documenting drought-related disasters: a global reassessment. The Journal of Environment & Development 16, 328. DOI: 10.1177/1070496507306222.

Foley J.A., Coe, M.T., Scheffer M. and Wang G. (2003). Regime shifts in the Sahara and Sahel: interactions between ecological and climatic systems in Northern Africa. Ecosystems 6, 524–539.

IPCC (2001). "Africa" (Chapter 10), in climate change 2001: impacts, adaptation and vulnerability, Cambridge University Press, Cambridge, pp. 487–532.

Mulugeta, G., Ayonghe, S., Daby, D., Dube, O.P., Gudyanga, F., Lucio, F. and Durrheim, R. (2007). Natural and human-induced hazards and disasters in sub-Saharan Africa. ICSU Regional Office for Africa Science Plan, p. 36.

Nicholson S.E. (2001). Climate and environmental change in Africa during the last two centuries. Climate Research 14, 123–144.

Scheuren, J.-M., Le Plain de Warux, D., Below, R., Guha-Sapir, D., Ponserre, S. (2007). Annual disaster statistical review. The numbers and trends 2007, Centre for Research on the Epidemiology of Disasters (CRED), Brussels.

Scholes R.J. and Biggs R. (eds) (2004). Ecosystem services in Southern Africa. A regional assessment. Council for Scientific and Industrial Research, Pretoria.

World Disaster Report (2001). http://www.ifrc.org/publicat/wdr2001/

Co-operation Plan on Hazards & Disasters Risk Reduction in Asia and the Pacific

Harsh Gupta

Abstract With the passage of time the impact of natural and human-induced environmental hazards and disasters continues to increase. ICSU is establishing a major new international initiative "Integrated Research for Disaster Reduction." Considering that the geographical area covered by the ICSU Regional Office for the Asia and Pacific (ROAP) accounts for more than one-half of the world population and about 80% of all losses due to natural hazards globally, a Science Plan to address three categories of hazards, namely earthquakes, floods and landslides has been developed. We realize that the Asia-Pacific region has a large number of islands and island countries, which are more vulnerable to natural hazards due to their geographical locations. Therefore, the issue of islands and natural hazards is specifically addressed.

The niche of the proposed Science Plan is to utilize the latest knowledge and best practices to address problems related with earthquakes, floods and landslides so to prevent hazards becoming disasters.

Keywords Integrated Research for Disaster Reduction · Hazard mapping

H. Gupta (✉)
National Geophysical Research Institute, (Council of Scientific and Industrial Research), Hyderabad – 500 007, India
e-mail: harshg123@gmail.com

This paper is based on ICSU Regional Office for Asia and Pacific (ROAP) Report: "Science Plan on Hazards and Disasters" prepared by a Planning Group with Harsh Gupta (Chair), Daniel Murdiyarso, Trieu Dinh Cao, Chamhuri Siwar, Seree Supharatid, Chen Dehui, Kyogi Sassa, Christel Rose, James Terry, and Mohd. Nordin Hasan

Introduction

The geographical area of Asia and the Pacific, accounts for more than one half of the global population and about 80% of all losses due to natural hazards globally (Fig. 1). In developing the Science Plan for an Asia-Pacific Hazards Research Program it was decided that initially earthquakes, floods and land-slides would be addressed.

The reason for this is that in the recent past, the Tangshan earthquake of 1976 in China claimed an estimated 242,000 human lives. The 2004 Sumatra earthquake of Mw 9.3 and the resultant tsunami claimed an estimated 250,000 human lives. Muzaffarabad earthquake of 2005 claimed 88,000 lives. The Wanchun earthquake of 2008 claimed 90,000 lives. Japanese seismologists estimate that if the 1923 Kanto earthquake, which heavily damaged Tokyo and Yokohama area causing some 120,000 deaths, was to repeat today, the economic losses could mount to US $ 1.2 trillion. Floods have caused havoc in the Asia-Pacific region. Floods and strong winds induced by two tropical storms caused 130,000 deaths in Bangladesh in the year 1991. Material damage due to summer floods in China during 1998 rose to US $ 20 billion. The 2008 Nargis storm and resultant floods claimed more than 120,000 lives in Myanmar. Landslides in Nepal during 2002 displaced some 266,000 people.

A unique feature of the Asia-Pacific region is the presence of a large number of islands and island countries, which are inherently more vulnerable to earthquakes, floods and landslides. At least 40 people died in the Solomon Islands after a tsunami swept ashore following the M 8.1 earthquake on 2 April 2007. Vanua Leve Island in Fiji witnessed worst ever recorded

T. Beer (ed.), *Geophysical Hazards*, International Year of Planet Earth,
DOI 10.1007/978-90-481-3236-2_7, © Springer Science+Business Media B.V. 2010

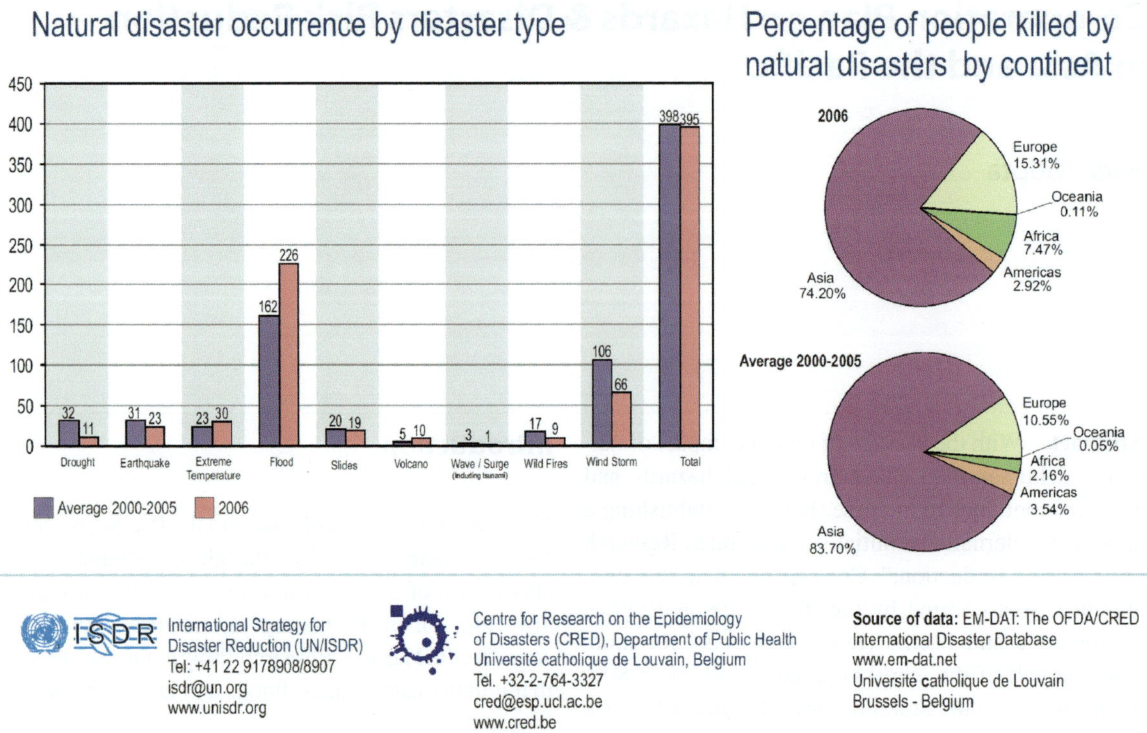

Fig. 1 Natural disaster occurrences by disaster type and percentages of people killed by natural disasters by continent (Source: UN/ISDR www.unisdr.org; CRED www.cred.be; and EMDAT www.em-dat.net)

floods as a consequence of Tropical Storm Ami on 12 January 2003, when 17 human lives were lost.

Aims and Objectives

The Science Plan envisages an integrated approach to the management of natural and human-induced environmental hazards and disasters to reduce the risk of resultant losses of life and property.

It aims to enhance the science of hazards and disaster management in Asia and Pacific and contribute to reducing the likelihood of hazards becoming disasters.

The Natural and Human-induced Environmental Hazards and Disasters Research Program seeks to accomplish these aims by undertaking coordinated multidisciplinary research leading to more effective societal responses to the risks associated with natural and human induced environmental hazards in the Asia-Pacific region. The legacy of the program would be an enhanced capacity to address hazards and make informed decisions on actions to reduce their impacts related to loss of lives and properties. There are several ongoing national and international efforts in the region, specifically addressing the problems of earthquakes, floods and landslides. The proposed program aims to identify niches and gaps in the current efforts and find a way to improve the situation, in collaboration with all concerned.

Earthquakes

Monitoring and Observation Systems

In the Asia-Pacific region there are well-equipped countries with modern seismographs. There are also several very poorly instrumented countries that do not have even the basic facilities to locate earthquakes. In addition to the seismographs, during the last couple of decades of the twentieth century, there has been a revolutionary development in estimating the accumulation of strains using global positioning system (GPS). Integrated global models of present day plate motions and plate boundary deformations have been prepared (for example Kreemer et al. 2003). Like seismographs, the installation of GPS in Asian and the Pacific countries is uneven. Therefore, one of the aims of the program is to identify such gaps and find ways and means to fill them.

Hazard Mapping

The Global Seismic Hazard Assessment Program (G-SHAP) was a flagship program of the International Decade of Natural Disaster Reduction (IDNDR). Under this program some 500 scientist from all over the world prepared the Global Seismic Hazard Map, which depicts the anticipated peak ground accelerations at the bed rock (Giardini 1999). Figures 2 and 3 depict them for the Asia-Pacific region. A major effort is required to estimate accelerations at the earth surface, particularly in the densely populated urban areas located in or near the zones of high seismicity to prepare micro-zonation maps.

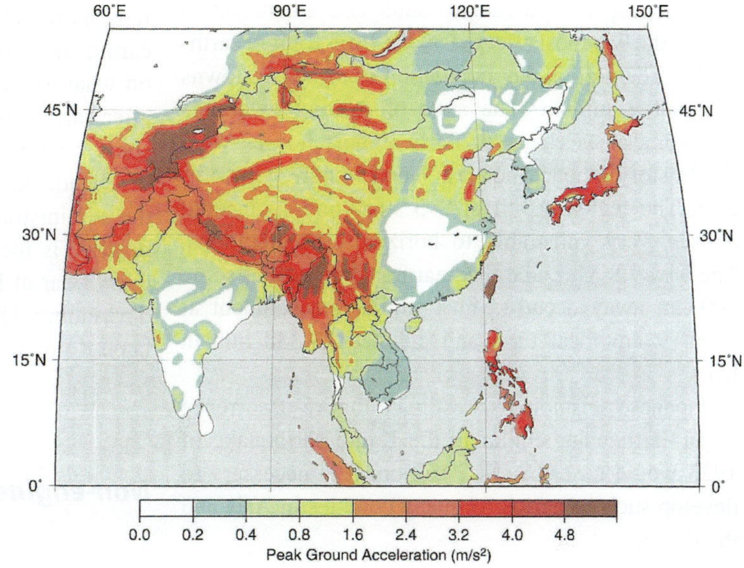

Fig. 2 Seismic hazard map of Asia depicting peak ground acceleration (PGA), given in m/s^2, with a 10% chance of excedence in 50 years. (www.seismo.eth.ch/GSHAP/)

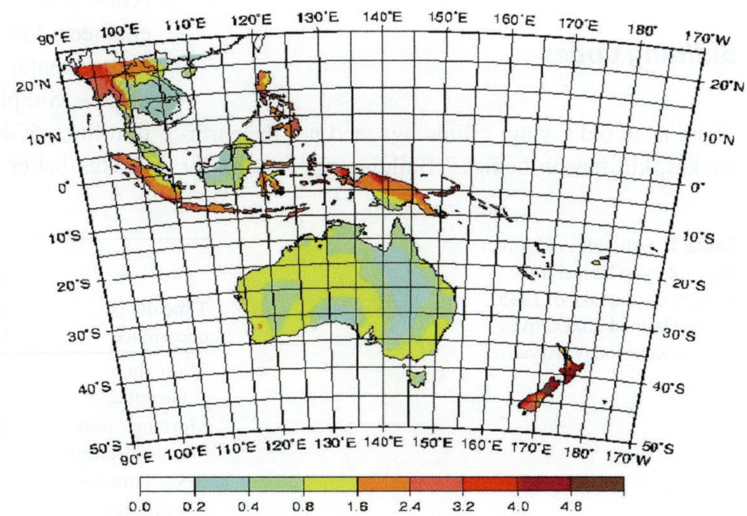

Fig. 3 Seismic hazard map of the Australia-South Pacific-Southeast Asia region depicting peak ground accelerations (PGA), given in m/s^2, with a 10% chance of exceedance in 50 years. (Giardini 1999)

Earthquake Scenarios

It is known that earthquakes re-occur at the same places. However, with the passage of time the population density and the style of buildings of the region changes. It is therefore helpful to develop scenarios of the consequences were an earlier earthquake to recur today. Arya (1992) estimated the number of human lives that would be lost if the Mw 7.9 Kangra earthquake of 1905 in the foothills of Himalaya was to reoccur today. The estimates are given in Table 1. In 1905 this earthquake claimed about 20,000 lives. However, over the last 100 years, the population density has increased about four fold. In 1905 about one-half million people lived in Dehradun and Dharamshala towns: the two most affected by the 1905 Kangra earthquake. The present day population of these two towns is about 2 million. The style of making dwellings and homes has changed. Earlier, most houses were single story wooden houses. Wood is now not available. The region is full of 2–3 story non-engineered houses, very vulnerable to horizontal accelerations. The Mw 7.8 Muzzafarabad earthquake, located some 200 km away occurred in a similar environment at about 9 am local time and claimed 73,338 human lives.

Japanese seismologists estimate that re-occurrence of an earthquake similar to the Kanto earthquake of 1923 would cost U.S. $1.2 trillion. It is necessary to develop such scenarios for many counties in Asia and the Pacific.

Building Codes

There is an old saying: "Buildings, and not the earthquakes, kill human beings." Building codes exist for most of the Asia-Pacific countries. What is lacking is their implementation. It has been observed that in the developed world the loss of human lives by earthquakes has decreased while the economic losses keep on rising. Whereas in the developing world, both are increasing (ASC 2006). So a major effort is required to assess the current status of building codes, their appropriate development and implementation.

Education and Awareness

It is extremely important to learn how to live with earthquakes. A lot of educational material is available on what to do and what not to do before, during and after earthquakes. Earthquake drills are carried out in some countries to remind citizens of the possibility of earthquakes and also to check the preparedness of the administrative machinery. A good example of this practice is the exercise conducted on the 15 January every year at Katmandu, Nepal in the memory of the devastating 15 January 1934 earthquake that devastated Katmandu.

Non-engineered Structures

A large percentage of the population in the Asia-Pacific countries lives in self-designed non-engineered structures. Such homes are very vulnerable to horizontal accelerations generated by the earthquakes. Simple inexpensive methods are available to retrofit such dwellings to make them withstand earthquakes better. It is recommended that such methods

Table 1 Estimates of human lives likely to be lost if the Kangra Earthquake of 1995, which claimed 20,000 lives, was to occur today (Arya 1992)

Time of occurrence	Deaths in collapsed house (%)	Deaths in part-collapsed house (%)	Total potential deaths
Midnight (sleeping)	40	20	344,000
Morning (awake and sleeping)	20	10	177,000
Noon Time (out working)	10	5	88,000

be taken up as a major initiative in earthquake prone rural areas.

Earthquake and Tsunami Warning Systems

Earthquakes

As of now, short-term earthquake forecasts are not available to cause effective evacuation and saving of human lives. Earthquakes generate both body and surface waves. However, it is known that most of the damage is done by the surface waves generated by the earthquakes (for example Richter 1958). These waves travel with a velocity of about 3 km s^{-1}, whereas the longitudinal body wave travels with a velocity of about 7 km s^{-1}. If the epicenter of a damaging earthquake is located some 300 km away from a major city, and there is a good network of seismic stations, it would take 25–30 s to locate the earthquake and estimate its magnitude. Whereas the surface waves would take about 100 s to reach the city. This gives a lead-time of about a minute to shut off electric supplies, gas lines, and passenger trains as well as to get people out of vulnerable spots.

Tsunami

Tsunami warning is a reality. The occurrence of a large earthquake in a tsunamigenic zone is one of the possible causes that may (or may not) generate a major tsunami. Local tsunamis can be generated by other geological processes such as failure of shelf slope, dissolution of gas-hydrates, meteorite impact etc. Travel times of tsunamis from various locations to coasts have been calculated. Once a tsunamigenic earthquake is detected, warning of the time of arrival and amplitude of a tsunami at specific locations is announced. This service has been available from the Pacific Tsunami Warning Centre for the past several decades for countries located on the coasts of the Pacific Ocean. However, there have been several false alarms. Consequent to the devastating tsunami caused by the 26 December 2004 Sumatra earthquake, efforts are under way by Australia, India, Indonesia and other countries to develop robust tsunami warning capabilities in the Indian Ocean.

Triggered Earthquakes

Artificial water reservoirs are built all over the world for flood control, irrigation and power generation. Under certain geological conditions, these reservoirs can trigger earthquakes. Triggered earthquakes exceeding magnitude 6 have occurred at least at four sites including two in the Asia-Pacific region. As of now, over 100 sites of triggered earthquakes are known globally (Gupta 2002, Fig. 4). Methods are now available to find safer sites. These should be taken into consideration while searching for suitable sites for artificial water reservoirs.

Floods

The Asia-Pacific countries have a major threat from floods. In one of the most affected countries, Bangladesh, as many as 80 million people are under the threat of floods. Over the years, several disastrous flood events, mostly induced by tropical cyclones or seasonal rains have occurred. Classical examples are the 130,000 deaths caused by floods in Bangladesh in 1991 (Chaudhary 1991), and the summer 1998 floods in China causing material damage worth US $ 30 billion (NCDC 1998). Floods also cause loss of top soil affecting fertility of the land. Figure 5 has images of flooding in Mumbai, India due to unprecedented rain of 994 mm in 24 h (Badami 2005).

Prevention and Mitigation

An important issue is public education based on the experience and flood related data collected by national and the international agencies. Mitigation also involves identification of vulnerable areas and better enforcement of building codes, construction regulations and evacuation procedures.

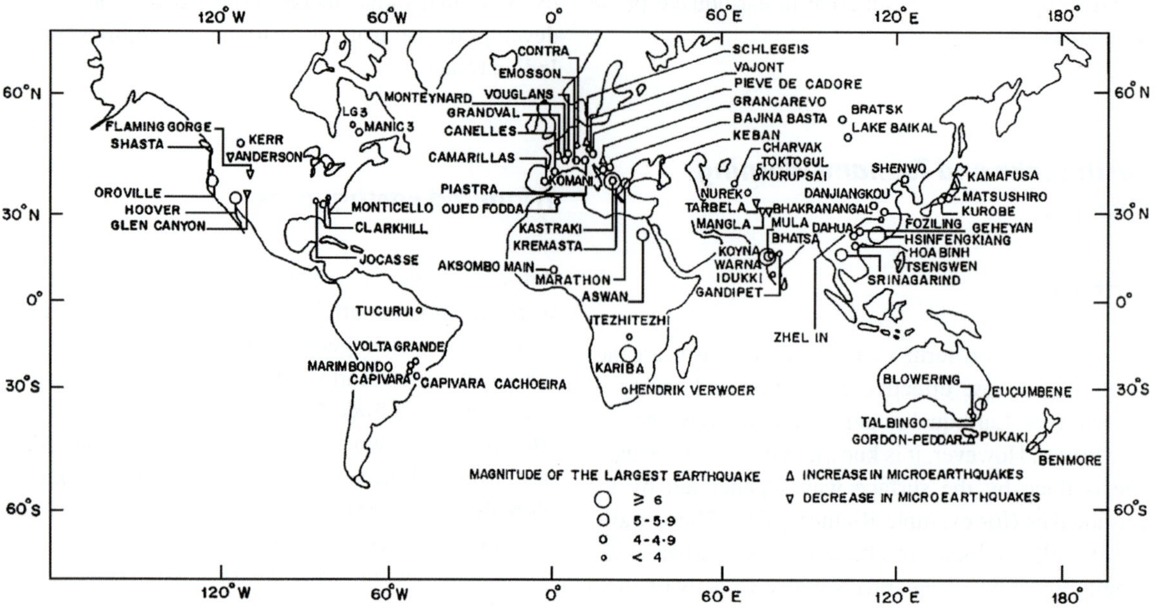

Fig. 4 Global distribution of known sites of earthquakes triggered by artificial water reservoirs. Earthquakes exceeding magnitude 6 occurred at Hsingfengkiang, China and at Koyna, India in the AP region (Gupta 2002)

Fig. 5 Images of disastrous flooding in Mumbai, India, after an unprecedented rainfall of 994 mm on 26–27 July 2005, causing 1,000 deaths and economic losses of US $ 750 million

Demonstration Studies and Linkages

Case studies of devastating floods are important to: (1) improve understanding of disaster processes, (2) to develop new disaster reduction strategies, (3) further improve flood prediction, and (4) enhance local and regional capabilities to better respond to such disasters. A few case studies, such as the Mumbai floods of 2005 and the seasonal floods in Bangladesh, would be instructive.

It is also proposed to take up demonstration projects to show the benefit of an integrated approach, new tools, and techniques to reduce the flood disaster. The key actions of the demonstration projects would be:

1. Flood monitoring in a targeted area to demonstrate the benefits of transition from research to application.
2. Ensemble prediction of flood hazards: to demonstrate the new ensemble flood forecast with a

prototype hydrometeorological ensemble forecast model driven by a multiglobal forecast model under the framework of the World Weather Research Programme of the World Meteorological Organization, for enabling risk reduction strategies in the countries most vulnerable to flood disasters.

3. Integrated flood management of a big river basin (e.g. the Yellow River in China): to demonstrate integrated basin flood management to reduce vulnerability while still preserving local ecosystems and biodiversity.

Floods are closely associated with extremes of the Earth's climate system and the effects of global warming, climate change and increases in heavy precipitation are now being felt in many parts of the Asia-Pacific region. As a consequence, flood-generation mechanisms are becoming more complex. Seasonal variation and geographic distribution of many flood-causing factors add to complexity. Flood disaster management will therefore benefit from new and strong linkages with existing research programmes on, landslides, health (water-borne diseases), economic and social vulnerability.

Improving Monitoring Systems

Improved monitoring of hydro-meteorological events is the key to reducing flood disasters. It is necessary to deploy modern automatic stations, Doppler radar, sodar, lidar, satellite based observing systems and other GPS-GIS platforms. It is desirable to collaborate with the existing global programs such as GEO (Group on Earth Observation) and GEOSS (Global Earth Observation System of Systems).

Predictability, Hazard Assessment and Regional Flood Warning Systems

Two approaches to flood forecasting in vogue include the empirical prediction approach based on statistical data, and numerical prediction based on the physical laws governing geophysical fluids. Flood predictability research should focus on quantification of uncertainties and construction of ensemble prediction systems. It is also important to improve the data quality used for various forecast components.

Most developed countries have sophisticated flood prediction and warning systems but this is not the case with the developing countries. It would be helpful to develop flood forecast models for big river basins and then to transfer them to regional and local scale scenarios in the Asia-Pacific region.

Major Research Activities

Geographical Flood Hazard Mapping

It is useful to investigate the past flood events in detail, and based on the knowledge gained from such studies develop the risk maps which should be used to optimize structural and non-structural flood control measures.

Investigation of Flood Mechanism

Floods are a part of the natural hydrological cycle, thus it is important to investigate regional influences on: weather systems responsible for heavy precipitation; hydrology of surface run off processes; and impact of urbanization and population growth on flood vulnerability and resultant management strategies.

Effects of Climate Change

Several climate models indicate an increase of temperature by 0.5–2 °C by 2030 in the Asia-Pacific region. Increased precipitation and rise in sea level are the other relevant issues.

Landslides

Landslides and the resultant disaster has been a topic discussed in various fields of science, engineering and administration. The accepted definition of landslide is "Movement of a mass of rock, debris or earth down a slope" (Cruden and Varnes 1996). Landslides are classified by the combination of types of movements (slide, flow, fall, topple and spread) and types of material (rock, debris, earth) such as rock slide, debris slide, debris flow, earth flow and rock fall (Terzaghi 1950). Topple is a rotational forward movement and spread is movement mostly on a flat ground. Topple and spreads are minor groups, and landslides belonging to these two categories do not occur frequently.

Landslide Causes

The mass movement of a landslide is initiated by "shear failure" of materials, when the shear stress is equal to or greater than the shear resistance on the slope. Therefore, landslides are caused by the increase of shear stress or decrease of shear resistance (and also a combination of these two factors).

Increase in shear stress can be due to earthquakes, tectonic movement and volcanic activity, as well as due the geological processes of erosion and deposition. Decrease in shear resistance can be due to pore water pressure changes, weathering, and ground water flow.

Climate change could also induce more landslides. For example climate change affects the pattern of rainfall and snow melt in many regions. Heavier and longer rainfalls may cause new landslides. However, a lot more is yet to learned on this issue.

Monitoring and Observation Systems

The velocity of landslide varies from one type of landslide to another and from region to region. Its knowledge for a specific locale is extremely useful in devising preventive measures. Knowledge of landslide frequency is crucial for probabilistic hazard evaluation. It is also important to know the cause of earlier landslides: was it an earthquake or a period of intense rain or a combination of the two? Did other anthropogenic factors such as deforestation or removal of earth play a role? The cause of a landslide trigger can change with the geological times.

Predictability and Hazard Assessment

Although landslides can be mitigated by engineering works, these should be cost-effective. Hence, low-cost technology engineering design should be available, especially in the developing world. Technology transfer is to be encouraged.

To develop an early-warning system for landslides is difficult. However, efforts in this direction need to be encouraged, especially in areas of high population density. False warnings are likely, but are a part of development.

Among elements to be considered in the system are the area that will be affected, and the possible loss of life and property. The early warning system should not only be restricted to mapping of vulnerable areas but also zoning of the potential hazards and disasters.

The system should be continuously upgraded, and public awareness increased in both urban and rural areas.

Prevention and Mitigation

Landslide may be prevented by stabilization of slope. There are three basic approaches:

1. Geometric methods, in which the geometry of the hillside slope is changed. Rock slope and soil slope geometric modifications will require different techniques.
2. Hydrogeological methods, in which an attempt is made to lower the groundwater level or to reduce the water content of the material.
3. Mechanical methods, in which attempts are made to increase the shear strength of the unstable mass or to introduce active external forces (e.g. anchors, rock or ground nailing) or passive external forces (e.g. structural wells, piles or reinforced ground) to oppose the destabilizing forces.

Linkages

Human Activity

The Department for International Development (DFID) Landslide Risk Assessment project has

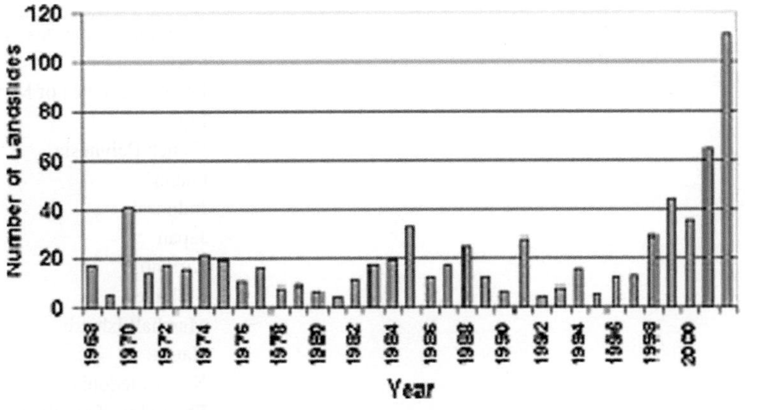

Fig. 6 Numbers of landslides per year in Nepal

attributed the rise in the annual number of landslides in Nepal since 1995 to anthropogenic activities (DFID 2005). This is illustrated in Fig. 6, based on a detailed landslide database compiled for Nepal for the period 1968–2002.

In recent years it has become increasingly apparent that humans are the key factor in the initiation of landslides.

Climate Change

The IPCC's Fourth Assessment Report has underlined the vulnerability of developing countries to climate change and its consequences, particularly in the Asia-Pacific region.

Landslide research and prevention should consider climate change and variability, especially in terms of rainfall regimes and their effect on landslides.

Cooperation with other international initiatives

Landslides are fully or partly targeted by other international initiatives. They include International Programme on Landslides (IPL) by ICL, UNESCO, ICSU, IUGS, IUGG, WMO, IFI, IGCP, IOGS, IYPE and others.

Special Vulnerability of Islands in the Asia-Pacific Region

Several of the Asia-Pacific nations are island nations, and many countries have large territories under islands. These areas are particularly more vulnerable to natural disasters (Bettencourt et al. 2002; Lawson 1993; Terry 2007; and WMO 2006). Table 2 provides a list of selected island nations and a few continental nations with off shore islands and significant island population. There are large island countries, such as Borneo and New Guinea with area exceeding 100,000 km^2. And there are small nations such as the Ha'apai group of central Tonga with area less than 10 km^2. Asia-Pacific is also unique as it contains all the five world's nations that are entirely atolls – Tuvalu, Kiribati, Tokelau, the Marshall Islands and the Maldives. Two broad categories of factors that make these areas vulnerable to natural hazards are physical and the human.

Physical Factors

Plate Boundary Associations

Most of the islands in the Asia-Pacific region owe their origin to plate boundary tectonics, and are situated on the "Pacific Ring of Fire" meaning high seismic and volcanic activity (Fig. 7). Volcanic activity can trigger a variety of landslides such as debris slide, earth flow and lahars.

Steep Lands

Volcanic islands have steep slopes and rugged topography and recently erupted volcanic material such as volcanic ash and pyroclastic flows, favoring landslides.

Table 2 Selected island nations of the Asia-Pacific region

Examples of nations or territories comprising islands only	
Cook Islands	New Zealand
Federated States of Micronesia	Niue
Fiji	Papua New Guinea
French Polynesia	Philippines
Guam	Singapore
Indonesia	Solomon Islands
Japan	Sri Lanka
Kiribati	Taiwan
Maldives	Tokelau
Marshall Islands	Tonga
Nauru	Tuvalu
New Caledonia	Vanuatu
Examples of continental/mainland nations with offshore islands and/or significant island populations	
Australia	Korea
China	Thailand
India	Malaysia

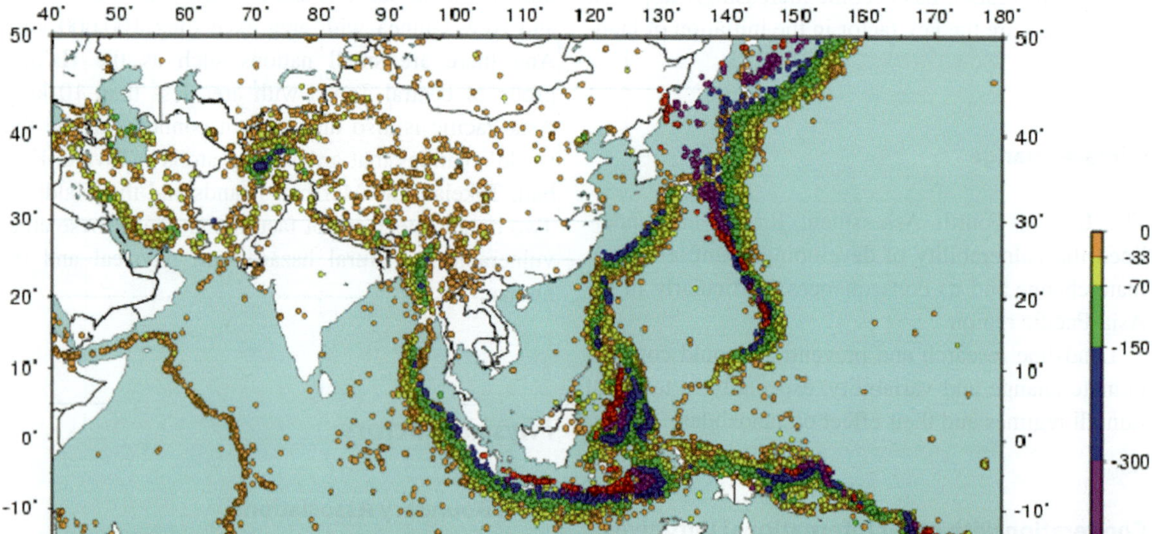

Fig. 7 Earthquakes associated with plate boundaries in the A-P region (USGS)

Wet climate and Tropical Cyclones

There is abundant rainfall exceeding 1,500 mm/yr, several islands having much more. Most of the precipitation is during the wet season or monsoons, resulting in large seasonal variations in the hydrological behavior of the rivers. Typhoons are frequent: the average number being 20 typhoons/yr in the North West Pacific and 10 in the South West Pacific.

Small Size and Low Elevation

Thousands of islands in the Asia-Pacific region are very small and/or low lying. Tsunami waves (generated by submarine earthquakes) and storm surges (generated by tropical storms) can completely engulf such places. Populations inhabiting such islands are therefore extremely vulnerable to sea flooding, as there is no high land providing opportunities for evacuation.

Human Factors

Remoteness and Inaccessibility

Many islands are in remote regions and the inhabitants are isolated. When natural hazards strike, the problems are multiplied due to:

- Hazard warnings systems are inadequate or non-existent.
- It is difficult for government authorities to make rapid assessments of the scale and severity of hazard impacts and the number of people affected and in need of assistance.
- There can be long delays in evacuating communities to safe locations.
- Emergency supplies such as food, medicines and temporary shelter cannot be easily or quickly transported to the afflicted areas.

Marginalization and Dependency on Local Resources

Island populations generally benefit less from national funding programmes and large-scale development projects. They also tend to be marginalised from the mainstream of economic activities in major conurbations. As a consequence, island communities are more dependant on natural resources and their local environments (forest products, freshwater resources, coral reefs, mangroves and coastal fisheries), as well as subsistence agriculture for their traditional livelihoods. This increases their vulnerability to hazards because in the event of a natural disaster, the health and productivity of these natural ecosystems are also badly affected, which can have a severe long-term impact on the food security of island populations in addition to the short-term impacts that are suffered immediately.

High Population Densities

Some small islands in the Asia-Pacific region are at great risk because of their extremely high population densities. Overcrowding has become a serious problem in many national capitals of island states because of the limited land area available for habitation, coupled with the "pull" factors influencing in-migration from outlying locations where access to employment and services are limited. For example on Betio the main islet of the capital of Kiribati, located in the south west of Tarawa atoll, 15,000 people live on an area less than 1 km^2 in size (Fig. 8).

Disaster Management and Capacity Building

Natural disasters have major socio-economic and health impact. These include human lives lost; spread of diseases; damage to property; loss of livelihood, jobs; and psychological stress.

Disaster risk management consists of a range of policies and practices developed to prevent, manage and reduce the impacts of disaster. These are:

Fig. 8 In Betio on Tarawa atoll (capital of Kiribati) 15,000 people live on just 0.8 km^2 of land. This overcrowding on a tiny low-lying islet greatly increases the vulnerability of the population to natural hazards

1. *Mitigation-prevention* – actions taken before or after a hazard event to reduce impacts on people and property;
2. *Preparedness* – policies and procedures designed to facilitate an effective response to a hazard event;
3. *Response* – actions taken immediately before, during and after a hazard event to protect people and property and to enhance recovery; and
4. *Recovery* – actions taken after a hazard event to restore critical systems and livelihoods and return a community to pre-disaster conditions.

Role of Social Sciences

Within the scope of disaster management, socio-economic research entails undertaking a coordinated research leading to more effective societal response to risks associated with natural and human induced environmental hazards, so as to make informed decisions on actions to reduce their impact (Bardet et al. 1995; del Ninno et al. 2001; Mekong River Commission 2006).

An important aspect of socio-economic studies would be to create a data base of the lessons learned from past disasters incorporating success stories, and case studies of impact analysis.

Improving the Quality of Decision Making

Quality of the information available and the manner in which it is used are factors that can influence decision-making. Cost benefit analysis has the potential to provide a decision-making tool for defining the problems and scoping the costs and benefits of alternative solutions.

Economic Valuation of Losses and Damages

It is not easy to estimate monetary value of a disaster because direct and indirect costs are involved. The direct costs include loss of physical and human capital, the cost of relief and clear up operations, rehabilitation, reconstruction, repair and maintenance. The indirect costs include reduced real estate value; loss of tax revenue; loss of industrial, agricultural and forest productivity; loss of tourist revenues; cost of preventive measures to mitigate future potential disaster; and consequences of diverting Government expenditure to relief efforts etc.

The focus on direct damage is largely due to the fact that there are difficulties in accounting for indirect and non-monetary damage and because economic studies of this nature are, not surprisingly, a low priority in the post-disaster recovery efforts.

Private Versus Public Costs

The direct and indirect costs as listed above may be further classified into private or public costs depending on who bears the burden of the costs or losses.

Macroeconomic Costs of Disasters

Several studies provide estimates of the macrolevel impacts of disasters. These include impacts on key macroeconomic variables (e.g. GDP growth), on particular sectors of the economy (e.g. agriculture), and total damage costs to physical and human capital at the national level. Macro-economic impacts of disaster impinge on national economic parameters such as GDP, employment, consumption, and business sentiments. Loss of potential incomes from tourism and investments may also occur. At the sectoral level, reduction in sectoral production (agriculture/industries/services) may occur, affecting employment, household incomes, consumption, expenditure, etc. through the industry-household linkages.

The magnitudes of these impacts depend on the sizes and durations of disasters, the structure of the economy, measures taken ex-ante to mitigate any impacts, the government's policy response to the shock, and the amounts and forms of external assistance. Most assessments of disaster impacts only focus on quantifying immediate direct damage in financial terms (i.e. the non market economic costs such as the value of lives lost, are not addressed).

The economic costs consist mainly of immediate damage assessments, in order to provide governments and donors with estimates of the amount of funds required to address emergency and reconstruction needs and by insurance companies for compensation. Long-term indirect costs in terms of flow of goods and services, reduced level of production and non-market impacts such as environmental damage and psychosocial effects are frequently omitted from such assessments (DFID 2005).

It is estimated by the World Bank that during 1990–2000, natural disasters resulted in damage that constituted between 2 and 15% of an exposed country's annual GDP.

Microeconomic Losses

The micro-economic impacts of disaster include effects on household resources and well-being due to loss or damage of household (e.g. house, appliances, utensils, assets, furniture, vehicles, etc.) and other assets (e.g. land, machinery, trees, etc.). The loss of crops and employment creates loss of income and security.

In the spatial and sectoral contexts in the rural and agricultural sector, there may be reduction of production of food and cash crops, affected through reduction of area and value of crops lost or damaged. There may also be reduction of livestock production due to livestock or cattle killed or injured. Public utilities and facilities may be also damaged.

For those who are working in the non-farm sector, there may be loss of non-farm employments and/or wages through loss of workdays retrenchment, displacement or wage deductions. In the urban/industrial sector, there may be loss of business assets, investments, opportunities and products. Urban houses may be partially or totally damaged. There may be damage to public amenities (e.g. schools, phones, playing fields, community centres, religious facilities, etc).

Role of the Private Sector and Civil Societies

The role of the private sector and civil societies is crucial in conducting vulnerability assessment and developing appropriate strategies to address calamities. Smart partnership strategies could, at local and regional levels, bring public and private sector experts and NGOs together with hazard researchers to develop vulnerability assessments and coping strategies, both pre-event mitigation plans and emergency response plans, and to provide input to establish government initiatives to evaluate and strengthen community resiliency nationwide.

The emergency response plans could serve to mobilize, within countries, government agencies and external donors and international programmes to provide the resources needed for such community-based efforts such as hazard maps, forecasts and outlooks, inventories of check lists of emergency plans, best practices of emergency plans, information on community profile, social and cultural vulnerabilities, and cost-sharing of implementation, etc.

NGOs and the private sector could also participate to enhance the "culture of prevention" which involves activities to provide avoidance of the adverse impacts of hazards and a means to minimize related environmental, technical and biological disasters. A case can be made for preventive measures, public awareness and education related to disaster risk reduction. This can lead to changed public attitude and behaviors, contributing to the culture of prevention. NGOs, with the support of the private sector, may organize community-based educational awareness programmes on disaster prevention and emergency relief.

Capacity Building

In the context of disaster management, capacity or capability has been defined as a combination of all the strengths and resources available within a community, nation or region that can reduce the level of risk, or the effects of a disaster. It includes physical, institutional, social and economic means as well as skilled personal or collective attributes such as leadership and management. Capacity building aims to develop human skills and societal infrastructures within a community, nation or region in order to reduce the level of risk and also to manage the aftermath and impacts of disaster on human, infrastructure and estate.

In order to integrate science for the benefit of society related to natural disasters, the knowledge and resources of the international science community need to be garnered to:

1. identify and address major issues of importance to science and society with respect to natural disaster;
2. facilitate interaction amongst scientists across all disciplines and from all countries;
3. promote the participation of all scientists in the international scientific endeavour; and
4. provide independent, authoritative advice to stimulate constructive dialogues among the scientific community and governments, civil society and the private sector.

Policy Relevance

For policy relevance, in the event of a natural disaster such as earthquake, landslide and flood, the most immediate capacity building concern is the establishment of a *Disaster Response and Management Committee*, whose tasks will be to coordinate and take immediate actions to assist and lessen the burden of the disaster victims. Several countries already have such a mechanism in place. The committee will need to establish procedures and coordinate relief and support activities at the federal, state, provincial/district and local levels. Support and relief may come from welfare agencies, private sector, religious and community organizations and NGOs, and also from foreign countries and organizations. The participation of the scientific community in this committee is essential to provide informed and authoritative advice to government in matters related to scientific and technical aspects of the disaster, risk monitoring systems, trauma counselling, disaster management, conducting environmental and socio-economic impact studies, etc.

Success Stories

At this stage it appears appropriate to cite a couple of success stories, among many, that show how the application of scientific and technological developments, interwoven with social fabric, improved the situation. One such example is that of mitigation of loss of human lives by cyclones on the Indian coasts. In 1977 about 20,000 lives were lost on the east coast of India during a super cyclone. Subsequently the entire coast of India was covered by radars and suitable arrangements were made to conveying the cyclone warnings (Fig. 9). As a result the number of lives lost was significantly reduced, to about 1,000 in 1996 and 27 in 2005 when the east coast of India was hit by cyclones similar to those of 1977.

Fig. 9 Cyclone monitoring radars on the Indian Coast

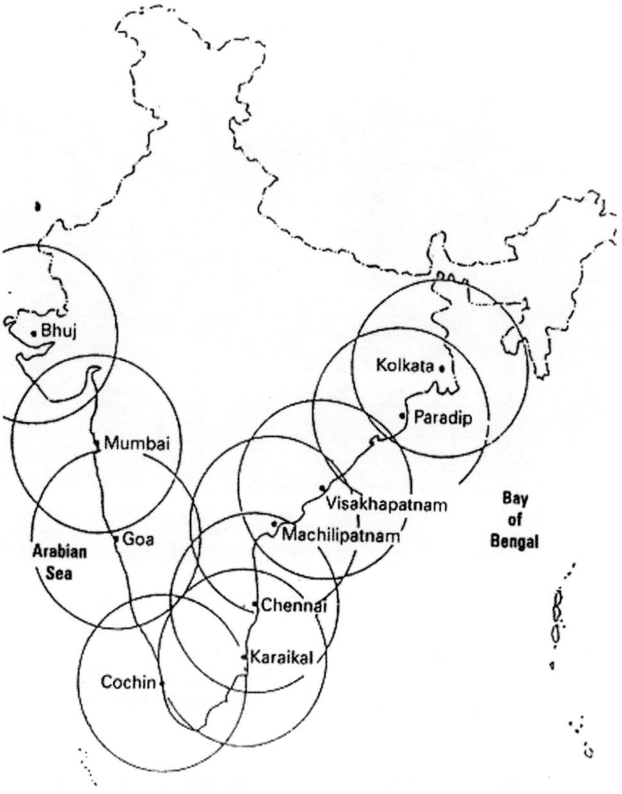

Another example is tsunami hazard mitigation. Soon after the occurrence of the devastating tsunami caused by the 26 December 2004 Sumatra earthquake, India took the initiative to set up an end to end system to mitigate tsunami and storm-surge hazard. It was pointed out that unlike the circum-pacific earthquake belt, where the entire belt is capable of generating a tsunamigenic earthquake, for the Indian Ocean, only Java-Sumatra to Andaman – Nicobar and Makaran Coast regions of Alpide belt can host tsunamigenic earthquake (Fig. 10). The mitigation system includes all the necessary elements: networking of seismic stations; deployment of ocean bottom pressure recorders; real time sea level monitoring stations; establishment of radar based monitoring stations for real time measurement of surface currents and waves; modeling for tsunamis and storm surges; generation of coastal inundation and vulnerability maps; operation of a tsunami and storm surges warning centre on 24×7 basis; capacity building and training of all the stakeholders and communication with the global community. This initiative was estimated to have a direct cost of US $ 30 million and was to be operative by August 2007. This has been achieved. The Indian National Centre for Ocean Information and Services (INCOIS), belonging to the Ministry of Earth Sciences (MoES), located at Hyderabad, is the nodal agency for this program. The system fared well during the occurrence of the 12/13 September 2007 tsunamigenic earthquakes (Table 3). One of the problems is the delay in estimating the size of large earthquakes. Empirical approaches are being developed to quickly estimate the size of the earthquakes occurring in Sumatra – Andaman zone of tsunamigenic earthquakes.

The Plan for Asia and the Pacific

We have so far summarized the vulnerability of the Asia-Pacific region to earthquake, flood and landslide hazards. The issues are very complex as the

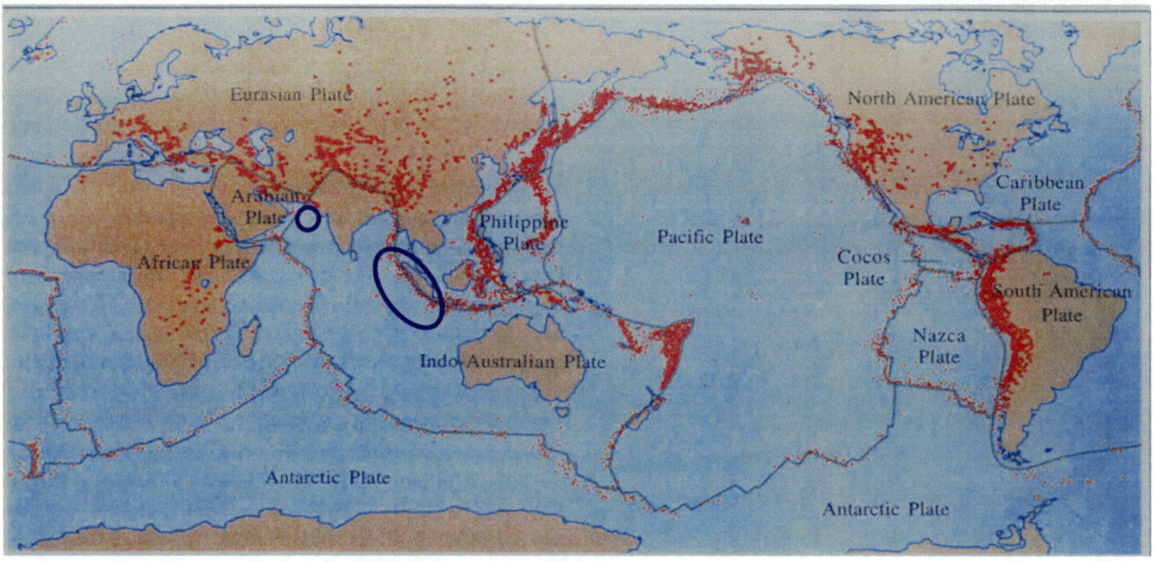

Fig. 10 Distribution of earthquakes and major plate boundaries. It may be noted that globally, more than 75% of earthquake energy is released in the circum-Pacific belt, about 20% in the Alpine-Himalayan belt, and remaining 5% through the mid-oceanic ridges and other Stable Continental Region earthquakes. For a tsunami to hit the Indian coast, it is necessary that a tsunamigenic earthquake occurs and its magnitude should be larger than M 7, and the possible locations of such events are enclosed in blue circle and ellipse

vulnerability to hazard changes from one location to another. In the following we present a basket of what is achievable.

Earthquakes

1. Several countries in the Asia-Pacific region do not have adequate earthquake recording and GPS monitoring of crustal deformation. Such countries need to be identified and a suitable mechanism to instrument them needs to be developed.
2. Develop scenarios of present day losses (social and economic) if earlier disastrous earthquakes recur. Also develop building codes for earthquake-prone countries where necessary.
3. Global Seismic Hazard Assessment Programme (GSHAP) was a flagship programme of IDNDR. Information on anticipated accelerations due to earthquakes at the bedrock for all the major cities in earthquake-prone areas is available. A major effort

Table 3 INCOIS generated a database of Model Scenarios considering various earthquake parameters. For the 12 September 2007 event scenario Ids 28.2 and 29.2 were picked from the scenario database. They were used to calculate the estimated travel time and run up heights at various coastal locations and water level sensors (Tide gauges & BPRs and tidal stations). Following are the estimated times of arrival and amplitude of tsunami and the observed ones at a few locations

Location	Estimated arrival time (h)	Estimated water level (cm)	Observed arrival time (h)	Observed water level (cm)
Padang	1751	80	1754	60
Coco's Island	1748	40	1748	50
Sahang	1903	20	1903	30
TB 3	1903	2	1913	1
TB10A	1931	1	1941	2
TB10	1930	2	1945	1
Port Blair	2010	10	2013	8
Chennai	2105	20	2110	18

is required to estimate the accelerations at the surface taking into consideration the local ground conditions and use them for micro-zonation of major cities.

4. A large percentage of the populations in the Asia-Pacific region live in self-designed non-engineered homes, which are very vulnerable to horizontal accelerations experienced during the earthquakes. This can be improved enormously by simple inexpensive retrofitting. This should be taken up as a major initiative with collaboration with UNESCO and other possible partners.
5. It is important to learn to live with earthquakes in earthquake prone areas. Several countries have developed and implemented earthquake drills. A lot of information is available as what to do before, during and after an earthquake. This needs to be appropriately spread to the public.
6. Artificial water reservoirs can trigger earthquakes. Ways and means are now available to find safer sites. Whenever a large artificial water reservoir is being proposed, necessary surveys should be conducted to find a suitable site.
7. Tsunami forecasts are now available for most AP region countries. However, countries need to develop facilities to make use of these advisories.
8. A few (four to five) case studies of earthquakes would be very helpful to understand why some earthquakes cause so much damage and loss of human lives.

Floods

1. An effort is required to set up flood prediction and warning systems in the ROAP region as has been done in many developed countries. This needs to be done in two steps: first, develop a big river basin flood warning system and second, transfer the big river basin flood warning system to a regional flood prediction and warning system.
2. It is crucial to communicate information related to flood prevention and mitigation to communities for their use. Equally important is to train the people to use this information.
3. It would be important to identify the vulnerability of countries to flooding, develop suitable building codes and construction regulations and develop a robust methodology to implement the same.
4. A few demonstration projects need to be selected. Two potential candidates are the Mumbai, India, flooding of 26–27 July 2005, and the flooding of Bangladesh during 1991.
5. The proposed flood-related work should be coordinated with other existing national and international programmes.

Landslides

1. Application of reliable landslide hazard and vulnerability zonation technology to identify landslide risk is necessary at country-to-community scales in the Asia-Pacific region. Adequate land-use planning to avoid landslide disasters should be promoted by this technology.
2. Cooperate with the existing landslide related efforts of several national and international agencies such as ICL, UNESCO, ICSU, IUGS, IUGG, WMO, IFI, IGCP, IOGS and IYPE.
3. The most cost-effective measure to mitigate landslide disasters is early warning and evacuation. Timely prediction and early warning technology suitable for natural and social conditions of the AP region should be developed.
4. Suitable methods to reduce landside causes or a landslide prevention technology is needed in areas such as cultural heritage sites and other locations of high-societal values where relocation and early warning are not possible or very difficult.
5. A few case studies of landslides would be very helpful to understand the parameters responsible for causing landslides.

Vulnerability of Islands

1. Archival of historical tropical cyclone records and documentation of changes observed reliably over that past 40 years or so to assess the future risk.
2. Analysis of EI Nino influence on cyclone parameters.

3. Improving the network of river hydrometric stations to better comprehend the flow characteristics of rivers of islands.
4. Examine the effect of climate change on river flow to project future trends.
5. Determining the thresholds (climatic, seismic, volcanic, geomorphic, etc.) triggering hill slope failure on islands.
6. Investigate the links between submarine landslides and local tsunami generation.
7. Development and implementation of appropriate and culturally sensitive options to natural hazards that will yield maximum benefit to all sectors of developing island economies.

Disaster Management and Capacity Building

It has been recognized that the available scientific knowledge is not being fully utilized in forecasting the hazards, estimating the risks, and getting the necessary outreach for the benefit of the public, particularly in a large part of the area covered by AP. It is important to stress:

1. disaster risk reduction and management;
2. the role of social sciences in understanding hazards and their mitigation;
3. the role of the private sector and civil societies;
4. building and enhancing regional capacity for disaster reduction;
5. building self-sustaining capacity at various levels for different hazards;
6. establishing continuity in capacity building.

Concluding Remarks

The science plan suggests to develop projects relevant to the problems identified. We hope to improve the situation in the Asia–Pacific region with support and collaboration of all concerned so that fewer hazard events develop into disasters.

Acknowledgement The present paper is based on the "Science Plan on Hazards and Disasters" developed by the ICSU Regional Office for Asia and Pacific (ROAP). For details please see http://www.icsu-asia-pacific.org/resourcecentre_hazard.html(website of ROAP). The author is grateful to UNESCO Holland for financial support to attend the IYPE Symposium held at Oslo, Norway on 7 and 8 August 2008, and to Tom Beer for invitation and suggestions to improve the manuscript. Ms. Uma Anuradha helped in the preparation of the mss.

References

Arya A S (1992) Possible effects of a major earthquake in Kangra region of Himachal Pradesh. *Current Science* 62(1 & 2): 251–256.

ASC (Asian Seismological Commission) (2006) Symposium on earthquake and tsunami disaster preparedness and mitigation, Final Report on The Sixth General Assembly ASC 2006, 7–10 November 2006, Bangkok, Thailand.

Badami S (2005) On Mumbai flooding of 26 July 2005 (www.karmayog.org/library/htm/libraryofarticles_319.html).

Bardet J P, Oka F, Sugito M et al. (1995) The Great Hanshin Earthquake Disaster, The January 17, 1995 South Hyogo Prefecture Earthquake, Preliminary Investigation Report, 5 February 1995.

Bettencourt S, Campbell J, de Wet N, et al. (2002) The impacts of climate change in Pacific island economies: policy and development implications. *Asia Pacific Journal on Environmental and Development* 9: 142–165.

Chaudhary A M (1991) Cyclone in Bangladesh. *Bangladesh Quarterly* 12(1): 7–12.

Cruden D M, Varnes D J (1996) Landslide types and processes. In: Turner A K Shuster R L (eds) Landslides: investigation and mitigation, Transportation Research Board, Special Report 247, pp. 36–75.

del Ninno C, Dorosh P A, Smith L C et al. (2001) The 1998 floods in Bangladesh: disaster impacts, household coping strategies and response, Research Report 122. International Food Policy Institute (IFPRI), Washington D.C.

DFID (Department For International Development) (2005) Natural disaster and disaster risk reduction measures. A desk review of costs and benefits, DFID, London, December 2005.

Giardini D (1999) The global seismic hazard assessment programme (GSHAP) 1992–1999 summary volume *Annali di Geofisica* 42(6): 957–1230.

Gupta H K (2002) A review of recent studies of triggered earthquakes by artificial water reservoirs with special emphasis on earthquakes in Koyna, India. *Earth Science Reviews* 58(3–4): 279–310.

Kreemer C, Holt W E and Maines A J (2003) An integrated global model of present day plate motions and plate boundary deformation. *Geophysical Journal International* 154: 8–34.

Lawson T (1993) Summary of the south-east Viti Levu landslide project – preliminary study. Fiji Mineral Resources Department, Suva.

Mekong River Commission (2006) Annual flood report 2005: flood management and mitigation programme, Vientiane, Lao PDR, July 2006.
NCDC (1998) Floods in China, summer 1998. (www.ncdc.noaa.gov/oa/reports/chinaflooding/chinaflooding.html).
Richter C F (1958) Elementary seismology. W.H. Freeman and Company, San Francisco and London, 768 pp.
Terry J P (2007) Tropical cyclones: climatology and impacts in the South Pacific. Springer, New York, 210 pp.
Terzaghi K (1950) Mechanism of landslides. In Engineering Geology (Berkel) Volume. Ed. da *The Geological Society of America*~ New York, 1950.
WMO (World Meteorological Organization) (2006) Statement on tropical cyclones and climate change. WMO international workshop on tropical cyclones, IWTC-6, San Jose, Costa Rica, November 2006, 13 pp.

Part III
Geophysical Risk and Sustainability: Climate and Climate Change

Closing the Gap Between Science and Practice to Reduce Human Losses in Hydro-Meteorological Disasters

Kuniyoshi Takeuchi

Abstract Future population increase is expected to be concentrated in urban areas. The resultant urban expansion further increases both potential hazard and vulnerability to hydro-meteorological disasters. This is unique to hydro-meteorological disasters (especially flooding) and not the case for solid Earth events such as earthquakes and volcanic eruptions where the occurrence is independent of urbanization. The basic reasons for disaster occurrence in developing countries, especially in urban areas, are poverty and governance. They are socio-economic and cultural problems which can be solved only by an integrated approach in which all the stakeholders and administrative sectors have to work together while science and technology play a supporting role.

For the reduction of human losses by natural disasters, early warning, evacuation and preparedness supported by accurate forecasts play a key role. Hydro-meteorological disasters are in a better position in this respect than other disasters as their forecasting technology is much advanced. Nevertheless the capability of forecasting is not well utilized in local practice for civil protection from hydro-meteorological hazards. This is the gap between science and practice. The gap can be filled within the framework of integrated flood management where the local ownership of flood forecasts should be promoted, as tried by ICHARM (one of UNESCO Category II centers), to make warning an integral part of the total community management against disasters.

Keywords Urbanization · Illegal settlers · Early warning and evacuation · Local ownership of flood forecasts

Introduction

The Bali Action Plan agreed upon at COP13 in Indonesia, December 2007 declared for the first time that adaptation to climate change is equally important as mitigation. It emphasizes the need for international cooperation especially for developing nations, the finances, capacity development, and most importantly "integration of adaptation actions into sectoral and national planning, specific projects and programmes, means to incentivize the implementation of adaptation actions" (UNFCCC 2007). It is obvious that the main barrier to mitigate disaster risk depends on social organization rather than on the extent of the knowledge available in human hands. Therefore the question is: how to mobilize actions based on best available knowledge and technology in practice? This is a very important challenge of "science for society."

This chapter will illustrate that the increase of flood disasters is mainly due to urbanization and related societal changes, especially problems of poverty and governance. Such societal problems can only be solved through an integrated approach where hydro-meteorological science and technology can play only a supportive role. For reducing human losses in disasters, however, hydro-meteorological prediction plays a key role that has been demonstrated in a number of countries including China and Bangladesh. But it is still not used as much as it could be in many develop-

K. Takeuchi (✉)
Public Works Research Institute (PWRI), International Center for Water Hazard and Risk Management (ICHARM) under the auspices of UNESCO, Tsukuba 305-8516, Japan
e-mail: kuni.t@pwri.go.jp

T. Beer (ed.), *Geophysical Hazards*, International Year of Planet Earth,
DOI 10.1007/978-90-481-3236-2_8, © Springer Science+Business Media B.V. 2010

ing countries. The paper emphasizes the need of local ownership of flood forecasts in order for the available technology to play its potential function to a full extent. As an example, the effort of ICHARM, one of the UNESCO Category II Centers, will be introduced.

Urbanization and Vulnerability to Natural Disasters

The world population is now 6.7 billion and expected to reach 9.2 billion by 2050 in medium variant (UN Population Division 2007) which is another 2.5 billion (equivalent to the world population in 1950) increase in 40 years and its consequences on the increase of needs in foods, energy, water, land and environments will be tremendous. The question is: where do those increased populations go? It is urban areas.

UN Population Division (2008) reports "during 2008, for the first time in history, the proportion of the population living in urban areas will reach 50%" and the urban population will be nearly doubled at 6.4 billion by 2050 (Fig. 1). It means that the rural population is nearly stabilized and all the additional population goes to urban areas. Since the concentration of population, properties and activities increases vulnerability to hazards, urbanization increases disaster risk of the world.

The concentration of valuable assets and activities such as lifelines (e.g., water, electricity and gas supply lines, communication lines, transportation lines), underground malls and centralized decision making and administrative function is a positive and bright side of urbanization, for which people gather. Although they are vulnerable to various hazards and need high level protection, they can afford the investment however costly it is. But there is the other side, i.e., overcrowded residences, areas of poor living conditions, development of high disaster risk areas etc. that are the negative side of urbanization and more difficult to justify investment. Urban migration is mainly a result of a gap between rural and urban conditions in economic opportunity and job availability. In the case of international migration, it is from poor countries to rich countries. The poor people who cannot sustain their lives in rural areas come into urban areas, in some cases across national borders, and occupy disaster prone areas where land price is cheap if not unauthorized. They join and increase the negative side of urbanization. The consequences are obvious, i.e., disaster risk increases.

Land Development Increases both Flood Hazard and Vulnerability

Development of land can, in general, increase both flood hazard and vulnerability if proper counter action is not taken. In case of earthquakes and volcanic erup-

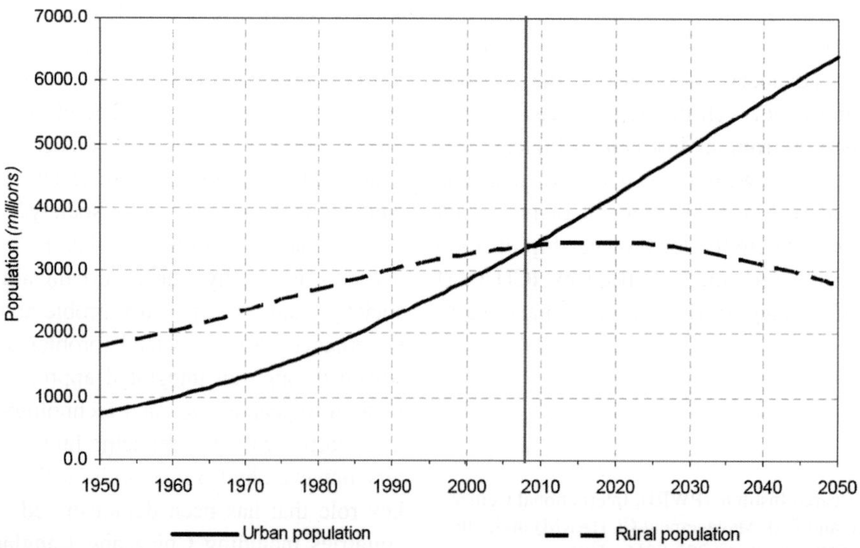

Fig. 1 Urban and rural populations of the world, 1950–2050 (UN Population Division 2008)

tions, land development does not affect hazard. It is therefore unique to hydro-meteorological disasters. Vulnerability increase as a result of land development is obvious with more people, more economic activities and more concentration of lifelines and properties. It is human exposure to hazardous areas which used to serve only for natural processes. As the principle of first come, first served is especially true with respect to land and water, the better lands in terms of productivity and safety are soon exhausted and eventually development expands to less productive and higher risk lands.

It is more severe in mega cities where new development takes place where the potential risk is high such as in low-lying areas and steep slopes. The poor and late comers tend to settle in hazardous areas and often form slums. Slums are formed in poor overcrowded areas and become highly vulnerable to disasters. "A slum household" is defined by UN-Habitat (2003) "as a group of individuals living under the same roof lacking one or more of the conditions below: Access to improved water, Access to improved sanitation facilities, Sufficient-living area, not overcrowded, Structural quality/durability of dwellings, Security of tenure." Under this definition, portions of Japan, France, UK, US and other developed countries also classify as slums.

Urbanization increases hazard, too. If it is not accompanied by proper mitigation works, any land development always increases flood hazard by decreasing infiltration capacity, storage capacity and flow resistances. Forests, bushes, grass land, wetlands, marshes, parks and any natural fields rich in flow retardation are converted to agricultural land or urbanized areas with roads, roofs, residential areas, manufacturing and commercial districts which are less flow resistant and retarding. Therefore, together with construction of levees, short cuts and other river improvement works, they make the flood concentration time shorter and flood peaks higher. Land subsidence due to over pumping of groundwater is also a serious factor in the increased frequency and magnitude of floods which is especially serious in urbanized deltas.

Such dual pressure to flood risk, from both the hazard side and vulnerability side, has been accelerated in developing countries by economic growth without proper counter actions or land use control. This is the heart of the urbanization problem in the world.

Illegal Settlers in River Space

A form of slum, inhabited by so called illegal settlers or squatters, in a river space is a common and very difficult problem in developing countries where poor people occupy river space without permission. A case in Central Jakarta is shown in Fig. 2. During floods, those houses in the river section are flushed away and the wreckage flows down into the highly populated central district of the city. At bridges or some narrow points, they get stuck and dam up rivers flooding surrounding areas.

Most people who lose houses during floods come back to the original place and rebuild their houses at the same site. This is because they have no other place to go to, their livelihood is in the neighborhood of their residence and their accustomed life style is there. It is extremely difficult for governments, national or local, to force settlers to move out from prohibited areas. Governments may be able to provide cheap apartments nearby or in the suburbs but find it very difficult to provide jobs for all to sustain their life. Without having a livelihood, people cannot move. Furthermore, as people cannot change their life style too much at once, if new employment requires a big change in their life style, people cannot accept the new settlement plan. Such life style includes human relationship in the neighborhoods, habits of dumping garbage and wastes directly into rivers, enjoying breezes over rivers, easiness of living, eating, bathing, chatting, playing, buying etc. etc. which are important components of their daily life and need a considerable effort and education to change. As a result, regardless of the obvious flood risk, they stay there. For the people living in such critical condition, it is a luxury to think and rationally plan their life in the long run, even for the next month or the next year. Watanabe (2003) describes this situation as "people waiting in the queue to death." He continues: "politicians tend to listen to the voice of the dead after disasters but not the voice of the living before disasters happen."

After the Mangahan Floodway was built in 1986 in Manila (See Fig. 3), the Philippines, thousands of people came into the 10 km long 300 m wide floodway and stayed there building illegal houses. When floods come, people just run away to outside of floodway. The governments try to provide early warning for the residents before it is too late. The Floodway was built in

Fig. 2 The illegal settlers over a river in Central Jakarta, Indonesia (25 May 2007)

order to divert the flood water of the Marikina River temporarily to Laguna de Bay Lake to reduce the flood discharge in the Pasig River and avoid flooding in the drainage network in Metro Manila. Since the completion of this floodway, no floods have occurred in Metro Manila at least by over banking of the Pasig River.

But unfortunately, a big discharge from the Floodway seems to amplify some flood problems in the low-lying shoreline villages of the Laguna de Bay Lake, especially when a Seiche or storm surge induced by a typhoon and the large discharge from the Floodway coincide (Felizardo 2006). People in the lake towns (especially Taytay town) located near the exit mouth of the Mangahan Floodway have complained about flood problems. They claim that there are higher maximum lake levels during floods and also long-term deposition of sediments at the lake bed near the mouth that increases the effects of the Seiche or storm surges (Liongson 2009).

In 1997, a counter project, West Mangahan Project, the 10 km long dike to protect the part of lowland, West Mangahan District of Metro Manila was announced as a partial solution to this problem. Then many squatters suddenly came in as soon as the announcement was made which delayed the completion of the project more than 3 years to 2007 (Shichijugari 2008).

The problems of illegal settlers are seen all over the world in developing countries. Poor people come, occupy and settle in the land between dikes and rivers or shore lines, within rivers including under bridges, low depressions, steep slopes, flood retardation ponds, planned dam inundation areas, planned road space, areas under public nuisance etc. etc. They are the most vulnerable areas for disasters and the first target of disaster victims. There is usually no legal framework to extend protection and sometimes even rescues as it is out of administrative scope and virtually no organized public supports except pressure of demolition. In the case of Cyclone Sidr in Bangladesh, November 2007, it is considered that among about 4,200 counted dead by the storm surge in the coastal zone, a large number of victims, squatters who lived outside of storm surge embankments might not have been included as their number is unidentifiable (JSCE 2008).

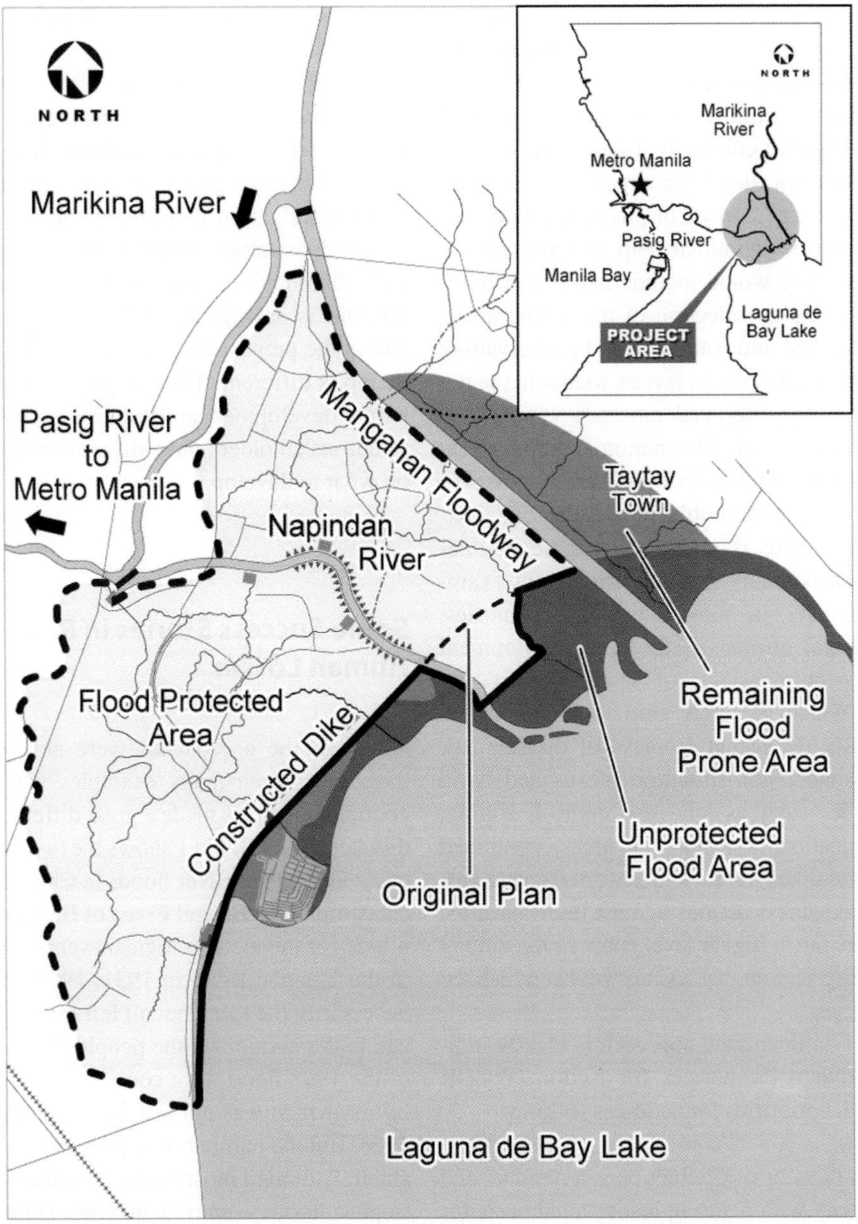

Fig. 3 A general plan of the Mangahan Floodway to protect Metro Manila and the West Marikina Dike to protect lowland shoreline villages of the Laguna de Bay Lake, in the Philippines. (Modified map based on Felizardo 2006)

Need for an Integrated Approach

As seen above, the main root causes of disasters in developing countries are poverty and governance. It is so at present and will be even more so in the future as urbanization accelerates. No disaster management or external aids for disaster mitigation can work without considering these fundamental causes. It needs an integrated approach where all related components should function in a holistic manner. It is especially important to break barriers between administrative sectors and promote interdisciplinary cooperation among engineers, scientists, administrators, sociologists, economists, psychologists, lawyers, politicians etc. Above all, capacity development including establishment of individual discipline with a spirit of mutual help and a "disaster culture" is the basis of disaster management. Disaster culture is a culture developed in a community from their long experiences of managing disasters. It may consist of physical way of living, human network of mutual help and various traditional events. They would include habits and traditions on such items as: selection of residential areas; form of houses; land and water use and conservation; maintenance of forests, ponds, levees, roads etc.; material, facilities and institutional emergency preparedness; human network of information dissemination; and especially the mutual help when any members of the community falls into trouble and suffers.

Unfortunately there is no ideal example in integrated approach but only partially successful examples. For example, in most developed countries, the long historical efforts of structural development resulted in the current level of safety against natural hazards, which is surely said to be a success story so far. But the recent increase of disaster risk requires many more non-structural means and basin wide approaches such as land use control, zoning, accepting inundation in residential areas, combined use of roads and dikes, rivers and sewers etc. and various local preparedness actions against multi hazards. They all require much higher level cooperation among all administrative sectors that has not yet been realized in many countries.

The need of an integrated approach is clearly indicated in the Hyogo Framework for Action (WCDR 2005). It lists the priorities for action as follows:

1. Ensure that disaster risk reduction is a national and a local priority with a strong institutional basis for implementation.
2. Identify, assess and monitor disaster risks and enhance early warning.
3. Use knowledge, innovation and education to build a culture of safety and resilience at all levels.
4. Reduce the underlying risk factors.
5. Strengthen disaster preparedness for effective response at all levels.

All the stated needs above, i.e., strong institutional basis, monitoring and early warning, knowledge and education, reduction of underlying risk and preparedness for effective response, require a strong intersectoral integrated approach.

It is important to note that all those priority actions are addressed to prevention and preparedness stages of the disaster management cycle (following occurrence of disaster: emergency response, rehabilitation, prevention/mitigation and preparedness). The integrated approach is rather simple in the emergency response and rehabilitation stages as the target actions are obvious, that is, salvage and recovery to original condition. But in the prevention and preparedness stages, the situation is different. Their target actions are an integral part of development and not necessarily obvious as different stakeholders have different views and interests on what to develop.

Some Success Stories in Reduction of Human Losses

Although the approaches were not quite integrated, there have been many examples where long lasting continuous efforts made a great difference and reduced disaster losses. Table 1 shows the two examples of such cases in Yangtze River floods in China and storm surge catastrophes in Bengal Coast of Bangladesh. The magnitudes of the meteorological events that caused floods in the Yangtze River in 1931, 1954 and 1998 are not necessarily the same but all left truly devastating damage to the nation and the people. The economic losses of the 1998 flood were considered to be over USD 30 billion which was not less than for those of 1931 and 1954. But the number of human losses was remarkably small. Although most literatures discussing how catastrophic floods occurred in 1998 blame deforestation and development of farmlands and residential areas, the decisive importance is the reduction of death tolls. This is due to early warning and systematized evacuation, relief and other preparedness for emergency response.

Similar changes in death tolls are seen in Bangladesh in 1970 (then East Pakistan), 1991 and 2007. The human losses were truly large in the first two cases but in the third case in 2007, by Cyclone Sidr, were considerably smaller in number. Cyclone Bhola hit the Ganges Delta on 12 November 1970, a category 3 cyclone, and the weakest in magnitude

Table 1 The chronological change in death tolls by floods in China and storm surges in Bangladesh with reference to Myanmar (after EMDAT, *Water Disasters Year Book in China, 2005)

China floods		Bangladesh storm surges		Myanmar Nargis	
Year	Death tolls	Year	Death tolls	Year	
1931	3,700,000 (400,000*)	1970	300,000	2008	138,000
1954	30,000	1991	139,000		
1998	3,700	2007	4,200		

of the three, but it took about 300,000 lives, one of the deadliest disasters in the history in the world. It is believed that information from ships in the Indian Ocean was disseminated and the majority of the people in the affected area were aware of the cyclone but only a small portion evacuated to safe structures and many were drowned. The aftermath responses were disorganized and very slow which increased the losses. The central government in West Pakistan was criticized triggering civil war on 26 March 1971 (the Independence Day of Bangladesh). This eventually expanded to the India–Pakistan War (3–16 December 1971) and ended in the independence of Bangladesh.

A very strong Cyclone hit near Chittagong on 29 April 1991 and took about 140,000 human lives. Most people reportedly had warnings of less than a few hours and some people could not leave or stayed behind not believing the danger to be as deadly as announced. But since some two million people evacuated, the death toll was smaller than that of 1970. In November 2007 Cyclone Sidr hit the Ganges Delta. Early warning announcements were made and reportedly more than three million people evacuated. In addition, the storm surge hit during low tide and in rural areas or small cities as compared with the 1991 cyclone that hit Chittagong, which drastically lessened the number of deaths.

The importance of warning, evacuation, sheltering and relief activities in saving human lives was very clearly, unfortunately too clearly, demonstrated by Cyclone Nargis in Myanmar in May 2008. The extent of local efforts to warn and evacuate the population is unknown but must have been minimal as such a strong cyclone is unexpected in Myanmar. Most regretfully, the military rulers of Myanmar refused international NGOs offers of help to rescue sufferers. Table 1 illustrates the success story of the reduction of death tolls in China and Bangladesh. Hopefully Myanmar deaths will eventually show similar reductions.

Importance of Local Ownership of Flood Forecasts

Early warning and preparedness are the keys to saving human lives. Importantly, precise information in advance of hazards is most valuable. Compared with solid earth hazards such as earthquakes and volcanic eruptions, hydro-meteorological hazards are much more predictable. Meteorological extremes are visually observable on site as well as remotely by the help of satellites, radars, telemeters etc. dependent on their local availability and the technology how to utilize them.

In order for forecasts to be useful in practice, various conditions need to be satisfied. They include the necessity that forecasts are:

1. Provided by a reliable authority that the local community can trust to assess local risk.
2. Provided with a decent accuracy (forecasters should not "cry wolf").
3. Provided with enough lead time and updated with a reasonable frequency.
4. Provided by popular channels (TV, radio, cell phones etc.) in absolute clarity (such as on Google earth).
5. Provided with necessary response instructions (e.g., evacuation routes, shelters etc.)

Among these conditions the reliable authority needs particular attention. If individual people do not feel that the forecast is about their own locality and some direct hazard is about to hit, they do not take concrete action such as evacuation. A key is whether the forecasters are familiar enough with the local details to include local information in the warnings.

It is usually impossible for forecasters to be aware of detailed local conditions because forecasting is a very specialized technology requiring experienced specialists. In the case of weather forecasts, in addition to nationwide large scope forecasts, local weather observatories make local analyses and provide downscaled information. It should be the same and even more necessary in the case of flood forecasts, because local hydrology is a very localized matter. Even if an upper basin receives much rainfall, if ground rocks are porous as in the Karst geology, river discharge will be small. If even a small natural dam is formed upstream in the main river course, downstream show little discharge and an outburst creates an abrupt flood discharge. In a highly urbanized basin, the concentration time is so short and no lead time is available for even a local forecaster.

The Toga River is a 7 km long urban river that originates from the Rokko mountains, passes through the midst of Kobe City and discharges out to the Seto Inland Sea, Japan. On 28 July 2008, a torrential rain caused a sudden flood discharge, increasing the water level from nearly 0 to 1.4 m within 10 min which killed four children playing in the river. No warning by a nationwide agency could be useful in such a local case. We do need local forecasters. This example may be too localized, but in any basin management, the local experts have to take responsibility as local forecasters who know local details both in natural and social conditions. This is why ICHARM (International Center for Water Hazard and Risk Management) under the auspices of UNESCO promotes the local ownership of flood forecasts.

ICHARM was established in Tsukuba, Japan, March 2006 as one of UNESCO Category II centers. Its aim is to serve as the global center of excellence to provide and assist implementation of best practicable strategy for water-related disaster risk reduction, focusing on flood-related disasters in the first phase. The concrete actions are research, training and information networking. A priority research area is, among others, the advanced early warning system that includes, starting from satellite observations, distributed hydrological models, inundation simulators and hazard mapping technology. This is coupled with training activities including a one year academic master's course in Disaster Management Policy jointly offered by Graduate Research Institute for Policy Studies (GRIPS) supported by Japan International Cooperation Agency (JICA).

ICHARM's central challenge is "localism," a principle that takes into account local diversity of natural, social and cultural conditions, being sensitive to local needs, priorities, development stage, etc., within the context of global and regional experiences and trends. Under such principle, ICHARM tries to promote local ownership of flood forecasts, by which local people can trust the source of forecasts and follow the administrative orders with confidence. ICHARM considers that the essential requirement in building local ownership of forecasts is the ownership of models and data. ICHARM provides models, along with associated training, to local experts and let them use the global data sets and their own local data by themselves. In this way, localities become self-supporting and self-confident in early warning operation. Table 2 lists the essential steps for this operation.

Table 2 The essential steps for promoting local ownership of flood forecasts

(i) ICHARM provides local experts an integrated flood analysis system (IFAS) developed on the basis of state-of-the-art technology together with technical training to learn how to use the model properly and effectively.
(ii) Using IFAS, local experts can make flood forecasts anywhere in the world including ungauged basins using satellite-based globally available data (rainfall, radiation, topography, land use, etc.).
(iii) Local experts will realize that flood forecasting only by globally available data is not accurate enough for local use and can improve further by the use of local ground-based observation data.
(iv) Local experts will promote their local measurements with their own incentives. The local observation data will be made globally available after local use.
(v) Local experts exercise early warning in their basin effectively and proactively with confidence based on their own knowledge and experiences of the basin. This is the aim of promoting the local ownership of flood forecasts.
(vi) ICHARM will assist and mutually learn with local experts throughout the process.

Flood forecasting and flood warning are different. In order to make use of advanced flood forecasts to save lives and properties, it is necessary for the forecasts to be interpreted for the local context and disseminated to the public with specific instruction for action by the mandated public authority. This is the task for which local experts' knowledge is indispensable and the local ownership of forecasts is necessary. The necessary information for flood warning differs basin by basin. In a very large basin where the upstream discharge information is indicative of the downstream phenomena some days later, satellite information or weather forecasts are less important while for basins prone to flash floods, satellite and numerical model estimates are more important.

ICHARM, as described in Table 2, advocates this concept and initiates training courses for needy nations. For widespread, genuine promotion in the world one needs more extensive mutual improvement activity directed at the local level. An example would be the effort of International Flood Network (IFNet), a non-profit organization to promote activities to reduce the negative impacts of floods all over the world established in 2003 in Tokyo joined by governments, municipalities, institutions, international agencies including WMO etc. IFNet globally provides extreme precipitation information freely through the internet at http://gfas.internationalfloodnetwork.org/gfas-web/ which will be validated with local observations by local experts. This is expected to help establish local ownership of forecasts.

Concluding Remarks

Global research has been very active for many decades since the International Geophysical Year in 1957–58 and, in climatological sciences, the start of the World Climate Research Program (WCRP) in 1980. Global warming and climate change analyses have been areas of remarkable success as evidenced by IPCC reports. The recent Global Earth Observation System of Systems (GEOSS) agreed in February 2005 by Group of Earth Observations (GEO) has given another thrust to global research. Supported by national satellites launching policy and fast computational technology, global science has shown rapid progress in observation, analyses and simulation of the Global phenomena. But once it comes down to local application of its outcomes, their use is limited.

The gap between global scientific achievements and local application needs considerable effort of downscaling. In hydro-meteorological field, considerable part of such effort is in the hands of hydrologists. Especially in the developing countries where ground observations are scarce and local extreme event forecasts have a high potential to reduce the number of human casualties at every extreme event, their practical use should be demonstrated in a concrete form, namely, as streamflow forecasts. The idea of promoting local ownership of flood forecasts is a response to such an urgent need.

The global research community repeatedly requests all nations to disclose their ground observation data. But in general most nations are reluctant and local communities do not see how disclosure of their data will be to their own benefit. The promotion of local ownership of flood forecasts is also an effort to correct such misunderstandings and deliver a visible benefit to local people. At the same time it is a certain effort to close the gap between science and practice.

Acknowledgements The author is indebted to Prof. Leonardo Q. Liongson, Director of University of Philippines-National Hydraulic Research Center for his providing precious information on Mangahan Floodway and Laguna de Bay. He is also grateful to ICHARM staff to modify the figures. The gratitude extends to Prof. Tom Beer, the editor of this book, for his patience, a number of suggestions and kind English editing.

References

Compilation Committee on Historical Floods in China (2005) 中国水灾年表 (Water Disasters Year Book in China)

Felizardo, J. C. (2006) Reducing flood disasters in the densely populated Mangahan floodway and the adjoining Laguna Lake shoreline in Metro Manila, Paper presented at the 9th International Riversymposium, Brisbane, Australia, 4–7 September 2006

JSCE (2008) Investigation report on 2007 Bangladesh cyclone Sidr storm surge disasters, JSCE Committee on Coastal Engineering and Committee on Hydraulic Engineering and Foundation of River & Watershed Environment Management

Liongson, L. (2009) Personal communication. A professor of the University of Philippines

Shichijugari, A. (2008, November) Twenty years overseas experiences in construction consulting firm, An Ozawa Award Memorial Paper, Kokkenkyo Joho, 792, 15–22

UNFCCC (2007) Revised draft decision -/CP.13, Ad Hoc Working Group on Long-term Cooperative Action under the Convention, Proposal by the President 14 December 2007, FCCC/CP/2007/L.7/Rev.1

UN-Habitat (2003) Slums of the world: the face of urban poverty in the new millennium? United Nations Human Settlements Programme, Working paper, HS/692/03E

UN Population Division (2007) World population prospects: the 2006 revision, Highlights, Working Paper No. ESA/P/WP.202

UN Population Division (2008) World urbanization prospects: the 2007 revision, UN DESA, Population Division, ESA/P/WP/205

Watanabe (2003) A draft proposal of a plan for disaster mitigation and preparedness applicable for developing countries, Annuals of Disaster Prevention Research Institute, Kyoto University, 46 B, 1–13

WCDR (2005) Hygo framework for action 2005–2015, Final Report of the World Conference on Disaster Reduction, 18–22 January 2005, Kobe, Hyogo, Japan (A/CONF.206/6)

The Role of Geosciences in the Mitigation of Natural Disasters: Five Case Studies

S.A.G. Leroy, S. Warny, H. Lahijani, E.L. Piovano, D. Fanetti, and A.R. Berger

Abstract Geoscientific data combined with historical documents on past natural hazard events and on the disasters that followed are essential to improve mitigation plans. It is only with this method that the full scale of potential rapid changes that are not covered by the instrumental record can be obtained. Therefore, the collection of these past data and their integration into planning should become one of the priorities of the Hyogo Framework of Actions. This paper analyses the following five case studies: global warming impact on the indigenous populations at high latitudes of Canada, hurricane impact on the southern coast of the USA as experienced in New Orleans, rapid level rise in several lakes of the Argentinian Pampas with emphasis on Laguna Mar Chiquita, the rapid sea level rise of the Caspian Sea as seen from Iran and the tsunami risk in a large Alpine lake of Northern Italy, Lake Como. In each area, the main natural hazard is part of a potential series of hazards that, if combined, could lead to a shift from disaster to catastrophe. The most successful cases of transfer of information between geoscientists and end-users are when the hazards and subsequent disasters are visible or when the messengers bearing the information are trusted by the local communities.

Keywords Geosciences · Natural hazard · Disaster · Mitigation · Canadian Arctic · New Orleans · Argentinian Pampas · Caspian Sea · Alpine lakes

Introduction

The rationale for this paper is to emphasize the role of geosciences in mitigating natural disasters and to highlight the need to bridge the gap from natural hazard study to human and societal responses (leading to mitigation).

A series of recent case studies, i.e. twentieth and twenty-first centuries, of natural hazards are presented and for each of them the link with past hazards, i.e. over the last millennia, is made to provide the full potential of hazard range. The chapter goes on to examine whether any lesson has been learned from the past or could have been learned. Finally, in the light of a positive answer to this question, the mechanisms of transfer of information are analysed.

The case studies and the contributors of this paper have been brought together following two past conference programmes, IGCP 490 on "The role of Holocene environmental catastrophes in human history" from 2003 to 2007 and ICSU Dark Nature on "Rapid Natural Change and Human Responses" in 2004 and 2005. The aims of these two programmes were to (1) refine the record of rapid (<100 years) environmental changes affecting physical environments and ecosystems during the last 11,500 years; (2) examine how past societies and communities reacted in the face of harmful change; and (3) explore the implications of rapid natural change for current environmental and public policies (Dark Nature[1]). The latter point proved to be very

S.A.G. Leroy (✉)
Institute for the Environment, Brunel University, Uxbridge (West London) UB8 3PH, UK
e-mail: suzanne.leroy@brunel.ac.uk

[1] Dark Nature website, 2008. Available at http://www.mun.ca/canqua/ICSU-DN/ [Accessed December 2008].

T. Beer (ed.), *Geophysical Hazards*, International Year of Planet Earth,
DOI 10.1007/978-90-481-3236-2_9, © Springer Science+Business Media B.V. 2010

challenging and is illustrated amongst the examples chosen for this paper.

What causes a catastrophe? Not all natural hazards (geological and hydro-meteorological) cause a disaster (loss of life and/or damage to environment), and not all disasters lead to a catastrophe. A catastrophe is the result of a combination of negative factors at the spatial, temporal and societal scales (Leroy 2006). The larger the area affected or the larger proportion of the settlement affected, the worse is the impact. Similarly, the sharper the onset and the longer the duration of the hazards, the worse is the impact. A society that is not willing to modify some of its rules (lack of flexibility and adaptability) or a society that is not ready (without mitigation plans), will also suffer from most severe impacts. The examples shown here are not real catastrophes but they may have been near misses, or they could potentially develop into one due to an accumulation of hazards.

Five diverse case studies have been selected throughout the world: two are linked to rising water levels (a lake and a sea), one is a geological hazard and two are linked to climate (global warming trend and a discrete event). Two are in North America (at low and high latitudes), one in South America, one in S-W Asia and one in Europe.

Global Warming at High Latitudes and Rapid Landscape Change in the Arctic

Introduction and Tableaux of Past Changes

"The Earth is faster now." This is how an aboriginal elder from northern Alaska describes changes she sees around her. Across the Arctic the effects of warming are clear (Furgal and Prowse 2008, Hassol 2004, Krupnik and Jolly 2002). The extent of sea ice is decreasing year by year, surface temperatures are rising, permafrost is melting, tundra soils degassing, and shallow coastlines retreating as waves batter soft sediments. Glaciers and ice sheets are retreating in many places, and ice patches on mountainsides are melting, to the delight of archaeologists who study the artefacts melting out after centuries of icebox preservation. Birds and insects previously unknown are now spotted in unfamiliar places, and some key land and marine species that are essential sources of food are under threat of extirpation. Some communities on shallow coasts are under increased threat of flooding. Roads, airport runways and buildings are subject to differential settlement.

In some places northern landscapes have changed little in thousands of years. Some aboriginal hearths and house sites, for example, have remained undisturbed near present-day beaches for 4–5000 years (McGhee 1996, 2007). However, elsewhere in the Arctic coasts have accreted or eroded, inland rivers have switched channels and migrated, glaciers have surged and retreated, slopes failed, and floods have occurred repeatedly (Pienitz et al. 2004, Overpeck et al. 1997). Studies of ice cores, permafrost ground temperatures, lake and coastal deposits and tree-ring records have shown that substantial landscape changes in the North American Arctic and Subarctic took place long before large-scale human influences on the environment occurred (Brown et al. 2000, Hewitt et al. 2000, Gajewski and Atkinson 2003). Variations in regional temperatures of as much as 10 °C occurred several times during the Quaternary (NRC 2002, Weart 2008), each leading to a longer period of steady temperatures. For example, Greenland ice cores bear evidence of a rapid warming of 5–10 °C in a decade or so, bringing to an end the Younger Dryas around 11,500 years ago. Though later climatic events may not have been as marked as this late glacial warming, it demonstrates just how dramatic natural climate change can on occasion be.

Rapid environmental changes also occur during periods of relatively stable climate. Volcanic eruptions, huge floods, destructive earthquakes, and, on a smaller scale, landslides and coastal storms can lead to marked changes in landscape conditions. About 1200 years ago volcanic ash from a vent near the Yukon-Alaska boundary blanketed the southern Yukon and adjacent parts of the Northwest Territories. This appears to have been the largest pyroclastic eruption anywhere in the past 2000 years (Lerbekmo 2008) and may well have had major ecological and human impacts (Workman 1979). Where thicker, the ash would probably have smothered much of the shrubby vegetation on which animals depended, leaving local inhabitants of the time – the Athapaskan people – without their usual source of game. Recent archaeological studies indicate a rapid change in hunting methods with the introduction of bow and arrow technology at the time of the eruption (Hare 2008). This sudden landscape change may have forced people to leave their traditional lands, some moving eastwards and others perhaps southwards to settle eventually in the southern USA, as the ancestors of the modern Navajo people (Moodie et al. 1992).

Among some Inuvialuit, the people of the western Arctic, a story is told of a once barren land transformed into one of lakes, rivers and plentiful fish and game. The cause was a sealskin bag full of water that grew and grew until the hunter who found it built a large raft and was able to escape when the bag burst, flooding the land and drowning the other people (Alunik et al. 2003). Reminiscent of the flood stories of other cultures, this suggests actual experience with widespread flooding.

The story of the Arctic peoples of North America – including the Dene, Inuit, Inuvialuit, Inupiat and their forebears - thus took place against a climatic and geological backdrop of varying stability. From time to time they faced major changes in landscape and ice cover, which must have challenged their way of life. Those who study cultural traditions, beliefs, attitudes and practices of northern peoples are working to extend the record of environmental change beyond the memories of those now living, or that of their parents and grandparents (Barber and Barber 2004, Cruikshank 2005, Berger and Liverman 2008).

The extent to which the various waves of migration across the Arctic over the long history of northern peoples (Fig. 1) were driven by environmental and climate changes, or were responses to new opportunities for trade or for hunting is still a matter of debate (Berner et al. 2005). McGhee (2007) suggests that it was the discovery of iron and copper that initiated migration into the eastern Arctic of ancestral Inuit around 1000 years ago, and that a few centuries later, trade opportunities with the prosperous Greenland Norse brought about a further expansion eastward. However, other movements may have been caused by the warmer climate during this period (the Medieval Warm Period), which brought open water near the coastal lands, restricting the sea-ice hunt and opening up maritime habitats. Further south, in northern Newfoundland, a period of warming sea temperatures around 1,100–1,500 years ago appears to have undermined aboriginal seal hunters and encouraged the northward movement of other indigenous people (Bell and Renouf 2008). Elsewhere in the Arctic, major landscape changes also took place during times of human occupation. Around 8000 years ago early people were living along the margin of the East Siberian Sea. Remains of their wooden houses can still be seen on what is now a remote island some 600 km north of the present mainland (Bauch and Kassens 2005, McGhee 2007).

Present: Responding to Climate and Landscape Change

Traditional and subsistence ways of life are becoming more difficult for many Northern peoples, and their long-standing ability to adapt to changing circumstances of weather and climate, is being seriously challenged by rapid climate warming, the effects of which are expected to become ever more severe (Einarsson et al. 2004). The long history of successful human survival in the harsh Arctic environment is witness to the extraordinary resilience of northern peoples, who have managed to keep their culture and traditions more or less intact in the face of environmental change. One important key to survival was their ability to move between sedentary and nomadic livelihoods and to maintain flexibility in where they lived and how they organized themselves socially (Berner et al. 2005). Another key has been their intimate knowledge of one's landscape based on observations of the seasons, the weather, the soil, and the plant and animal life. Indigenous people and others, who live close to the land, at least beyond towns and cities, possess a special understanding of the natural world around them.

The way in which today's Arctic inhabitants are themselves tracking and adapting to the challenging climate-induced changes in their environment (Ashford and Castleden 2001, Nunavut Department of the Environment 2005, Ford 2008) may be relevant to other rural societies facing change. Local observations of changing environments have provided knowledge essential to circum-Arctic nations as they press for action on climate change. Typical of the kinds of changes now being seen across the Arctic (Climate Change Impacts and Adaptation Network 2007[2]) are

[2] Climate Change Impacts and Adaptation Network, 2007. C-CIARN Nunavut- State-of-play report. Available at http://www.c-ciarn.ca/pdf/c-ciarn-nunavut_e.pdf [Accessed October 2008].

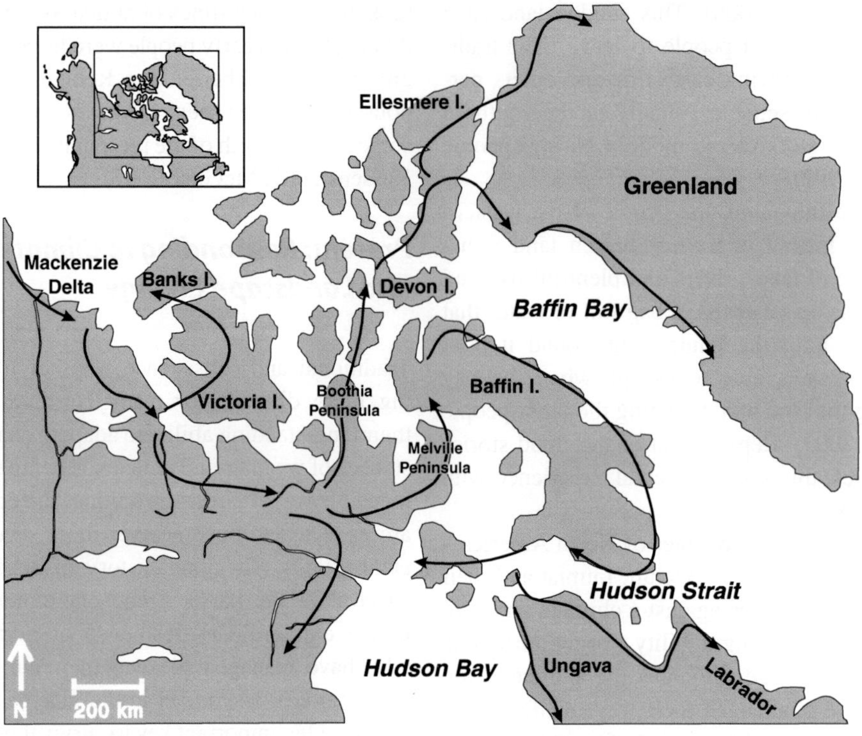

Fig. 1 Migration routes for the Tuniit (Palaeo-Eskimo) peoples of North America, during their long eastward migration over the past several thousand years. Modified after McGhee (1996) Maps 5.2 and 5.3. See also McGhee (2007)

those noted by the Inuvialuit of Banks Island in the Western Canadian Arctic:

- Multiyear ice no longer comes close to Sachs Harbour in summer, resulting in rougher seas and more dangerous travelling on ice.
- Open water is now closer to land in winter, and it is now harder to predict springtime ice break-up, weather and storms.
- Permafrost is no longer solid in places, and lakes are draining into the sea as permafrost thaws and ground slumps occur (Ashford and Castleden 2001).

Lessons

There are two important roles for the earth sciences in this time of profound change: assessing what is taking place now in the northern landscape, and unravelling the geological record of the recent past, even if the IPCC projects future changes of much greater magnitude. Tracking, describing, analyzing and communicating the physical changes taking place in contemporary terrestrial and marine environments is of prime importance, for the landscape provides the background against which ecological and social changes are now taking place. Garneau and Alt (2000) and Furgal and Prowse (2007) provide convenient summaries of these conditions. Of equal importance is the investigation through a wide range of palaeoenvironmental approaches of the geological changes of the past. Unravelling the record of past climates can yield valuable information about the extent and causes of change, whether abrupt or gradual. It also contributes to an understanding of past human development, including the origins of cultural traditions (Cruikshank 2005).

To strengthen communications with the public, the Geological Survey of Canada has produced a series of climate change and "geoscape" posters to illustrate how Canadian landscapes are shaped by geological processes past and present. These are aimed at schools and the general public. For example, the geoscape poster for Whitehorse, the small, subarctic capital city of the Yukon Territory (Turner et al. 2003) illustrates clearly how water and energy resources can be conserved, natural hazards and disasters reduced, and natural ecosystems protected. A climate change poster for Nunavut (the Inuit homeland in the eastern Canadian Arctic) explains how glacier fluctuations and sea levels are related, and how changes in the permafrost active layer affect infrastructure (Climate Change Impacts and Adaptation 2001). The north is also covered by a new Canadian earth science programme to promote private and public collaboration on adaptation issues and to develop tools to help in adapting to a changing climate (Adaptation website, Natural Resources Canada[3]).

In the Arctic, the past century saw one of the most rapid and most successful adaptations to external forces. This took place when many of the northern peoples of Canada and elsewhere were obliged by government policy to abandon their nomadic ways and settle in towns and villages where there are strict controls on resource use and management, land use and ownership, and where market-drive economic activities now dominate. In two or three generations, the way of life of the Inuit and other northern peoples has changed substantially, as they strive to build a new society and exercise more control over their lands.

It might be suggested that the ways in which prehistoric nomads reacted to rapid climate and landscape change could provide helpful insights as to how societies and individuals facing even faster change in the twenty-first century might adapt. However, comparing a contemporary, largely sedentary, technologically advanced society with far smaller numbers of hunters and fishers of past centuries and millennia seems rather far-fetched. In the face of the current changes now coming from climate warming, loss of traditional biodiversity and widespread air-borne pollution from southern sources, the new way of life means that traditional adaptation strategies are no longer possible, and possibilities for flexibility much reduced. Scientific research, especially from the earth sciences, is providing some of the essential understanding of the ecosystem and landscape changes involved. However, it remains to be seen how the peoples of the circum-Arctic will adapt, as access to their hunting and fishing grounds become more difficult, and the social, cultural and economic consequences of changes to their way of life loom ever larger.

Hurricanes in the Tropical Atlantic: New Orleans

Introduction and Past Records of Hurricanes

The Louisiana coast is located on the Gulf of Mexico, between latitudes 29 and 30 °N. Because of its location, this region is prone to tropical cyclones that develop in the Atlantic, i.e. hurricanes. Hurricanes are low pressure, intense tropical weather systems connected with strong thunderstorms and have a well-defined counter-clockwise circulation of winds near the earth's surface, and sustained wind speed of 119 km/h and higher. These complex systems feed on heat released from rising hot and humid air masses. Hurricanes develop each year in the Atlantic Ocean and the Gulf of Mexico from June to November. Although the timing is stable, the intensity of hurricane activity in the Gulf varies, as does the number of hurricanes that make landfall in urban areas. According to the U.S. National Oceanographic and Atmospheric Administration,[4] the 2008 Hurricane season set records. For the first time since hurricane monitoring started, six consecutive tropical cyclones made landfall on the U.S. mainland, and this is also the first Atlantic season to have hurricanes in five consecutive months. The majority of hurricanes entering the Gulf of Mexico are Category 3 and lower. One of the threatened geographic areas, the Northern coast of the Gulf of Mexico, hosts unique coastal ecosystems and major U.S. cities. In Louisiana, New Orleans is a metropolitan port city of historical significance with a thriving and colourful tourism industry. To its south, swamps accounting from 40% of the nation's wetlands create a natural barrier between New Orleans and the

[3] Adaptation Website, Natural Resources Canada. Available at http://adaptation.nrcan.gc.ca/index_e.php/ [Accessed 5 November 2008].

[4] National Oceanographic and Atmospheric Administration, 2008. Available at http://www.noaanews.noaa.gov/stories2008/20081126_hurricaneseason.html [Accessed December 2008].

Gulf of Mexico. Economically, this region is the heart of Louisiana owing not only to tourism, but also to fisheries and the oil and gas exploration and refining infrastructure. Twenty-five percent of all the oil and gas used in America and 80% of the nation's offshore oil and gas travel through Louisiana's wetland; more than 95% of all marine species living in the Gulf of Mexico spend all or part of their life cycle in Louisiana's wetlands; 30% of the nation's fisheries catch comes from offshore Louisiana; and 75% of all waterfowl breed exclusively in the wetlands (America's Wetland Campaign to Save Coastal Louisiana's website[5]).

Storms from 1944 till present are recorded in details by the U.S. National Hurricane Center and their impact can be evaluated. Since the 1800s, storms known from historical documents can also be taken into account when analyzing risks. Gulf coast hurricanes have changed the face and economic strength of historic cities for decades. For instance, the "Great Storm" that devastated Galveston, Texas, in 1900 is known from survivors' memoirs and various books (Greene and Kelly 2000, Bixel and Turner 2000). From these writings, it is evident that at the turn of the century, Galveston was the major port city in the United States. But this storm's high death toll (over 8,000) led city officials to encourage the transfer of the population and infrastructure inland, in Houston, Texas.

As controversy is fuelling the debate on the relationship between global warming and frequency of hurricanes worldwide, it is urgent to analyze the frequency and intensity of great storms further back in time, by understanding their sedimentary records (Leroy and Niemi 2009) and integrate the findings into future mitigation plans and assess future risks. To document the frequency and intensity of past hurricanes, detailed sedimentary records are now analyzed and compared to that associated with modern storm deposits. For instance, Horton et al. (2009) and Park et al. (2009) analyzed fresh hurricane-induced storm deposits to improve the recognition of older hurricane sedimentary signatures. Horton et al. (2009) collected hurricanes Katrina and Rita storm surge deposits before they were removed or naturally eroded. They estimated the surge to be of 7.5 m in Alabama with inland extents greater than 700 m. They noted that the surge sedimentary unit had a thickness ranging from 7 to 13 cm, was coarser than pre-storm surge units, and had lower organic content and a virtual absence of foraminifera tests. Park et al. (2009) conducted a similar study, but focused on storm surge sediments from a lake in San Salvador Island, Bahamas. They used a multi-proxy approach (e.g. grain sizes and composition, microfossil assemblages and geochemical analyses) to define the geological signature of storm deposits in this region.

These modern deposits can then be compared to the sedimentary records of past storms. Palaeotempestology, the study of the signature left by hurricanes in sedimentary archives, such as sand layers in the fine-grained sedimentary record of lakes or coastal lagoons, is a key step in this direction. For instance, Liu and Fearn (2000) proposed on the basis of coastal lake and marsh sediment studies that, in the past, Category 4 or 5 hurricanes hit the U.S. Gulf coast only about once every 300–600 years on average at any one point. In a similar study, Donnelly and Woodruff (2007) analysed the frequency of hurricanes that affected a Caribbean lagoon within the last 5,000 years and found periods of more intense hurricane activity and periods of quiescence, which they linked to the El Niño-Southern Oscillation and the West African monsoon. McCloskey and Keller (2009) studied a 5,000-year sedimentary record of hurricane strikes on the central coast of Belize and concluded that major hurricanes have struck the Belize coast on average once every decade for the past 500 years, and that two periods of hyperactivity occurred between 4,500 and 2,500 years BP. This type of hurricane hyperactivity might have induced significant stress on the Maya civilization (McCloskey and Keller 2009) just as it does on modern civilization living in the path of frequent hurricanes today.

A Recent Disaster: Hurricane Katrina

When on 29 August 2005, Hurricane Katrina made landfall between two of the Gulf's largest cities, New Orleans and Mobile, as a Category 3 hurricane with winds at 209 km/h, storm surge up to 9.1 m, and rainfall up to 30.5 cm, hundreds of thousands of families were affected (Figs 2 and 3). Approximately 5.8 million people in three states experienced hurricane-force

[5] America's Wetland Campaign to Save Coastal Louisiana. Available at http://www.americaswetland.com/ [Accessed August 2008].

Fig. 2 (**a**) View of the Bart family residence at the corner of Annette St. and Sumpter St., two blocks from the London Avenue Canal (New Orleans), the day following Katrina's landfall. Note the top of the truck in front of the second house. Photograph by Michelle Dejoie Manning. (**b**) Same view, 2 months later. After waters receded, locals were allowed to return to their home to remove debris such as these refrigerators seen in the background. Note the watermark on the same truck. Photograph by Christine Alexander

Fig. 3 Inside the Bart's residence, all walls, electrical systems, all belongings, and 50 years of memories have been destroyed by water and mold. Photograph by Christine Alexander

winds (Gabe et al. 2005) and most of the major damage was located in a 160-km radius from the landfall area. But the bulk of the devastation was caused by flooding associated with breaches of New Orleans levees and floodwalls that left most of the city under several metres of water. Because of the extent in geographic impact, intensity, death toll and damages to properties and the environment, Katrina was one of the most catastrophic hurricanes in U.S. history. The Louisiana Recovery Authority[6] assembled a one-year anniversary

[6] Louisiana Recovery Authority Anniversary Report, 2006. Available at http://lra.louisiana.gov/index.cfm?md=sitesearch&tmp=home&keyword=hurricane+katrina+anniversary+data [Accessed August 2008].

report in 2006, listing the full impact of the storm in Louisiana only, the tolls are staggering.

In addition to the obvious cost in human life, structural and economical losses, the cultural landscape of Louisiana has taken a major hit. The loss of houses or work, and the rising cost of insurance made it impossible for entire families to come back to their hometown. Among these permanently displaced families are musicians, artists, and some of the last stronghold of the Cajun/Creole French heritage. Many Mexican immigrants moved from Texas to Louisiana to help with the rebuilding effort, further enhancing the cultural shift of the region. According to an August 2008 report from the Greater New Orleans Community Data Center,[7] 16 of 50 New Orleans neighbourhoods that flooded following Katrina have less than half the households present in June 2005.

Transfer of Information

To its credit, Louisiana had an excellent evacuation plan in place. It operated in partnership with the National Hurricane Center (NHC). The NHC broadcasts 5-day forecast for all hurricanes on local and national television channels. Hence, the public and officials were aware of the storm approach. But the 3-day forecast zone of hurricane Katrina extended from Texas to Florida, and three days before the storm made landfall, the main evacuation routes were already saturated with persons evacuating four states (Florida, Alabama, Mississippi and Louisiana). On 28 August 2005, hurricane Katrina was upgraded to a Category-5 hurricane. At that time, the mayor of New Orleans ordered the first-ever mandatory evacuation of the city. Numerous smaller cities along the storm's predicted path were also issued mandatory evacuation orders. Following this announcement, the Louisiana State Police implemented the Contraflow Lane Reversal on three major interstates, i.e. I-10, I-55 and I-59, allowing 1 million persons to evacuate successfully and find refuge in shelters, university dormitories, and with friends and families out of state. Yet, a reported 150,000 persons were unable to evacuate. One of the reasons was that 112,000 households were without private vehicles (Cutter et al. 2006). Others chose to remain because they did not believe the storm constituted a serious threat, while others were too poor or physically unable to evacuate. The high death toll from Katrina can be attributed in part to the fact that hurricanes have been affecting Louisiana for years without long-lasting consequences, and therefore, many families underestimated Katrina's threat. Today, the state is working on improving its evacuation plans. For instance, evacuation buses have been added to assist families without transportation.

The devastation brought by Katrina also jump-started new collaborations between scientists. For instance, at the LSU Center for Computation and Technology, the tragedy mobilized groups working on high performance computing, sensor networks and visualization, GIS, remote sensing and coastal and atmospheric modelling. One of their missions is to provide data to Public Health governmental agencies to better prepare for storms through visualization and modelling.

The greatest challenge remains storm protection. The region desperately needs better protection for the population living in the low-lying parts of town. Early French settlements in New Orleans in the 1700s were strictly restricted to the high natural levees of the Mississippi River (Fig. 4). After the construction of pumps, rainwater drainage canals and levee systems, many of the marshes and swamp forests (visible in the 1723 map of Fig. 4) were drained and converted to residential areas. Today, these neighbourhoods host the majority of the Greater New Orleans population and most are located below sea level.

The levees were not high or strong enough and some of the failures were linked to the poor integration of known sedimentological data into their design. In some cases, the levees were not armoured; therefore lacking protection against erosion from overflowing water. This problem was addressed by the Corp of Engineers, which used reversed "T" shaped walls in the reconstruction process. Another issue concerns the subsurface sediment on which the levees were built. Nelson and Leclair (2006) reported that some of the levees are anchored in or close to peat deposits. Peat is an unstable sediment that was originally formed in swamp forests that made up the region's environment before the French settlements. Another puzzling discovery was the several metres of sand deposit at

[7] Greater New Orleans Community Data Center, 2008. New data reveals 16 New Orleans neighborhoods have less than half their pre-Katrina households. Available at http://www.gnocdc.org/ [Accessed November 2008]

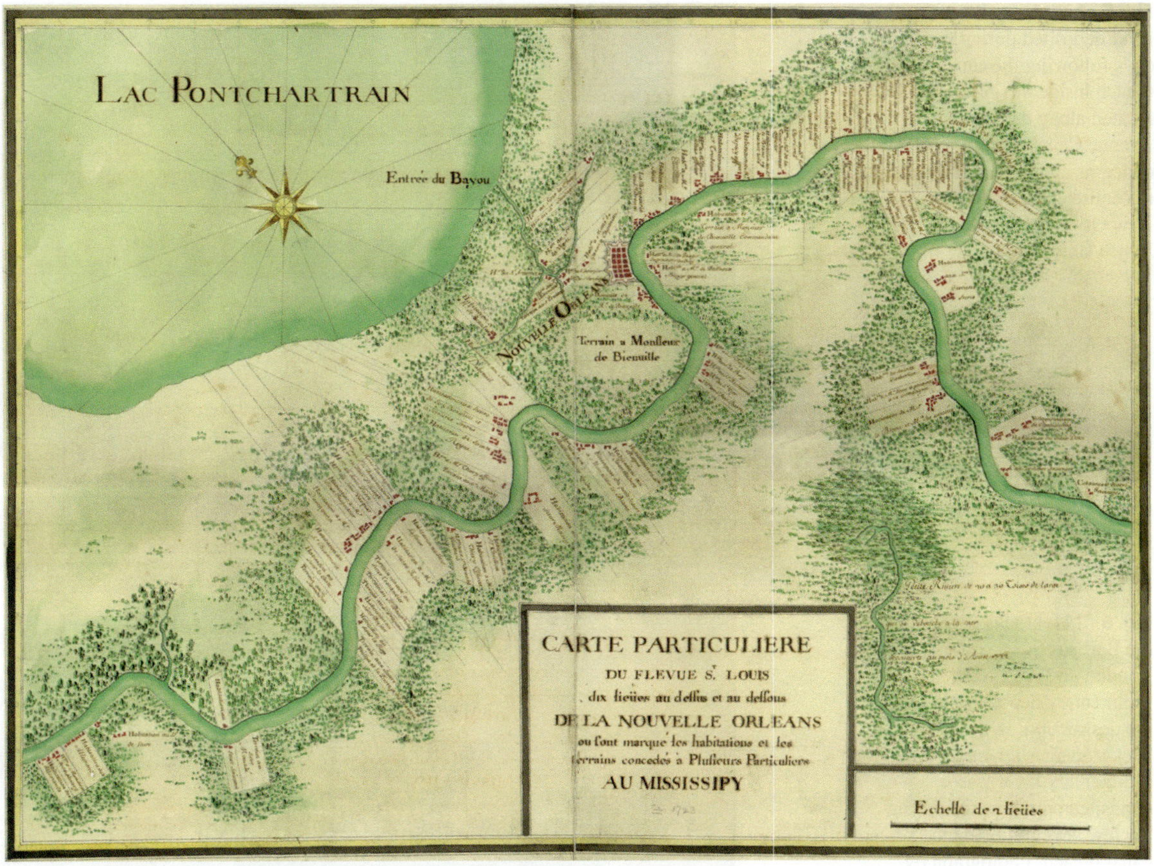

Fig. 4 A particular map of the St. Louis River (now Mississippi River) ten leagues above and below New Orleans, on which are marked the homes and lands granted to some private individuals along the Mississippi (ca. 1723). Note that most of the areas along the rivers are natural swamp. Houses were built only on the high banks of the river. *Cartes Marines* from the Newberry Library (call number: Ayer ms map 30, no. 80)

the site of one of the breaches (Fig. 5). An astounding 26,380 m^3 of sediment were deposited in the city in the days that followed Hurricane Katrina (Fig. 6). This means a deposition ranging between 0.3 and 1.8 m in two days close to the levee breaches (Nelson and Leclair 2006). Since no sands were used to build the levee, Nelson and Leclair (2006) argue that this sand originated from the underlying Pine Island Trend beach deposits. This unit was deposited some 4,000 years ago as sea level rose (Coleman et al. 1998) and is found beneath all the drainage canals. Nelson and Leclair (2006) point that this raises one of the most significant concerns about the future stability of the New Orleans drainage-canal levee system.

In addition to immediate concern of levee stability, the larger-scale hurricane protection system needs to be revisited. van Heerden and Bryan (2007) proposed a three-layer defense plan. The first layer should be the improvement of existing levees, and construction of new levees or floodwalls in front of all major population centres. A second layer should protect this inner layer. An obvious natural barrier is a healthy swamp; and Louisiana's wetlands could be replenished by flooding from the Mississippi River. Finally, a third layer should protect the wetlands from erosion by strengthening barrier islands.

When Mechanisms of Economy and Mitigation are Shaping a City

One of the broader implications of hurricanes is the economic impact on hurricane-affected regions. Physical geography influences economic performance (Diamond 1997). The geographical tropics contain the poorest countries in the world showing that geography

Fig. 5 An impressive sand layer deposited during the few days following the canal breach inside this residence located along the London Avenue Canal raises important concerns about the stability of the entire canal system in New Orleans. Photograph by Renée Hetherington

Fig. 6 Two blocks away from the Bart's residence, along the London Avenue Canal, sedimentary deposits add to the destruction. In the background, a 1.1-metre thick sedimentary layer was deposited on this porch during the few days following the canal breach. Photograph by Klaus Arpe

is a key factor in the distribution of wealth. Sachs et al. (2001) stated that the geography/wealth relationship is due to three components: tropical countries face higher rates of infectious disease, lower agricultural productivity and have more difficulties transporting foods and goods. One factor that should also be considered when assessing needs of tropical countries such as those in Central America and southern U.S. is the frequency and intensity of hurricanes. The economic impact of hurricanes has been underestimated in the geographical analysis of wealth distribution and considering current prediction on global warming, hurricanes' economic impact should become increasingly significant. Horowitz[8] predicts that a 1°C temperature increase across all countries would yield a 3.8% decrease in world GDP.

[8] Horowitz, J.K., in press. The Income-Temperature Relationship in a Cross-Section of Countries and its Implications for Predicting the Effects of Global Warming. Environmental and Resource Economics. Available at http://faculty.arec.umd.edu/jhorowitz/ [June 2009].

Katrina's cost are estimated to be several billion US dollars, but the reconstruction effort is mostly federally funded – a support system that other hurricane-prone countries such as the Dominican Republic, Cuba or Haiti do not have. But even for a large economic powers such as the U.S., hurricanes have some devastating trickling effects, such as the increase of insurance premiums, on each citizen and specifically those in hurricane-affected area. Elsner and Kara (1999) reported that hurricanes accounted for 62% of all catastrophic insurance losses. Since 2004 and the intense Florida and Louisiana hurricane seasons, storm-related payouts almost erased earnings of some of the biggest publicly-traded auto and home insurers. As a consequence, rates of insurance have increased significantly forcing some businesses to relocate out of state. In some of the most vulnerable areas, insurance companies cancelled policies. The only option for many homeowners to find insurance is via the Federal Government. According to officials at the U.S. Department of Homeland Security's FEMA[9], financial assistance to state and local governments to help repair or replace infrastructure throughout the Gulf coast damaged by Hurricanes Katrina and by hurricane Rita (a Category 3 hurricane that made landfall in Louisiana just 24 days after Katrina devastated the region) has topped the 10 billion USD mark. Public officials in southern states, have been fighting insurance companies for years over rising rates and companies "non-renewed" policies for hurricane-battered places like Florida and Louisiana. Vitello (2007),[10] the president of the Insurance Information Institute, said "Considering what happened between 2003 and 2005, and considering that the best meteorological minds are telling us that for the next 15–20 years hurricane activity will be heavier than normal, if we didn't do something to reduce our exposure, we'd be out of business."

These types of views bring out another issue. As CO_2 concentration keeps increasing in the atmosphere, and as temperature responds with a general warming trend (0.6°C in the past 100 years) (Mann and Kump 2008), there are little doubts that sea level will keep rising, threatening Louisiana's wetland, and increasing flood risks. Because hurricane feeds on warm and moist water, it seems logical to assume that the frequency and strength of hurricanes will increase as well, but current data do not prove this conclusively and there is still no consensus regarding links between hurricane activity and climate change (Arpe and Leroy 2009). Meanwhile, New Orleans' face has changed forever with demographic data showing a smaller and whiter population (Stewart and Donovan 2008) than pre-Katrina.

Rapid Sea Level Rise: Caspian Sea

Geographical Setting

At the border of Asia and Europe, the Caspian Sea is the world's largest isolated basin (Fig. 7). Its catchment basin covers an area of 3.5 million km^2, which is mainly located in five littoral states, namely Iran, Turkmenistan, Kazakhstan, Russia and Azerbaijan. The catchment also covers small sections of Turkey, Armenia and Georgia (Rodionov 1994).

The Caspian Sea is a semi-elliptical basin oriented in a N-S trend with a length of about 1,200 km and a mean width of about 310 km. The sea surface area and water volume (at the level of –27.5 m relative to world sea level) is around 376,400 km^2 and 78,100 km^3 respectively (Terziev 1992). On the basis of its bottom morphology, the Caspian Sea can be divided into three distinct sub-basins, deepening southward to a maximum depth of 1,025 m. The Southern Caspian basin that has the main volume of water is separated from the middle basin by the Apsheron sill at a depth of 150 m (Kosarev 1975). The total length of the Caspian coast is around 4,460 km (Caspian Environmental program 2002[11]).

The general atmospheric circulation over the Caspian Sea and its drainage basin, which comes from the north Atlantic and the Arctic and their interac-

[9] FEMA, 2005. Available at http://www.fema.gov/news/newsrelease.fema?id=20359 [Accessed December 2008].

[10] Vitello, P., 2007. Hurricane Fears Cost Homeowners Coverage. Available at http://www.nytimes.com/2007/10/16/nyregion/16insurance.html [Accessed December 2008].

[11] Caspian Environmental Program, 2002. Caspian Demographic Profile. CRTC "Human Sustainable Development & Health" Ashgabat, Turkmenistan. By CRTC SHD&H. February 2002. Available at http://www.caspianenvironment.org/scripts/print.pl?word=report_miscell2.htm [Accessed December 2008].

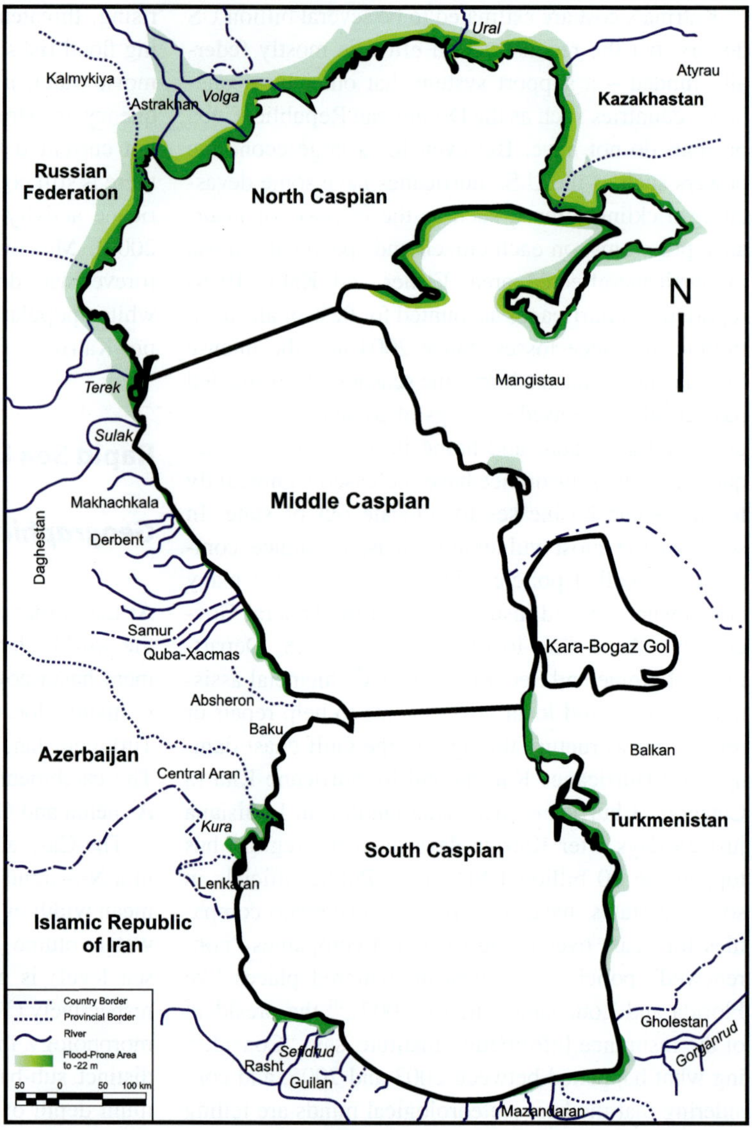

Fig. 7 Projected inundation area around the Caspian Sea in case of sea level rise up to –22 m that experienced in the late Holocene and historic periods. Compiled and modified after Caspian Environment Program 2002

tions with regional and local circulations, determines the hydrometeorological regime of the Caspian Sea (Rodionov 1994).

Past and the Future

Historical literature and ancient coastal remains have recently contributed to unravel the sea level fluctuations for the past two millennia successfully (Apolov and Fedorova 1956, Kroonenberg et al. 2007, Leroy et al. 2007). Using geological and geomorphological investigations at a quasi-annual resolution, a likely 65 year cycle has been highlighted (Kroonenberg et al. 2000, Giralt et al. 2003). Since instrumental measurements, the sea level has fluctuated with an amplitude of 3.8 m (Voropaev 1996). The rate of the last sea level rise that began in 1979 is hundred times faster than that in the world oceans (Kroonenberg et al. 2000). Rapid sea level changes had different, often deep, impacts on the Caspian shores (Kaplin and Selivanov 1995, Voropaev et al. 1998, Kroonenberg et al. 2000), depending on coastal slope and composition. The gentle slope of the north coast and some parts of east coast are prone to inundation with sea-level rise and emergence with sea level fall.

Atmospheric pressure, wind, wave, riverine influx, temperature-salinity, impulsive energy (such as earthquake, submarine landslide and mud volcano), and the position of the Sun and the Moon, all contribute to changing the Caspian Sea level. Wind waves and wind-induced sea level changes (storm surge) are very important in short-term sea level fluctuations (Terziev 1992, Kazancı et al. 2004). The northerly and southerly winds are prevailing wind patterns that produce corresponding waves, which reach their maximal elevation in the middle and northwest parts of the south basin (Koshinskii 1975). The highest wave is recorded in Azerbaijan coastal waters at a height of 12 m (Terziev 1992). Sea level rise due to storm surge occurs at various scales in the three sub-basins. The maximum wind-induced sea level rise is around 1 m in the middle and south basins, while in the north Caspian with its shallow waters and a gently sloping coastal area, a 4 m rise in water has been recorded in a storm in 1957 (Terziev 1992). In addition to positive storm surge, negative storm surges are also important for the north Caspian coast. They displace coastal waters towards the sea and cause ports to become unusable. The impulsive energy in the Caspian Sea that may have a potential to generate tsunami waves could have three sources: submarine landslides from the south and western shelf of the middle Caspian basin, mud volcano explosion in the south Caspian and earthquake in the middle Caspian Sea. The first two sources are poorly studied, while seismogenic tsunamis have historical records in the Caspian coasts. A possible return period for tsunamis with a height of 3 m is suggested as 60 years (Dotsenko et al. 2002).

Long-term sea level fluctuations depend on the sea-water balance (difference between river influx to the sea and evaporation over the sea surface) and geological processes. Geological processes such as subsidence, uplifting, spreading, and sedimentation, can cause a change in the Caspian Sea water volume over the long term (Federov 1996, Rychagov 1997). Geological processes over the Caspian catchment basin can affect the water balance, such as watershed and river course switching to another basin (Varushchenko et al. 1987, Leroy et al. 2006).

Water balance is the determining factor for Caspian Sea level change since isolation from the Paratethys 5 Ma ago. The main input elements of Caspian Sea water are river influx, precipitation over the sea surface and groundwater seepage. The loss of water depends on evaporation from the sea surface and from Kara Bogaz Gol Bay, which acts as an evaporative basin (Giralt et al. 2003). The rate of the last sea level rise since 1979 was around 14 cm/a reaching its maximum value in 1995 (Voropaev 1996). The sea level rise is attributed to an increase in Volga discharge (Frolov 2003). Intense water consumption from the mid twentieth century especially in the Volga basin has dropped the sea level by about one metre (Zonn 1996). Without this human impact, the water level would be 1 m higher than the present stand.

The prediction of Caspian Sea level changes has been the subject of many research projects during the past century. They have attempted to forecast the Caspian Sea level on the basis of stochastic models of water balance, palaeodata of sea level changes, cyclicity of sea level changes, and correlation between sea level change and the factors that can contribute to its water balance (Shiklomanov et al. 1995, Frolov 2003). However long-term forecasting of the Caspian water balance and consequences for its level have remained unsuccessful. None of the studies predicted the sea level rise in 1979 and the fall in 1995. The Caspian Sea, including its closed catchment basin, has an area of around 1% of the earth's surface. The hydrometeorological cycle in such a huge area is globally affected (Voropaev 1996). The main problem of forecasting is the mechanism by which controlling factors act on the elements of water balance. The transfer of humidity to the Caspian basin and the physical processes in and over the Caspian Sea, which determine the rate of evaporation, are critical points that are not yet clearly understood (Kosarev and Tuzhilkin 1997, Arpe and Leroy 2007).

Some short-term sea level changes have a predictable character. Warning systems for forecasting wind waves and storm surges are relatively successful in coastal states in reducing the risk. For tsunamis, there is no system for early warning and some potential sources are as yet poorly known.

Present: Impact of Sea Level Rise on the Coastal Community

Human activities of circum-Caspian nations are distributed unevenly in the region. Hazardous sea level change in the Caspian region became important from the late nineteenth century with the development of industry, oil exploitation and marine transportation (Komorov 1996, Lahijani 2001). Natural extreme events such as the tsunami in 1895 due to an earthquake (Dotsenko et al. 2002), the storm surge of 1952 and the high wave of 1957 had disastrous effects. During a four-day surge (10–13 November 1952) in the north and north-west Caspian Sea caused by a strong storm, water level in places rose up to 4.2 m, then penetrated 25–35 km inland and flooded 17,000 km^2 including five islands. The rate of rising water level was 20 cm/h. The return period of this type of surge is estimated around 150–200 years. During the storm of 20–21 November 1957, the Caspian west coast, mainly including Azerbaijan and Dagestan, was hit by high waves. The waves destroyed oil structures and were associated with loss of human life (Terziev 1992). The sea level rise of 1979–1995 inundated 8,300 km^2 of coastal territory (Zonn 1996). Most economic loss occurred on the western and southern coasts, which are densely populated. Human activities in the coastal area accelerated after the collapse of the Soviet Union in 1991. Petroleum-related industry and transportation are the main branches of economic activity. The Caspian coast, with a mostly arid climate and a smaller area of sub-tropical climate on the southern coast (from Lenkoran in Azerbaijan to Gorgan in Iran) and with its lack of rivers on the east coast, has a limited attraction for development. Therefore human settlements and infrastructure are unevenly distributed along the coast. At present the population settled in the Caspian catchment basin is estimated to be around 80 millions, most of whom live in Russia (73%), Iran (13%) and in Azerbaijan (10%) (Lahijani 2001). Around 3.9 million people, or approximately 50% of the Azerbaijan population, live in the coastal zone in the four regions of Lenkoran, Central Aran, Apsheron and Quba-Xacmas. The Iranian regions of Guilan, Mazandaran, and Golestan border the Caspian coast with a population of 6.3 million persons that is 9% of Iran's population. The population in a narrow, 2 km wide, shore strip zone on the Iranian coast is around 370,000 that increases to 1.63 million within the first 10 km and 3.70 million within 50 km (Pak and Farajzadeh 2007). The total population in the 13 coastal regions of littoral states is around 15 million (Caspian Environmental Program 2002[12]).

The Iranian Coast with a population density of 235.9 per km^2 (Pak and Farajzadeh 2007) has the highest density in the Caspian periphery. The Iranian Caspian coast with sandy to gravelly beaches and a forested landscape with a relative proximity to large cities attracts domestic tourism during the main vacations when the population doubles. Except for the easternmost part of the coast, each metre of coastal area is occupied by homes (Fig. 8), towns, agricultural fields, fisheries and resort facilities. The main infrastructure on the Caspian shores consists of ports, railroads and pipelines on the western and northern coasts and of oil fields that are mainly developed on the Azerbaijan and Kazakhstan coasts.

Natural hazards, including long-term sea level change, storm surge, tsunami and wind waves, form a potential threat to the Caspian coast. However, the whole Caspian Sea rarely experiences brief hazardous sea level changes simultaneously. The North Caspian with its shallow waters (on average 5 m deep) does not allow the generation of high waves. In contrast the area is liable to severe surges, both positive and negative. The north sub-basin is frozen in winter, when there is no opportunity for brief sea level changes. The middle Caspian receives wind waves, surges and tsunami waves. The west coast, mainly in the middle sub-basin and the north west of the south sub-basin, is the most hazard-prone coast. The Apsheron sill prevents the development of tsunami waves towards the south basin. The South Caspian coasts are exposed to medium-range storms and surges.

As humans intensively occupied a narrow shore zone during the twentieth century, a sea-level rise now means a disaster for the coastal area. All the coastal countries are in a developing stage of their economies. The three newly formed circum-Caspian states (Azerbaijan, Kazakhstan and Turkmenistan) mainly rely on oil exploitation from the Caspian Sea and its peripheral fields.

A rise of sea level by more than 1 m will inundate the five littoral states to various extents, which is at a minimum in Iran with 300 km^2, but in Kazakhstan it is up to 6,300 km^2. This range of sea level

[12] Caspian Environmental Program, 2002. Caspian Demographic Profile. CRTC "Human Sustainable Development & Health" Ashgabat, Turkmenistan. By CRTC SHD&H. February 2002. Available at http://www.caspianenvironment.org/scripts/print.pl?word=report_miscell2.htm [Accessed December 2008].

Fig. 8 Sea wall along the Mazandaran coast, at Babolsar, Mazanderan University's residential campus. The sea wall protected the campus against sea level rise and coastal retreat. Photograph by Majid Shah-Hosseini (2007)

rise is probable during a storm surge. A rise in sea level to the level that was experienced in late Holocene and historical periods at around 22 m below world sea level (Rychagov 1997, Kroonenberg et al. 2007, Lahijani et al. 2009) would inundate many population centres and infrastructure (Fig. 7). Many coastal structures were designed for extreme events around the mid-twentieth century. Now the sea level is 2 m higher than during that period and therefore the hazard could overwhelm the structures. Long-term sea-level rise remains the main potential threat for the Caspian coast. A population of more than half million in the Astrakhan region including the southern part of Astrakhan city is at high risk. The Azerbaijan coast is more vulnerable to sea level rise. Its economy depends on Caspian resources, 50% of its population including the capital and other large cities are located on coastal areas and their main infrastructure is coast based. Around 140,000 persons in Azerbaijan live in high-risk areas including some parts of the capital Baku and in Lenkoran. A steeply sloping coast prevents deep penetration of flooding into the Iranian coastal zone. Nevertheless, sea level rise could impact on low-lying areas with high-density populations. Anzali (a port in the Guilan province), the east part of Mazandaran and the south of Golestan are facing a high rate of risk with a population of 150,000. The east and northeast coasts with scattered populations would have a lower risk in future sea level rise. Main roads and railways in Astrakhan-Atyrau and some oil structures in Kazakhstan and Turkmenistan are located in a high-risk area. People in these two east littoral states that settled in high-risk areas is however estimated to be less than 20,000 persons (Caspian Environmental Program 2001[13]).

Lessons: Adaptation and Mitigation Measures

The economic loss of the last sea level rise is born partially by the littoral states, as overall measures that could be planned to reduce or restrict vulnerability are

[13] Caspian Environmental Program, 2001. Caspian Sea Potential Inundation and Impacts on Human and Natural Environment. December 2001. Ashgabad/Turkmenistan. By Prof. Dr. Frank Schrader, Germany. Available at http://www.caspianenvironment.org/scripts/print.pl?word=report_miscell2.htm [Accessed December 2008].

based on the resilience of people or the reliability of protective devices (Smith 1996). Both approaches are in use in the developing littoral states. Building dams and dikes to protect coastal settlements and infrastructures against inundation and reinforcing ports for new water stand provide a feeling of reliability when applying technology and engineering design. To accept hazardous sea level rise as a part of life demonstrates a lower capacity to face natural extreme events. Adaptive methods could be applied by the littoral states themselves, in that they could avoid, or protect themselves against sea level hazard (Soroos 2000).

At the level of the whole Caspian Sea, mitigation measures are implemented by individual states, as regional cooperation does not work. Stabilizing the Caspian Sea level was of great concern during the Soviet period, when sea level was declining, but due to ecological problems (such as joining the separate basins of Caspian Sea, Aral Sea, Black Sea, west Siberia, Barents Sea and Kara Sea (Voropaev and Ratkovich 1985)), it has been abandoned. As the Caspian problems are shared between coastal states, all the states should promote initiatives about Caspian environment regardless of any legal debate (Oliounine 2003).

The Iranian coastal zone of the Caspian Sea is governed by four laws: (1) The law of "coastal lands and coastal freed lands ratified in 1975." Under this law, a 60 m wide strip of the Caspian coastal zone from the maximum level in 1962 and the land up to 150 cm of that level, which emerged due to sea level fall, are declared as state-owned territories. Some exceptions and details are mentioned in the Law for private and state activities. Ministry of Agriculture, Department of Forestry and Pasture is authorized for the execution of the Law, but the related executive regulations are still not prepared. Moreover this law is over-ruled by the Caspian Sea rise since 1962, as the land is not under water. (2) The law of "fair distribution of water ratified in 1982." Under this Law all coastal zones of seas, gulfs, bays, lagoons and river banks are state owned and any construction and land use are permitted only by the Ministry of Energy, Water Department. This Law mainly concentrates on water uses and sand excavating. The two above-mentioned laws are in conflict over some points. (3) The "Code of port and shipping organization ratified in 1969." Any port construction and maritime activities need the authorization of the Ministry of Road and Transportation, Port and Shipping Organization. (4) A national law of "land and housing" and regulation of each municipality apply to urban construction, including the Caspian coastal zone.

During the past three years, the National Government tried to free 60 m coastal zone for public access. The main part of the Mazandaran coastal zone is occupied by governmental agencies. Some of them prepared a public access agreement to the coastal zone over the past few years. In spite of different Laws for management of the Caspian coastal zone, jurisdictional difficulties remained until now. In practice the two first Laws have been abandoned, because they conflict with each other, no regulations were prepared after their ratification and the limits of governing territories are not clear (Moghimi 1997). Moreover some parts of the coastal zone have been under seawater since the sea level rise of 1979. The two later Laws are in force for construction of buildings in urban areas and ports. Even in Caspian coastal cities, the threat of sea level rise is not considered for urban development. Regulation for construction in rural areas is not rigid. Valued coastal biospheres that are mentioned in the Ramsar Convention[14] are however being monitored and preserved by the Department of Environment.

The circum-Caspian nations, which are enjoying its resources, need more awareness of the Caspian environment and the behaviour of its water level. The attention of the international and regional scientific communities to Caspian-related researches has increased during the past two decades. Despite of good progress in research, intensive human activities along the Caspian shores in vulnerable areas show that scientific information is not well received both by local people and decision makers. The transfer of knowledge to policy makers and stakeholders should be focussed on overall ecosystem services, which are more valuable than short-term economic benefits. Different levels of cooperation may be recommended for the Caspian issues, but public demand and awareness is vital for the improvement of the present situations (Pak and Farajzadeh 2007).

14 Ramsar Convention on Wetlands. Available at http://www.ramsar.org/ [Accessed December 2008].

Rapid Lake Level Rise in the Pampean Plains of Argentina

General Setting

The present-day South American climate distribution is closely linked to the particularly tapered shape and the topography of the continent as well as to global circulation cells, ocean currents and the proximity of large bodies of water (Cerveny 1998). The Andes Cordillera largely controls the atmospheric circulation of Southern South America fostering a tropical-extratropical air mass exchange especially along their eastern side (Garreaud et al. in press). East of the Andes, a vast lowland area from Colombia and Venezuela to the Argentinian Pampas is an outstanding South American geomorphological feature. In Argentina, the Pampean plain is a flat and low area with altitudes between 80 and 400 m a.s.l. that extends to the south to ca 40°S, characterized by widespread Late Pleistocene-Holocene loess deposits (Zárate 2003).

The twentieth century climate variability shows contrasting hydroclimatic patterns on the Pampean plains. The region was affected by long drought intervals throughout the first three quarters of the twentieth century followed, since the 1970s, by an abrupt hydroclimatic shift to a humid phase. High-resolution palaeoclimatic reconstructions to fully appreciate the climate variability are still scarce across the Pampean plains. Middle latitudes palaeoenvironmental reconstructions proposed so far were exclusively based on geomorphic-stratigraphic data (Iriondo 1999, Kröhling and Iriondo 1999) and historical proxies (Cioccale 1999). However, during the last few years there has been a noticeable increase in high-resolution studies based on palaeolimnological record of Pampean lake systems (Piovano et al. 2002, 2004a, 2004b, 2006a, 2009, Córdoba et al. 2006, Palamedi 2006). The initiation of the program PALEOPAMPAS[15] now provides the opportunity to analyze past climate variability at larger time-windows and from a more regional perspective.

[15] PALEO-PAMPAS "Paleolimnological reconstructions across the Pampean Plains," 2007. Available at http://www.cicterra-conicet.gov.ar/areas-limno-programa.htm [Accessed December 2008].

Laguna Mar Chiquita: A Sensor of Past and Present-Day Hydroclimatic Changes in the Pampas

Laguna Mar Chiquita (30° 54′ S – 62° 51′ W) is a highly variable and shallow (10–12 m maximum water depth) hypersaline endorheic lake containing a detailed record of past hydroclimatic changes from Late Glacial times until the Little Ice Age and the late twentieth century (Piovano and Leroy 2005a, Piovano et al. 2006b, Piovano et al. 2006a, Piovano et al. 2009). Variability in precipitation and river discharge during the twentieth century has triggered noticeable sharp lake-level fluctuations, across the Argentinian Pampas (Fig. 9), where Laguna Mar Chiquita is a sensitive hydroclimatic indicator for middle latitudes in South America (Piovano et al. 2002, Pasquini et al. 2006, Piovano et al. 2004a). Historical and instrumental data show that during dry intervals (i.e. prior to the 1970s) the lake surface was reduced to $\sim$1,000 km^2, from an extent up to $\sim$6,000 km^2 during periods with a positive hydrological balance (Fig. 10). At highstands, like today, it is not only the largest saline lake in South America but also one of the world's largest. Well-dated short cores provide a calibration of the lake's sedimentary, isotope and biological response to the last 100 years of documented lake levels changes, which yield a well-constrained multiproxy model for the basin (Piovano et al. 2002, 2004a, 2004b, Varandas da Silva et al. 2008). A semi-quantitative estimation of palaeolake-levels using the carbon isotope composition of organic matter as a hydrological proxy (Piovano et al. 2004a) shows for the last millennia a pattern of alternating lake highstands and lowstands (Piovano et al. 2009). Palaeohydrological reconstruction for the period coeval with the Medieval Climatic Anomaly (Villalba 1994) indicates a wet phase (by 1,060 cal. yr BP) with lake level magnitudes equivalent to the present-day highstand. Conversely, the palaeohydrological proxies for the cold period corresponding to the Little Ice Age indicate very dry conditions with the occurrence of short-lived humid pulses, especially during the second half of the nineteenth century. With the exception of a few short-term lake-level rises, the lowstand recorded after the Medieval Climatic Anomaly is dominated by the conspicuous hydroclimatic shift that took place in South Eastern South American (SESA) during the last quarter of the twentieth century (Fig. 9). The wet spell,

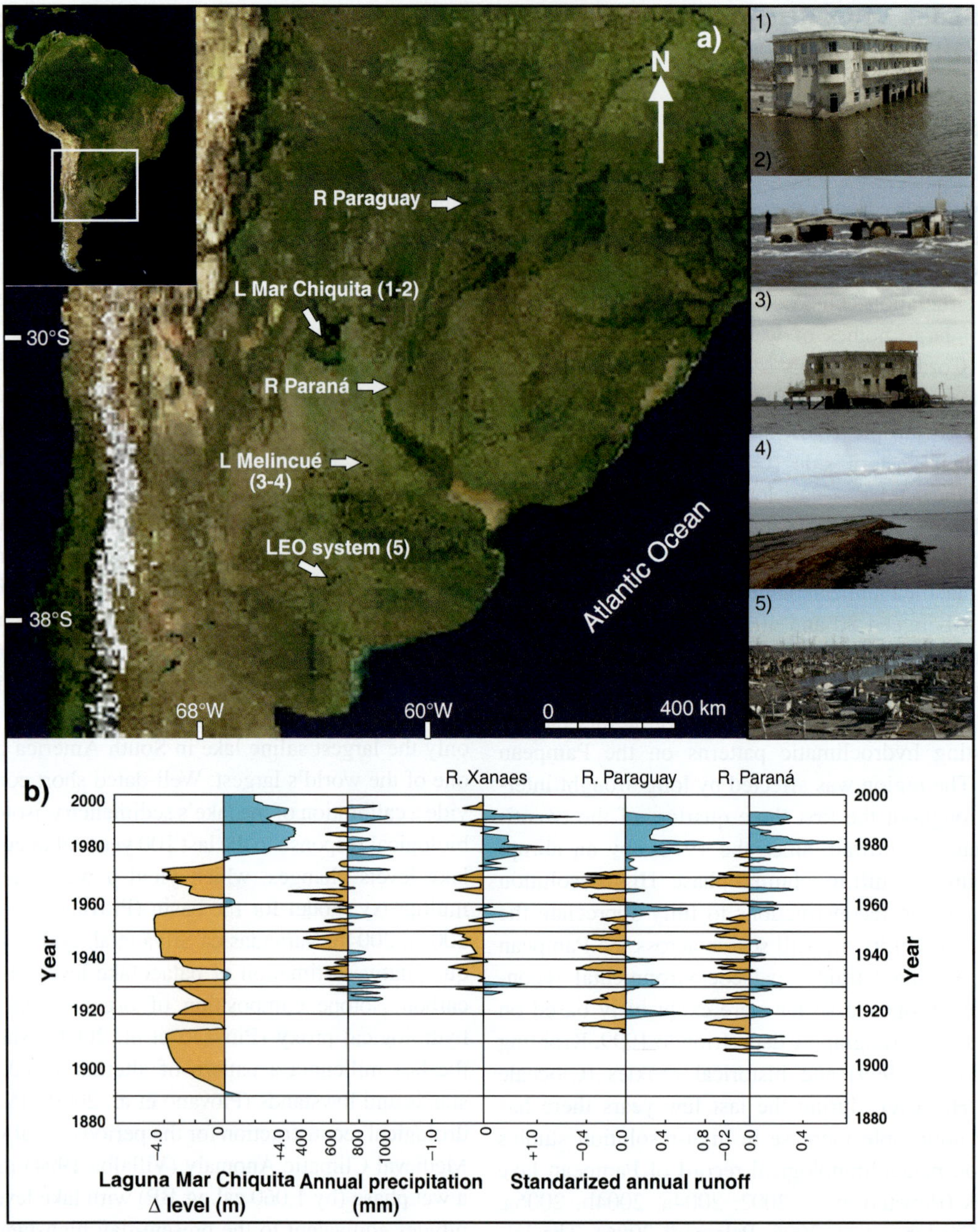

Fig. 9 (**a**) Satellite image of Southern South America of the area marked with a rectangle in the image of the left upper corner. White arrows indicate Río Paraguay and Paraná gauging stations used in graphs of Fig. 9b, Laguna Mar Chiquita, Laguna Melincué and Lagunas Encadenadas del Oeste de Buenos Aires (LEO system). Pictures on the right illustrate the consequences of water-level increase in the Pampean lakes; 1-2: Laguna Mar Chiquita (Pictures taken in year 2004); 3-4: Laguna Melincué (year 2005) and 5: Laguna Epecuén in LEO system. (**b**) Lake-level curve for Laguna Mar Chiquita. The interval AD 1890–1967 was reconstructed from historical data (dashed line). Instrumental records started in AD 1967. Δ lake level = 0 is an intermediate lake-level stage that matches the AD 1977 shoreline elevation (66.5 m a.s.l.). Positive values represent highstands (light blue areas), and negative values indicate lowstands (orange areas). Annual precipitation for the AD 1925–96 interval. Values above average are in light blue and below average in orange. Standardized runoff of rivers Xanaes (within the Laguna Mar Chiquita basin) and Ríos Paraguay and Paraná from Río de La Plata Basin. Discharges above and below the mean annual runoff are in light blue and orange respectively

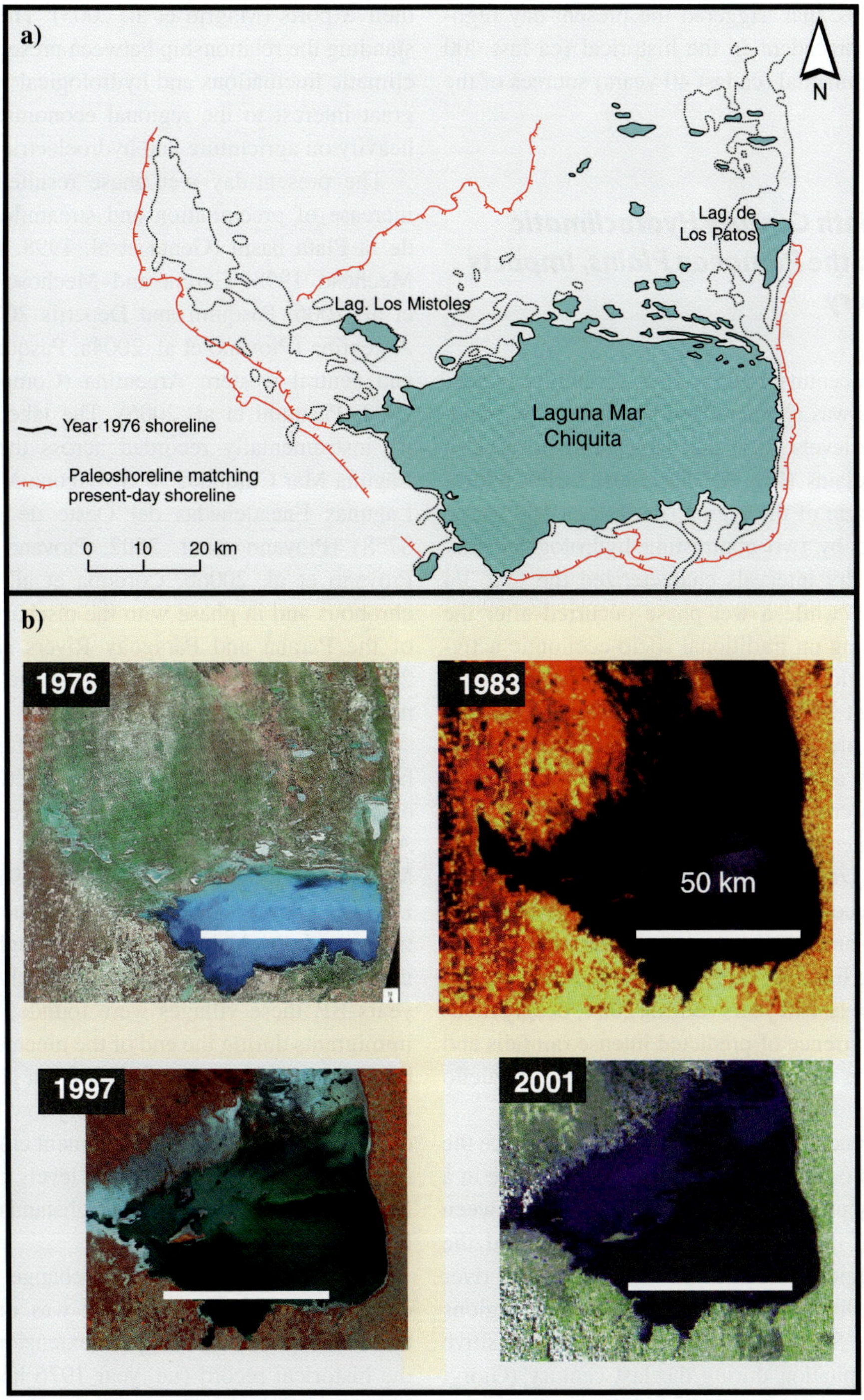

Fig. 10 (**a**) Laguna Mar Chiquita extension and palaeoshorelines mapped in a satellite image of year 1976 (see Fig. 10b). The present-day size of the lake matches with the position of palaeoshorelines. (**b**) Satellite images showing the lake-surface variations between 1976 and 2001. Satellite images were obtained at http://conae.gov.ar excepting the image of 1976 (taken from Bucher et al. 2006)

since the 1970s, that triggered the present-day highstand has no precedent in the historical (ca last 400 years) or instrumental (ca last 40 years) sources of the region.

The Twentieth Century Hydroclimatic Changes in the Pampean Plains, Impacts and Recovery

The twentieth century hydrological variability in central Argentina was characterized by distinctive fluctuations of lake levels, river discharges and surfaces of flooded low plains (Fig. 9). The most recent hydroclimatic scenario of the region (i.e. last ca 100 years) is represented by two contrasting hydrological situations. Long dry intervals characterized the first 3/4 of the century while a wet phase occurred after the 1970s impacting on traditional socio-economic activities in the region. In addition to the economic damage, the current wet interval in the Argentinean Pampas expanded the area devoted to crops (mainly soya) that in turn deeply affected the quality of non-marketable natural resources and processes (e.g. erosion control, fresh water supply, biodiversity, biogeochemical cycles) (Viglizzo and Frank 2006). Episodes of heavy rainfall are becoming more frequent. For instance the frequency of precipitation events exceeding 100 mm in Central and Eastern Argentina have increased threefold during the last 40 years (Barros 2004) deeply modifying the recurrence of predicted intense rainfalls and thus triggering unusual erosive processes (Argüello et al. 2006).

The most recent hydroclimatic change (i.e. since the 1970s) is also occurring at a sub-continental scale in a wide and very productive region of the SESA between 22 and 40°S, including Uruguay, Paraguay and the subtropical regions of Argentina and Brazil (see river runoff in Fig. 9b). Amongst all sub-continental regions of the world, SESA has shown the largest positive trend in precipitation during the last century (Giorgi 2002). The increase in annual precipitation in the last 40 years has been more than 10% over most of the region, but in some places it has been higher than 30% (Castañeda and Barros 1994, Minetti et al. 2003). Agricultural products from this region provide sustenance for the majority of the population of these countries (>200 millions), and constitute a large fraction of their exports (Magrin et al. 2005). Therefore, understanding the relationship between present-day and past climatic fluctuations and hydrological variability is of great interest to the regional economies that depend heavily on agriculture and hydroelectricity.

The present-day wet phase resulted in a general increase of precipitation and streamflows in the Río de la Plata basin (Genta et al. 1998, Robertson and Mechoso 1998, García and Mechoso 2005, Barros et al. 2006, Pasquini and Depetris 2007), in central Argentina (Piovano et al. 2004a, Pasquini et al. 2006) and central western Argentina (Compagnucci et al. 2002, Pasquini et al. 2006). The lake level variability instrumentally recorded across the Pampas (i.e. Laguna Mar Chiquita, 30°S; Laguna Melincué, 33°S; Lagunas Encadenadas del Oeste de Buenos Aires, 37°S) (Piovano et al. 2002, Piovano et al. 2004a, Piovano et al. 2006b, Córdoba et al. 2006) is synchronous and in phase with the discharge fluctuations of the Paraná and Paraguay Rivers (Piovano et al. 2004a) pointing toward a large-scale climatic phenomenon affecting SESA (Fig. 9a).

The wet spell that started after the 1970s has affected the socio-economic activities of several lakeshore villages across the central plains of Argentina (e.g. Miramar in Laguna Mar Chiquita, Melincué in Laguna Melincué, Carhué and Guaminí in Lagunas Encadenadas del Oeste de Buenos Aires; pictures on the right in Fig. 9a). Although the area was initially occupied by Indians probably since 10,000 years BP, these villages were founded by Europeans immigrants during the end of the nineteenth century or beginning of the twentieth century, a period matching the end of the Little Ice Age. Very low lake levels and extensive droughts forced settlement close to the lakes, usually below the topographic levels of geomorphological evidences of former highstands (Piovano and Leroy 2005b).

Although the hydrological change started in the early 1970s or even before, it was only after 1977 that the Laguna Mar Chiquita extension went beyond the historical record (i.e. year 1976 lake shoreline in Fig. 10a) producing drastic economic and social consequences. For instance, the number of inhabitants in Miramar fell from ca 5,000 to 1,600 as people were forced to move away from the rising lakeshore during the period 1977–1985. An area of 120,000 m^2 of buildings (including 90% of hotels) was flooded or destroyed. This new scenario strongly disrupted the tourism-based local economy that was flourishing during former low lake-levels years due to the therapeutic properties of hypersaline waters (Piovano and Leroy 2005b).

The new hydroclimatic conditions have rendered obsolete a great part of the infrastructure related to water management, since it was designed for a different climate. Most of the infrastructure was, and still is, designed with the implicit assumption of a stationary climate, reflecting the lack of awareness of the technical community about the regional climate trends and their hydrological consequences (Barros et al. 2006). Particularly, the inhabitants of Miramar seem to have recovered from the past traumatic experience and now are adapted to higher, but always fluctuating, lake level scenarios. In fact, the new lake situation is widely considered as a positive factor for improving the regional development based on the tourism industry. This activity is additionally promoted since the lake ecosystem became a protected site by the Ramsar Convention on Wetlands.[16]

Sublacustrine Landslides and Tsunamis in a Large Alpine Lake

Past Landslides and Tsunamis in Lake Como

Lake Como (198 m a.s.l.) is located in Northern Italy (Fig. 11). It is the deepest lake of the Alps (425 m), and it has a particular lambda shape that allows it to be divided easily into three main branches: the northern Alto Lario, the southwestern Como branch and the southeastern Lecco branch. The deepest sector of the lake is the Como branch, with an extensive area at a depth of 400 m (Fig. 12a, b). This branch has a typical fjord morphology, deep and narrow, a length of 27.8 km, and the peculiarity of being hydrologically closed. In fact the western end of the lake, where Como is located, is surrounded by hills (altitude of 336–469 m a.s.l.). The only effluent of the lake, River Adda, flows out from the eastern branch (Fig. 11). In both the southern lake branches (Como and Lecco), the presence of several turbiditic deposits were defined, mapped and characterized, resulting from the combination of a bathymetric survey (multibeam Simrad 3,000) with a high-resolution seismic reflection study (single-channel 3.5 kHz sub-bottom profiler) and a coring campaign (gravity corer) (Fanetti 2004, Fanetti et al. 2008). In particular two deposits with a significant thickness (> 1.5 m), a volume (10^6 m^3) and with a basin-scale distribution were characterized in the Como branch, the shallower one named Megaturbidite 1 (MT1) and the deeper one Megaturbidite 2 (MT2) (Fig. 12b). The estimated ages of these turbiditic deposits, extrapolated from mean sedimentation rates based on radiocarbon (^{14}C) and radionuclide (^{137}Cs) analyses, are around the mid-12th (MT1) and early 6th (MT2) centuries AD.

The multibeam data together with the acoustic-facies distributions and the volumes of these two major sedimentary deposits, MT1 ($\sim 3\times10^6$ m^3) and MT2 ($\sim10.5\times10^6$ m^3), indicate that they resulted from large slides that occurred at the northern tip of the Como branch, along the steep slopes of a sub-lacustrine plateau. In fact, at the beginning of the western branch, a bathymetric sill exists (Fig. 12a), at 140 m water depth, with two morphological scarps, on both the NW and NE flanks. Moreover the volume of the material deposited in the western branch is comparable with the sediments missing from the NW slope of the sill (Fanetti et al. 2008).

Dangerous tsunami-like waves (seiches) can be generated by large sub-aqueous landslides leading to such megaturbidites in this fjord-like basin. Possible trigger mechanisms leading to these catastrophic events in the Como branch include a combination of steep-slope overloading, with significant lake-level fluctuations related to Holocene climate change and/or earthquake shaking. In particular, the MT1 event may have been caused by an earthquake with an estimated magnitude of 6.2, which occurred in AD 1222 at Brescia (Guidoboni 1986), a city located $\sim$ 90 km from the plateau (Fig. 12a); while for the older event (MT2) historical documents (Fanetti et al. 2008) report a catastrophic alluvial flood event in October AD 585, ravaging northern and central Italy.

Evidence for the repetition of the geohazard is based on the Lake Como sedimentary archive, which is limited to the shallower portion of the sediments (maximum investigation depth $\sim$ 16 m in the Como branch)

[16] Ramsar Convention on Wetlands. Available at http://www.ramsar.org/ [Accessed December 2008].

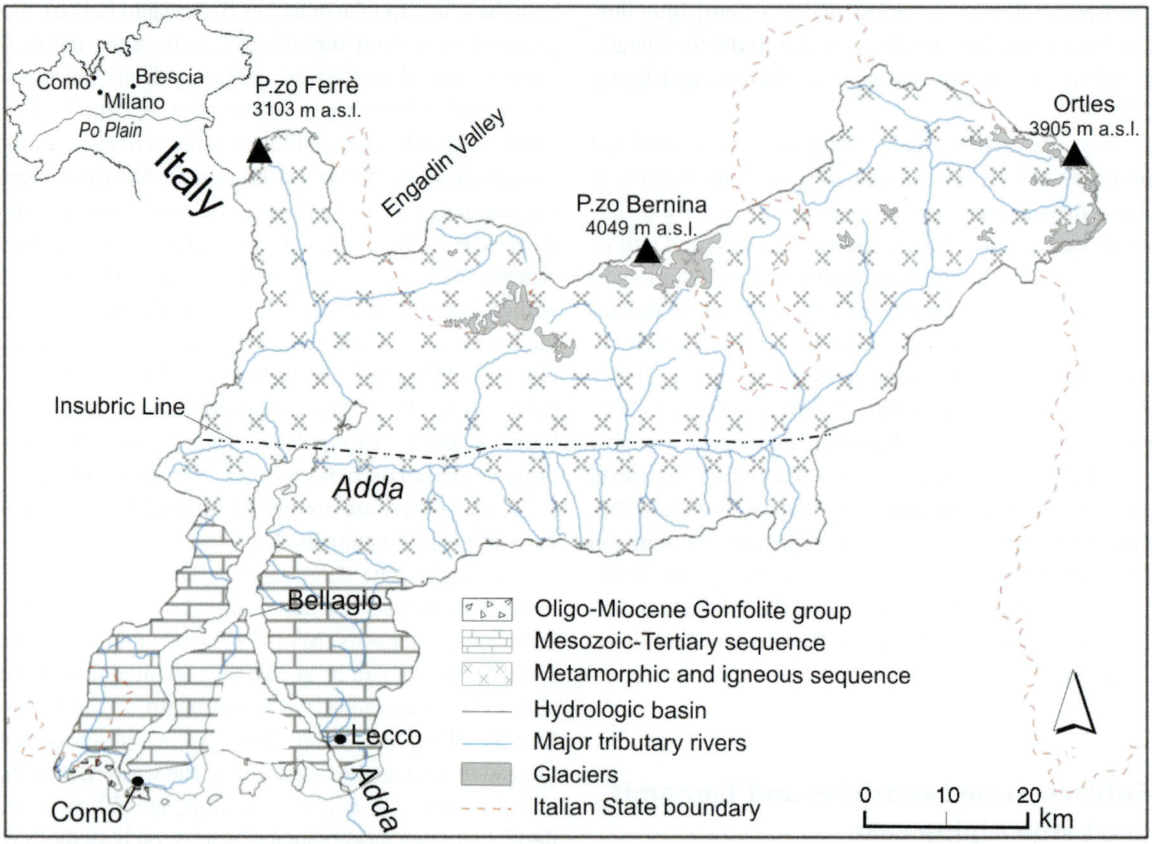

Fig. 11 Geological and regional setting of Lake Como catchment area

where two significant catastrophic events have already been noted. Moreover it has to be highlighted that, in the historical documents of Como town (Fanetti et al. 2008), sudden floods or lake surface movements have also been recorded that for the knowledge of the time were enigmatic and not attributed to a natural event (heavy rains, sub-aerial landslides,...). Now we can assume that such phenomena could have been the result of past anomalous waves or ancient tsunami, linked to sublacustrine landslides, exactly the same events found in the sedimentary record.

The two main turbiditic deposits in the Como branch that occurred in the mid-12th (MT1) and early 6th (MT2) centuries had a time interval of about six centuries (Fanetti et al. 2008). In the eight centuries since the last catastrophic event, no other large-scale sublacustrine landslide has occurred. Different land practices that have reduced erosion and sedimentary input to the lacustrine basin could be the reason of this longer quiescent period (Fanetti and Vezzoli 2007).

Another type of geohazards that is significant in the Lake Como area, is the subsidence of the town of Como. During the period 1950–1975, because of deep-water withdrawal from wells, the Como area was affected by a human-induced accelerated subsidence with a velocity of 10–20 mm per year, i.e. one order of magnitude higher than the natural rate (2.5 mm/a). Nowadays the subsidence rate has returned to its natural trend (Comerci et al. 2007).

The Present Geohazards Potential

Tsunami events are typically characterized by an instantaneous onset, especially in lacustrine environments, given that in such settings also amplification will occur even with a moderate phenomenon. According to the triggering factor the speed could however be slightly different: either earthquakes, which will take people by surprise as they are still unpredictable, or

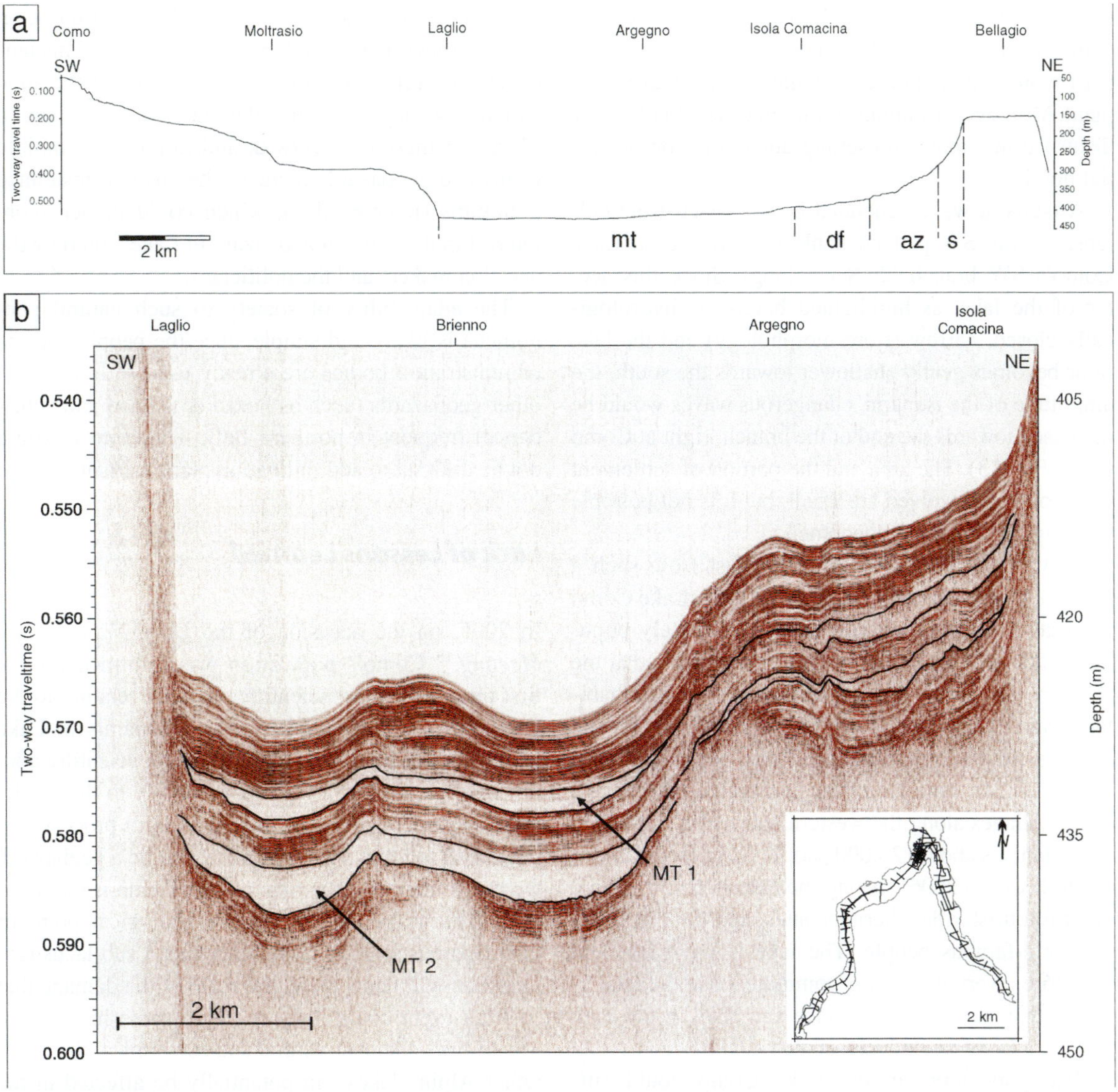

Fig. 12 (**a**) Topographic profile of the Como branch obtained by the longitudinal 3.5 kHz seismic line showing the Bellagio plateau, the deep over 400 m basin and the shallower transect towards Como town. Dotted lines indicate the extension of the four stages of the large slope failures associated with MT1 and MT2 in the deep basin: the large slide scars (s), the ~2.4-km-long accumulation zone (az), the ~4.8-km-long debris-flow deposits (df) and the ~5.5-km-long megaturbidites (mt). (**b**) Detailed seismic section of the basin fill where are located the large ponded megaturbidites (MT1 and MT2) and the well-stratified hemipelagic sedimentation draping the basin morphologies

long rainy periods which can be predicted and could cause progressive alarm in the community. Because the speed of a tsunami on Lake Como, the onset of the hazard would be rapid, while the duration of the phenomenon would be relatively short. Indicatively, similarly documented events in alpine type lakes (e.g. Lake Lucerne; Schnellmann et al. 2006) reach their maximum amplitude within a couple of hours. Following a tsunami disaster, a short bad period is expected that involves both human society and the biosphere: this period would not be longer than the food storage capacity because Lake Como is in a highly developed area (relatively easy to rescue); but significant and persistant damage to the drinkable lacustrine water is expected (most of the drinking water of the area comes from the lake). Also, the lacustrine fauna (e.g. fish) would suffer because of the sediment displacement. Moreover communication routes would be easily disturbed in a mountain setting due to the consequential flood.

A tsunami wave generated by a sublacustrine collapse on the SW plateau flank will involve all Lake Como's SW branch (27.8 km long). Since this sector of the lake, as highlighted before, is hydrologically closed, narrow (fjord morphology) and the lake floor becomes gently shallower towards the south, the amplitude of the tsunami's dangerous waves would be increased towards the end of the branch, right at Como town (Fig. 13). The area and the portion of settlement that would be involved are small, but it is not possible to escape such a rapid phenomenon.

Nowadays Como would recover fast from such a disaster but only at the cost of many lives. Lake Como is located in the Lombardy Region, a densely populated and urbanized area. Two towns are located at the southern tips of the lake (Como town, 83,600 inhabitants, and Lecco town, 45,500 inhabitants) and several villages are set along the shores, right at the lake level (Fig. 12a). An estimate of the average people present in the villages along the western shore plus the Como inhabitants is about 120,000 people (Gruppo di Lavoro Lago di Como 2006). Along the lakeshores, there are several tourist villas, heritage sites, and the dwellings of world-famous people. The lake is navigable and therefore, every day, many commuters use the lake as a way to reach their workplace and also, in the high season, a lot of tourists cruise on the lake.

The population around Lake Como could suffer from a modern disaster because tsunami waves, according with their intensity, could: (1) generate serious damage to the houses near the shore; (2) flood part of Como town, and of the other settlements, close to the lake; (3) induce shore instability due to the rapid change in the water level; (4) remix the lacustrine sediments and therefore compromise both the drinkable water sources and the fishing economy; and (5) interrupt strategic life lines. The tsunami threat would also seriously compromise the ever-growing tourist economy.

The possibility of job loss is real in a tsunami scenario. In fact psychological fear alone could induce a lot of tourists to chose other holiday resorts, as in the case of the Twin Towers act of terrorism (9/11), when most of American tourists skipped Europe in their journeys as reported in the newspapers. This tendency can seriously damage the economy of the area, which nowadays also includes the influx of visitors. Moreover there is worldwide indication that possible or defined geohazards increase the costs of insurance premiums (Beer et al.[17]), which could induce more interest in the tsunami geohazard in Lake Como by the decision makers and the politicians.

The adaptability of society to such natural phenomena is likely and simple since the people and the administration bodies are already used to living with other geohazards (such as landslides, flood and subsidence) frequent in northern Italy. Therefore it would not be difficult to add mitigation plans for tsunamis.

Lack of Lessons Learned

In 2005, on the occasion of the *Dark Nature Final Meeting*,[18] Como's population was informed for the first time of the new scientific database relating to the lake and of the potential threats. Tsunamis in Lake Como are not freak phenomena as both scientific documentation and historical reports testify to their existence in the recent time. Since any earth process that poses risk to human life can be said to be a geohazard, we can declare that the risk in Lake Como is medium-high. This is because, from the geological point of view, there is still the probability that a sub-lacustrine landslide will happen and the scale of the damage that such an event could induce is high (both in terms of loss of lives and in structural and economical damage). Other Alpine lakes can potentially be affected in the same way.

In truth, the main recognized vulnerabilities of the area do not include the possibility of tsunamis as real hazards. Consequently no mitigation or evacua-

[17] Beer, T., Bobrowski, P., Canuti, P., Cutter, S. & Marsh, S., 2004. Hazards, minimising risk, maximising awareness. Planet Earth in our hands, Earth Sciences for Society. IYPE brochure, p. 16. Available at http://www.esfs.org [Accessed December 2008].

[18] Michetti, A.M., Pasquarè, A.F., Haldorsen S. & Leroy, S.A.G., (Eds) 2005. *Final Meeting, Dark Nature – Rapid Change and Human Responses*. 6–10 September 2005, Villa Olmo, Como, Italy. Available at http://atlas-conferences.com/cgi-bin/abstract/caqy-01 [Accessed May 2009].

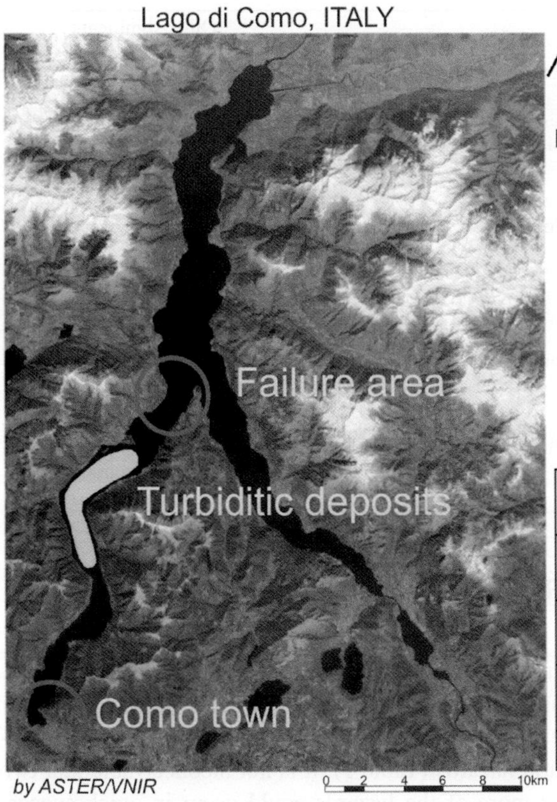

Failure source	Sublacustrine plateau area (Bellagio)
Volume of the deposits	MT1=3.5*10^6 m^3 MT2=10.5*10^6 m^3
Landslides estimaded age	MT1: 1280±120 AD MT2: 800±200 AD
Distance from the source area to Como town	~ 27.8 km
Como town inhabitants	83600

Fig. 13 Satellite image of the lake Como with the failure area, the turbiditic deposits and the location of Como town. On the right a cartoon of the lake to explain the direction (arrows) of an anomalous wave generated by a sublacustrine collapse of the plateau western flank. In the table the physiographic characteristics of the MT1 and MT2, and the Como town inhabitants

tion plans exist, and information and education about tsunami hazards is absent as well.

Up to now the transfer of information from scientists to end users or to the media, or to education, has only taken place once through the *Dark Nature Final Meeting* – 2005. The transfer of information occurred at the administrative level, involving all the private and public organisations that deal with environmental management of the Lake Como area, and the scientific staff of the Università degli Studi dell'Insubria, Chemical and Environmental Science Department.

Discussion

Lessons Learned from the Past: Understanding and Communicating

The main lesson from these five examples is the importance of understanding the record of rapid landscape change and natural hazards. It is after all the palaeoenvironmental record in the sediment of the Caspian Sea, Mar Chiquita, Lake Como, Gulf of Mexico and the Arctic that has shown that such hazards are not one-time affairs, not necessarily human-induced, and that they must be taken into account in development planning and policy. In the case of new world countries, geoscientific analyses at high time resolution are critical to complement the short instrumental and written records which started only very recently, such as in Argentina. It is only through earth science that the background rates, trends and cyclicity of natural hazards may be established. Where these cannot be worked out, those responsible for disaster management should at least recognize that bad things have happened in the past, and that their people/societies/cultures have survived through them and in some cases may not have fully recovered from them (Diamond 2005).

The second lesson is that the earth science community needs to work much harder to get the

Table 1 List of actual hazards in the five regions analysed in this chapter

Region	Main hazard	Other significant hazards
Arctic	Rapid climatic change at high latitude	Loss of sea ice and permafrost, landslides
Coast of Louisiana	Hurricanes	Delta erosion, delta subsidence, wetland loss, sea-level rise
Caspian Sea Coast	Rapid sea level rise	Rapid sea level drop, tsunamis, earthquakes, wind surges (and retreat of water by wind)
Argentinian Pampas	Rapid lake level rise	Rapid lake level drop, soil erosion, hydroclimatic event
Alpine lakes in Italy	Underwater landslides and tsunamis	Earthquakes without tsunami, floods, on-land landslides

palaeo-record accepted and understood by the general public (Liverman 2008). Liverman (2008) suggests that the most common barriers to communication are: the medium of communication itself chosen by the scientists, the technical language of science, the lack of experience and training in dealing by the media, the inability to tailor communication to a variety of audiences, the lack of understanding of the decision-making structures and the lack of incentive to engage outside the scientific community. Moreover geoscience communication cannot be done by working in scientific isolation but needs to be an interdisciplinary effort that takes cultural factors into account (Hewitt 2008[19]). Trying to establish a direct link between geoscience information and policies would, in many countries, require a good deal of further study. For example Sieh (2006) after studying the causes and the effects of the Sumatran earthquake and tsunami of 2004 suggests the following chain of events from basic science to people's everyday lives: basic science (understanding), emergency response preparedness, warning capability, public education and finally infrastructure. He argues that, in developing countries, many lives can be saved by educating the population at risk via posters, meeting with local officials and lectures to ordinary citizens.

[19] Hewitt, K., 2008. Culture and Risk: Understanding the Socio-Cultural Settings that Influence Risk From Natural Hazards. Global E-Conference. ICIMOD and the Mountain Forum. Available at http://www.mtnforum.org/rs/ec.cfm [Accessed December 2008].

Ratchet Effect of Hazard Accumulation

In each of the five case studies, the main natural hazard was analysed for its extent and its impact. However each region potentially may suffer from a range of natural hazards (Table 1). It is not inconceivable that two or more of the hazards could occur in quick succession (before total societal recovery) or even at the same time. For example, in 2005, the hurricane-damaged city of New Orleans was hit again one month later by hurricane Rita; and in 2008, hurricane Gustav added wind-damage to structures that were being rebuilt. Following an accumulation of disasters, it becomes more difficult or impossible to return to previous conditions: this is known as the ratchet effect (Chambers 1989, Ford et al. 2006). Each time there is a new disaster, the capacity for society to recover decreases and it may reach a point when there is a societal collapse.

After a disaster, there are two ways to recover: (1) return to past conditions, but retain the same susceptibility/vulnerability to hazards, (2) adapt and modify society (sometimes in depth) in order to increase resilience (Leroy 2006). The second situation is of course by far more preferable; though there are situations where the best solution would be for people to move away from a disaster area, such as one buried by ash deposits or lava flows, or inundated by sea-level rise. However our modern world is closely linked to technology and infrastructure that cannot be easily transported, decreasing therefore the likelihood of migration. There was little adaptation after the

1906 earthquake in San Francisco, at least not until recently – and yet this was rebuilt as one of the most attractive US cities. Many people tend to return to their traditional homelands after an eruption. It may not be preferable from a management viewpoint, but from a cultural standpoint it is perhaps easiest to move back into the threatened area, as people have done from time immemorial (e.g. Berger 2007) and rely on resilience.

Transfer of Information

In each of the five case studies, the transfer of information to the local communities from the geoscientists has been done with varying success. In one extreme case there is no interest at all and in another one the local community has integrated the geoscientific data very well. Table 2 summaries the type of recovery and the lessons learned in the five case studies.

Global Warming at High Latitudes, Yukon

New scientific research in the Arctic, as elsewhere, underpins moves to influence public policy on climate change and its human drivers. The work of government geological survey departments and of academic researchers is communicated to the public and to policy makers through authoritative reports (e.g. Furgal and Prowse 2008), and a multitude of journal articles (e.g. Berger and Liverman 2008). Indeed, based on observations such as those above, the Manitoba Government recently developed a school guide to help teachers and students understand climate change and its impact on Arctic communities (Manitoba Education and Youth 2003). This shows how collaboration between scientists and aboriginal peoples can help to attain a better understanding of the world around them (see also Ashford and Castleden 2001, Krupnik and Jolly 2002, Berger and Liverman 2008).

Hurricanes in New Orleans

Outstanding plans were born from the lesson learned. Hurricane prediction, evacuation and recovery efforts are managed very effectively. Thousands of families were safely evacuated just preceding the landfall of Hurricane Katrina, and procedures were improved to help the population that remained in the city in 2005. All plans were extremely effective during the succeeding hurricane season as seen with the crisis management during Hurricane Gustav and Hurricane Ike in 2008.

With warmer sea-surface temperature and the associated - more than probable -intensification of hurricane seasons, in a region that is losing wetlands, the protection issue is the only key element that still needs improvement. Some of the elements of the three-layer protection system (strengthening of barrier islands,

Table 2 Lessons learned in the five case studies

Hazard	Effect on people	Recovery	Learning
Underwater landslide & tsunami (Alpine lakes)	Shaking, flooding, loss of drinking water	Yes	No, too long ago. Return to previous unsafe conditions along the lake
Hurricane (New Orleans, Galveston)	Wind damage, flooding	Yes, but not total	More technology or displacement of economic centre
Global warming/climatic change (northern high latitudes)	Shift of resources	Yes, but not total	Population movement or more technology. Governmental information and support
Rapid sea level rise (Caspian)	Flooding	Yes, moving away from the coast	Yes, for the governmental sector. No, for the private one
Rapid lake level rise (Argentina)	Flooding	Yes, moving away from the shores	Yes, participatory approach

swamp replenishment, and better levee system) proposed by scientists are in progress and other components are considered, but a full protection system is not yet in place. To implement such as a system, it will take continued collaboration between scientists and engineers, several years of new infrastructure construction, and major federal funding. It is clear to politicians, scientists and citizens that basic building laws limiting construction to certain geographic areas or heights in the city will only solve short-term problems and the focus will need to stay on protection now that prediction, evacuation and recovery efforts are adequate.

Rapid Sea Level Rise in the Iranian Coast of the Caspian Sea

Iran's Caspian coast is facing three major problems: high density population in the coastal zone, inundation of this zone due to sea–level rise, and weakness of the existing laws and lack of much required laws for regulating the relationship among the stakeholders (Pak and Farajzadeh 2007). These problems have increased the governmental level of interest and it has initiated now the integrated coastal zone management (ICZM) program. In this program, basic information has been gathered for establishing a realistic strategy. Moreover, the Caspian Sea has attracted more scholarly attention over the past two decades, with an increase in research contribution at the international level. Before passing the ICZM plan through bureaucratic procedures, the government began to free a 60 metre coastal zone for public access. The enhancement of public awareness and of inter-sectorial cooperation is crucial for management of the coastal zone. The level of scientific knowledge and the quality of institution related to the Caspian Sea are relatively good, but the share of the private sector remains negligible. Moreover, the transfer of scientific knowledge to the wider public requires more effort, which is needed if the laws and managerial procedures are to be respected.

Rapid Lake Level Rise in the Argentinian Pampa

After demolishing that part of the village that was affected by the 1970s lake-level rise (i.e. buildings under lake-water and ruins), new regulations on urban development were established by land-use decision-makers together with local officials. The implementations of the urban regulations are based on the combined knowledge of society, hydrologists and palaeoclimatologists about the short- and long-term changing nature of the lake. Regulations take into account prior lake-level variability and prohibit the building below the maximum possible lake-level. The necessity for constraining urban growth was seen as a crucial task for the whole of society. In this sense, the ICSU conference organized during the year 2005 in Miramar (Piovano and Leroy 2005b) played an important role in confronting not only local officials but also society with the importance of planning actions based on understanding how to diminish "environmental risks," when a society is vulnerable in face of the environmental unevenness.

Palaeolimnological data point towards the need to reinforce palaeoclimatic research at mid-latitudes in South America to fully appreciate natural climatic variability beyond the instrumental record and to plan future strategies leading to sustainable development.

Underwater Landslides and Tsunamis in Large Alpine Lakes, Italy

Unfortunately the potential recurrence time of the hazard appears to the politicians and to the end-user to be too far-off. All the evidence is below the lake, buried underneath the lacustrine sediments, or lost in ancient historical observations. The absence of evident and clear signals of the possible threat makes it easy to ignore this as a serious geohazard. The time scale is too large: the problems that are urgent for the local political class have a time scale of years and not of centuries.

Therefore no lessons for public policy have been learned from the past tsunami events which occurred in Lake Como in historical time.

In brief, it is clear that the information transfer from the earth scientists to the end-users works best when the hazards are frequent and visible and when the messengers are trusted by the local community. The latter may lead then to a participatory approach to the mitigation of disasters. An analysis of the role of culture in disaster management clearly emphasises the need to give a large role to the most vulnerable groups in the decision–making process relating to them in order to ensure success for the mitigation measures (Hewitt 2008).

Geosciences and the Hyogo Framework of Action

The contribution of geosciences to the list of priorities of Hyogo Framework for Action[20] and to the conclusions of the review of the "Yokohama Strategy for a Safer World"[21] (adopted by the United Nations in 1994) is that good mitigation decisions will be taken only if the full scale of hazards, disasters and risks, which is certainly not covered by instrumental records only, is known. Therefore it is necessary to look at the past millennia and include in any framework of action information on hazards, disasters and catastrophes obtained from historical documents, archaeological, palaeoenvironmental as well as geological data, all at a time scale relevant to society (Leroy 2006).

Conclusions

The incorporation of geoscientific data into mitigation plans has been examined in the case of five recent examples of disasters. A first lesson learned is that geosciences are able to provide high-time-resolution information on past hazards that are directly relevant to society.

The second one is the difficulty that many geoscientists meet when they try to communicate to the wider public, as often they are not trained for this and their efforts are not valued by their employers. Amongst the five examples analysed here, in one case there is no interest by the wide public as the hazard is perceived as too improbable because too infrequent. In another case, the local community has fully integrated in their building plans the information provided by the geoscientist owing to their trust in him.

In conclusion, we strongly recommend that any framework of action to mitigate natural hazard disaster integrates geoscientific information on hazards in their plans. It is crucial to be aware of the full range of potential hazards and of their frequency.

[20] Hyogo Framework for Action, 2005. Available at http://www.unisdr.org/wcdr/intergover/official-doc/L-docs/Hyogo-framework-for-action-english.pdf [Accessed December 2008].

[21] "Yokohama Strategy for a Safer World", 1994. Available at http://www.undp.org/cpr/disred/documents/miscellanous/yokohamastrategy.pdf [Accessed December 2008].

Acknowledgements The first author is grateful to Tom Beer (CSIRO) to have been invited to give a talk on environmental catastrophes at the IGC33 (International Geological Congress) in Oslo. The talk was part of a 2-day megasession on hazards as a contribution to the International Year of Planet Earth (IYPE). SL would also like to thank the sponsors of the ICSU Dark Nature (via IUGS) and IGCP 490 conference programmes. Some circum-Caspian demographic data and the inundation map used in this publication originate from the Caspian Environment Programme which is thanked for all its work beneficial to the coastal states. ARB acknowledges the assistance of Robert McGhee and Katherine Trumper. Mike Turner (Brunel University) has kindly revised the English of a draft of the manuscript. We are grateful to Klaus Arpe (Max Plank) for fruitful comments on earlier versions of this paper.

References

Alunik, I., Kolausok, E.D. & Morrison, D., 2003. *Across time and tundra. The Inuvialuit of the Western Arctic*. Vancouver: Raincoast Books.

Apolov, B.A. & Fedorova, E.I., 1956. *Investigation of Caspian Sea level fluctuation*. Moscow: Report of IOAN 15, pp. 72–228 [in Russian].

Argüello, G., Dasso, C. & Sanabria, J., 2006. Effects of intense rainfalls and their recurrence: case study in Corralito ravine, Córdoba Province, Argentina. *Quaternary International* 1, 140–146.

Arpe, K. & Leroy, S.A.G., 2007. The Caspian Sea level forced by the atmospheric circulation as observed and modelled. *Quaternary International* 173–174, 144–152.

Arpe, K. & Leroy, S.A.G., 2009. Impacts from SSTs, ENSO, stratospheric QBO and global warming on hurricanes over the North Atlantic. *Quaternary International* 195, 4–14.

Ashford, G. & Castleden, J., 2001. *Inuit observations on climate change*. Final report. Winnipeg: International Institute for Sustainable Development.

Barber, E.W. & Barber, P.T., 2004. *When they severed earth from sky: how the human mind shapes myth*. Princeton, NJ: Princeton University Press.

Barros, V., 2004. *Tendencias climáticas en la Argentina: precipitación*. Proyecto Agenda Ambiental regional-Mejora de la Gobernabilidad para el Desarrollo Sustenable. PNUD Arg./03/001. Fundación Torcuato Di Tella y Secretaria de Medio Ambiente y Desarrollo Sustentable.

Barros, V., Clarke, R. & Silva Dias, P., 2006. *El cambio climático en la cuenca del Plata*. 1st edn. Buenos Aires, Argentina: Consejo Nacional de Investigaciones Científicas y Técnicas.

Bauch, H.A. & Kassens, H., 2005. Arctic Siberian shelf environments – an introduction. *Global and Planetary Change* 48, 1–8.

Bell, T. & Renouf, M.A.P., 2008. The domino effect: cultural change and environmental change in Newfoundland 1500-1000 cal BP. *Northern Review* 28, 72–94.

Berger, A.R., 2007. Rapid geological change challenges concepts of sustainability. *Geoscience Canada* 34, 81–90.

Berger, A.R. & Liverman, D.G., 2008. Special collection: rapid landscape change, and human response in the Arctic and Subarctic. *Northern Review* 28, 8–160.

Berner, J., Callaghan, T.V., Fox, S. et al., 2005. *Arctic climate impact assessment: scientific report*. Cambridge: Cambridge University Press.

Bixel, P.B. & Turner, E.H., 2000. *Galveston and the 1900 storm*. Austin, TX: University of Texas Press.

Brown, J., Hinkel, K.M. & Nelson, F.E., 2000. The circumpolar active layer monitoring program (CALM): research designs and initial results. *Polar Geography* 24 (3), 165–258.

Bucher, E.H., 2006. *Bañados del río Dulce y Laguna de Mar Chiquita*. Córdoba, Argentina: Academia Nacional de Ciencias.

Castañeda, E. & Barros, V., 1994. Las tendencias de la precipitación en el Cono Sur de América al este de los Andes. *Meteorológica* 19, 23–32.

Cerveny, R.S., 1998. Present climates of South America. In: J.E. Hobbs, J.A. Lindesay & H.A. Bridgman, eds. *Climates of the southern continents: present, past and future*. Hoboken, NJ: John Wiley.

Chambers, R., 1989. Vulnerability, coping and policy. *Institute of Development Studies Bulletin* 20 (2), 1–7.

Cioccale, M., 1999. Climatic fluctuation in the central region of Argentina in the last 1000 years. *Quaternary International* 62, 35–47.

Climate Change Impacts and Adaptation, 2001. *Degrees of variation: climate change in Nunavut*. Geological Survey of Canada, Miscellaneous Report 71.

Coleman, J.M., Roberts, H.H. & Stone, G.W., 1998. Mississippi River delta: an overview. *Journal of Coastal Research* 14, 698–717.

Comerci, V., Capelletti, S., Michetti, A.M., Rossi, S., Serva, L. & Vittori, E., 2007. Land subsidence and Late Glacial environmental evolution of the Como urban area (Northern Italy). *Quaternary International* 173–174, 67–86.

Compagnucci, R.H., Agosta, E.A. & Vargas, W.M., 2002. Climatic change and quasi-oscillations in central-west Argentina summer precipitation: main features and coherent behavior with southern African region. *Climate Dynamics* 18, 421.

Córdoba, F., Piovano, E. & Pasquini, A., 2006. The 20th century limnological and rainfall variation across the Pampean plains of central Argentina. *Reconstructing Past Regional Climate Variations in South America over the late Holocene: A new PAGES initiative*. International Symposium. 4–7 October 2006, Malargue, Argentina.

Cruikshank, J., 2005. *Do glaciers listen? Local knowledge, colonial encounters and social imagination*. Vancouver: University of British Colombia Press.

Cutter, S.L., Emrich, C.T., Mitchell, J.T., Boruff, B.J., Schmidtlein, M.T., Burton, C.G. & Melton, G., 2006. The long road home: race, class and recovery from Hurricane Katrina. *Environment* 48, 9–20.

Diamond, J., 1997. *Guns, germs, and steel: the fates of human societies*. New York: W.W. Norton & Company.

Diamond, J., 2005. *Collapse: how societies choose to fail or succeed*. New York: Viking Press.

Donnelly, J.P. & Woodruff, J.D., 2007. Intense hurricane activity over the past 5,000 years controlled by El Niño and the West African monsoon. *Nature* 447, 465–468.

Dotsenko, S.F., Kuzin, I.P., Levin, B.V. & Solovieva, O.N., 2002. Tsunamis in the Caspian Sea: historical events, regional seismicity and numerical modelling. *Petropavlovsk-Kamchatsky Tsunami Workshop*, 10–15 September 2002.

Einarsson, N., Larsen, J.N., Nilsson, A. & Young, O.R., 2004. *Arctic human development report*. Akureyri, Iceland: Stefansson Arctic Institute.

Elsner, J.B. & Kara, A.B., 1999. *Hurricanes of the North Atlantic: climate and society*. New York: Oxford University Press.

Fanetti, D., 2004. *Holocene evolution of the Lake Como western branch: definition of the limnogeological, geophysical and geomorphological characteristics of an Alpine lake*. Unpublished PhD thesis, Università degli Studi dell'Insubria.

Fanetti, D., Anselmetti, F.S., Chapron, E., Sturm, M. & Vezzoli, L., 2008. Megaturbidite deposits in the Holocene basin fill of Lake Como (Southern Alps, Italy). *Palaeogeography, Palaeoclimatology, Palaeoecology* 259 (2–3), 323–340.

Fanetti, D. & Vezzoli, L., 2007. Sediment input and evolution of lacustrine deltas: the Breggia and Greggio rivers case study (Lake Como, Italy). *Quaternary International* 173–174, 113–124.

Federov, P.V., 1996. Ponto-Caspian basin as a stratotype of Pleistocene of Europe. *Stratigraphy and Geological Correlation* 4 (6), 99–104 [in Russian].

Ford, J.D., 2008. Climate society and natural hazards: changing hazard exposure in two Nunavut communities. *Northern Review* 28, 51–71.

Ford, J.D., Smit, B. & Wandel, J., 2006. Vulnerability to climatic change in the Arctic: a case study from Arctic Bay, Canada. *Global Environmental Change* 16 (2), 145–160.

Frolov, A.V., 2003. *Modelling of the long-term fluctuations of the Caspian Sea level: theory and applications*. Moscow: GOES Publisher [in Russian].

Furgal, C. & Prowse, T.D., 2008. Northern Canada. In: D.S. Lemmen, F.J. Warren, J. Lacroix & E. Bush, eds. *From impacts to adaptation: Canada in a changing climate 2007*. Ottawa: Earth Sciences Sector, Natural Resources Canada, pp. 57–118.

Gabe, T., Falk, G., McCarty, F. et al., 2005. *Hurricane Katrina: social-demographic characteristics of impacted areas*. Congressional Research Service report for Congress. The Library of Congress RL33141.

Gajewski, K. & Atkinson, D.A., 2003. Climatic change in northern Canada. *Environmental Reviews* 11, 69–102.

García, N.O. & Mechoso, C.R., 2005. Variability in the discharge of South American rivers and in climate. *Hydrological Sciences* 50, 459–478.

Garneau, M. & Alt, B.T., 2000. *Environmental response to climate change in the Canadian High Arctic*. Ottawa: Geological Survey of Canada Bulletin p. 529.

Garreaud, R., Vuille, M., Compagnucci, R. & Marengo, J., in press. Present-day South America climate. *Palaeogeography, Palaeoclimatology, Palaeoecology*, Available online 9 September 2008.

Genta, J.L., Perez Iribarren, G. & Mechoso, C., 1998. A recent increasing trend in the streamflow of rivers in southeastern South America. *Journal of Climate* 11, 2858–2862.

Giorgi, F., 2002. Variability and trends of sub-continental scale surface climate in the twentieth century. Part I: observations. *Climate Dynamics* 18, 675–691.

Giralt, S., Julià, R., Leroy, S.A.G. & Gasse, F., 2003. Cyclic water level oscillations of the KaraBogaz Gol – Caspian Sea system. *Earth and Planetary Science Letters* 212 (1–2), 225–239.

Greene, C.E. & Kelly, S.H., 2000. *Through a night of horrors: voices from the 1900 Galveston storm.* College Station, TX: Texas A&M University Press, p. 206.

Gruppo di Lavoro Lago di Como, 2006. *Progetto PLINIUS. Criticità e azioni per il recupero della qualità delle acque del Lario.* Centro Volta Como, Italy.

Guidoboni, E., 1986. The earthquake of December 25, 1222: analysis of a myth. *Geologia Applicata e Idrogeologia* 21, 413–424.

Hare, G., 2008. The effect of the White River Ash on the archaeological record: a view from the Yukon Alpine. *Northern Review* 28, 136.

Hassol, S.J., 2004. *Impacts of a warming Arctic. Arctic climate impact assessment.* Cambridge: Cambridge University Press.

Hewitt, K., Byrne, M.-L., English, M. & Young, G., 2000. *Landscapes of transition: landform assemblages and transformations in cold regions.* New York: Kluwer Academic/Plenum Publishers.

Horton, B.P., Rossi, V. & Hawkes, A.D., 2009. The sedimentary record of the 2005 hurricane season from the Mississippi and Alabama coastlines. *Quaternary International* 195 (1–2), 15–30.

Iriondo, M., 1999. Climatic changes in the South American plains: records of a continent-scale oscillation. *Quaternary International* 57–58, 93–112.

Kaplin, P.A. & Selivanov, A.O., 1995. Recent coastal evolution of the Caspian Sea as a natural model for coastal responses to the possible acceleration of global sea-level rise. *Marine Geology* 124, 161–175.

Kazancı, N., Gulbabazadeh, T., Leroy, S.A.G. & Ileri, O., 2004. Sedimentary and environmental characteristics of the Gilan-Mazenderan plain, northern Iran: influence of long- and short-term Caspian water level fluctuations on geomorphology. *Journal of Marine Systems* 46 (1–4), 145–168.

Komorov, I.K., 1996. *Rebirth of Volga: pace of rescuing Russia.* Novgorod, Russia: Ecologia [in Russian].

Kosarev, A. N., 1975. Hydrology of the Caspian and Aral seas, Moscow State University, p. 372 [in Russian]

Kosarev, A. N. & Tuzhilkin, V.S., 1997. Annual cycle of variability in thermohaline field of Caspian Sea. *Water Resources* 24 (4), 463–467.

Koshinskii, S.D., 1975. *Characteristics of the strong waves over the Soviet seas: part one: the Caspian Sea.* Leningrad: Gidrometeoizdat [in Russian].

Kröhling, D.M. & Iriondo, M., 1999. Upper quaternary palaeoclimates of the Mar Chiquita area, North Pampa, Argentina. *Quaternary International* 57–58, 149–163.

Kroonenberg, S.B., Abdurakhmanov, M., Badyukova, E.N., van der Borg, K., Kalashnikov, A., Kasimov, N.S., Rychagov, G.I., Svitoch, A.A., Vonhof, H.B. & Wesselingh, F.P., 2007. Solar-forced 2600 BP and Little Ice Age highstands of the Caspian Sea. *Quaternary International* 173–174, 137–143.

Kroonenberg, S.B., Badyukova, E.N., Storms, J.E.A. & Ignatov, E.I., Kasimov, N.S., 2000. A full sea-level cycle in 65 years: barrier dynamics along Caspian shores. *Sedimentary Geology* 134, 257–274.

Krupnik, I. & Jolly, D., 2002. *The earth is faster now: indigenous observations of Arctic environmental change.* Fairbanks: Arctic Research Consortium of the US.

Lahijani, H., 2001. *The role of circum-Caspian states in the Caspian pollution.* Tehran: Ministry of Energy, Department of Water Research, Internal research report [in Persian].

Lahijani, H., Rahimpour Bonab, H., Tavakoli, V. & Hosseindoost, M., 2009. Evidence for late Holocene highstands in central Guilan – East Mazanderan, south Caspian coast, Iran. *Quaternary International* 197, 55–71.

Lerbekmo, J.F., 2008. The White River Ash: largest Holocene Plinian tephra. *Canadian Journal of Earth Sciences* 45, 693–700.

Leroy, S.A.G., 2006. From natural hazard to environmental catastrophe, past and present. *Quaternary International* 158, 4–12.

Leroy, S.A.G., Marret, F., Gibert, E., Chalié, F., Reyss J.-L. & Arpe, K., 2007. River inflow and salinity changes in the Caspian Sea during the last 5500 years. *Quaternary Science Reviews* 26, 3359–3383

Leroy, S.A.G., Marret, F., Giralt, S. & Bulatov, S.A., 2006. Natural and anthropogenic rapid changes in the Kara-Bogaz Gol over the last two centuries reconstructed from palynological analyses and a comparison to instrumental records. *Quaternary International* 150, 52–70.

Leroy, S.A.G. & Niemi, T.M., 2009. Editorial: hurricanes and typhoons: from the field records to the forecast. *Quaternary International* 195: 1–3.

Liu, K.B. & Fearn, M.L., 2000. Holocene history of catastrophic hurricane landfalls along the Gulf of Mexico coast reconstructed from coastal lake and marsh sediments. In: Z.H. Ning & K. Abdollahi, eds. *Current stresses and potential vulnerabilities: implications of global change for the Gulf Coast region of the United States.* Baton Rouge: Franklin Press, pp. 38–47.

Liverman, D.E., 2008. Environmental geoscience: communication challenges. In: D.G.E. Liverman, C. Pereira & B. Marker, eds. *Communicating environmental geoscience.* London: Geological Society London, Special Publications 305, pp. 197–209.

Magrin, G.O., Travasso, M.I. & Rodriguez, G.R., 2005. Changes in climate and crop production during the 20th century in Argentina. *Climatic Change* 72, 229.

Manitoba Education and Youth, 2003. *A teacher's guide for the video Sila alangotok – Inuit observations on climate change.* A resource for senior 2 science. Winnipeg: Manitoba Education and Youth, School Programs Division.

Mann, M.E. & Kump, L.R., 2008. *Dire predictions: understanding global warming.* Intergovernmental Panel on Climate Change. New York: Dorling Kindersley.

McCloskey, T.A. & Keller, G., 2009. 5000 year sedimentary record of hurricane strikes on the central coast of Belize. *Quaternary International* 195 (1–2), 53–68.

McGhee, R., 1996. *Ancient people of the Arctic.* Vancouver: University of British Columbia Press.

McGhee, R., 2007. *The last imaginary place. A human history of the Arctic world.* Chicago: University of Chicago Press.

Minetti, J., Poblete, A., Acuña, L. & Casagrande, G., 2003. Nonlinear trends and low frequency oscillations in annual precipitation over Argentina and Chile. 1931–1999. *Atmósfera* 16, 119–135.

Moghimi, S., 1997. *Report on laws of managing and protecting Iran's coast.* Tehran: Ministry of Energy, Internal report [in Persian].

Moodie, D.W., Catchpole, A.J.W. & Abel, K., 1992. Northern Athapaskan oral traditions and the White River volcano. *Ethnohistory* 39, 148–171.

Nelson, S.A. & Leclair, S.F., 2006. Katrina's unique splay deposits in a New Orleans neighborhood. *GSA Today* 16 (9), 4–10.

NRC, 2002. *Abrupt climate change: inevitable surprises.* Washington DC: Committee on Abrupt Climate Change, National Academy Press.

Nunavut Department of Environment, 2005. *Inuit Qaujimajatuqangit of climate change in Nunavut: a sample of Inuit experiences of recent climate and environmental changes in Clyde River, Pond Inlet, Resolute Bay, Grise Fiord, Nunavut.* Iqaluit, Government of Nunavut, Department of Environment, Environmental Protection Division.

Oliounine, I., 2003. Leadership seminar on Caspian Sea and its Deltas region sustainable development and regional security 27–30 May 2003, Astrakhan, Russian Federation. *Ocean and Coastal Management* 46 (8), 797–806.

Overpeck, J., Hughen, K., Hardy, D. et al., 1997. Arctic environmental change of the last four centuries. *Science* 278, 1251–1256.

Pak, A. & Farajzadeh, M., 2007. Iran's integrated coastal management plan. Persian Gulf, Oman Sea, and southern Caspian Sea coastlines. *Ocean and Coastal Management* 50, 754–773.

Palamedi, S., 2006. *El Registro del cambio climático en la Región Pampeana Argentina, la Laguna Melincué.* Unpublished Thesis. Facultad de Ciencias Exactas, Físicas y Naturales, Universidad Nacional de Córdoba.

Park, L.E., Siewers, F.D., Metzger, T. & Sipahioglu, S., 2009. After the hurricane hits: recovery and response to large storm events in a saline lake, San Salvador Island, Bahamas. *Quaternary International* 195 (1–2), 98–105.

Pasquini, A. & Depetris, P.J., 2007. Discharge trends and flow dynamics of South American rivers draining the southern Atlantic seaboard: an overview. *Journal of Hydrology* 333, 385–399.

Pasquini, A., Lecomte, K., Piovano, E. & Depetris, P.J., 2006. Recent rainfall and runoff variability in central Argentina. *Quaternary International* 158, 127–139.

Pienitz, R., Douglas, M.S.V. & Smol, J.P., 2004. *Long-term environmental change in Arctic and Antarctic lakes.* New York: Springer.

Piovano, E.L., Ariztegui, D., Bernasconi, S.M. & McKenzie, J.A., 2004a. Stable isotope record of hydrological changes in subtropical Laguna Mar Chiquita (Argentina) over the last 230 years. *The Holocene* 14, 525–535.

Piovano, E.L., Ariztegui, D., Cioccale, M., Córdoba, F. & Zanor G., 2006a. *Reconstrucciones paleolimnológicas desde el Ultimo Máximo Glacial en el sur de Sudamérica: ¿megasistemas en antifase hidrologica?* III Congreso Argentino de Cuaternario y Geomorfología, Tomo II: 659–669.

Piovano, E., Ariztegui, D., Córdoba, F., Cioccale, M. & Sylvestre, F., 2009. Hydrological variability in South America below the tropic of Capricorn (Pampas and eastern Patagonia, Argentina) during the last 13.0 ka. In: F. Vimeux, F. Sylvestre & M. Khodri, eds. *Past climate variability from the Last Glacial Maximum to the Holocene in South America and Surrounding regions (Focus on local and large scale teleconnexions).* Springer – Developments in Paleoenvironmental Research Series (DPER).

Piovano, E.L., Damatto Moreira, S. & Ariztegui, D., 2002. Recent environmental changes in Laguna Mar Chiquita (central Argentina): a sedimentary model for a highly variable saline lake. *Sedimentology* 49, 1371–1384.

Piovano, E.L., Larizatti, F.E., Favaro, D., Oliveira, S.M., Damatto, S.R., Mazzilli, B. & Ariztegui, D., 2004b. Geochemical response of a closed-lake basin to 20th century recurring droughts/wet intervals in the subtropical Pampean plains of South America. *Journal of Limnology* 63, 21–32.

Piovano, E.L. & Leroy, S.A.G., 2005a. Holocene environmental catastrophes in South America: from the Lowlands to the Andes. *Episodes* 28 (4), 296.

Piovano, E.L. & Leroy, S.A.G., 2005b. *Holocene Environmental catastrophes in South America: From the Lowland to the Andes.* Abstract Volume and field guide. Miramar-Córdoba, 11–17 March 2005. Third Joint Meeting of ICSU Dark Nature and IGC, p. 490.

Piovano, E.L., Villalba, R. & Leroy, S.A.G., 2006b. Holocene environmental catastrophes in South America: from the Lowlands to the Andes. *Quaternary International* 158, 1–3.

Robertson, A.W. & Mechoso, C.R, 1998. Interannual and decadal cycles in river flows of southeastern South America. *Journal of Climate* 11, 2570–2581.

Rodionov, S.N., 1994. *Global and regional climate interaction: the Caspian Sea experience.* Water science and technology library 11, Baton Rouge: Kluwer Academic Press.

Rychagov, G.I., 1997. Holocene oscillations of the Caspian Sea and forecasts based on paleogeographical reconstructions. *Quaternary International* 41–42, 167–172.

Sachs, J.D., Mellinger, A.D. & Gallup, J.L., 2001. The geography of poverty and wealth. *Scientific American* 3, 70–75.

Schnellmann, M., Anselmetti, F.S., Giardini, D. & McKenzie, J.A., 2006. 15,000 years of mass-movement history in Lake Lucerne: implications for seismic and tsunami hazard. *Eclogea Geologicae Helvetiae* 99, 409–428.

Shiklomanov, I.A., Georgievsky, V.Yu. & Kopaliani, Z.D., 1995. Water balance of the Caspian Sea and reasons of water level rise in the Caspian Sea. In: *UNESCO – THP – IOC – IAEA Workshop on Sea Level Rise and the Multidisciplinary Studies of Environmental Processes in the Caspian Sea Region.* Paris, France, 19–12 May 1995, IOC, UNESCO.

Sieh, K., 2006. Sumatran megathrust earthquakes: from science to saving lives. *Philosophical Transactions of the Royal Society A* 364, 1947–1963.

Smith, K., 1996. *Environmental hazards: assessing and reducing disaster.* Routledge, London.

Soroos, M., 2000. Environmental change and human security in the Caspian region: threats, vulnerability and response strategies. In: W. Ascher & N. Miravitskaya, eds. *The Caspian Sea: a quest for environmental security.* NATO Series, 2. Environment Security 67, pp. 13–28.

Stewart, I. & Donovan, K., 2008. Natural hazards. In: S. Buckingham & M. Turner, eds. *Understanding environmental issues*. London: Sage, pp. 207–234.

Terziev, S.F., 1992. *Hydrometeorology and hydrochemistry of seas*. Vol. 6, the Caspian Sea, No 1. Hydrometeorological Conditions. Leningrad: Gidrometeoizdat [in Russian].

Turner, R.J.W., Mougeot, C.M., Roots, C.F., Clague, J.J. & Franklin, R., 2003. *Geoscape Whitehorse – geoscience for a Yukon community*. Ottawa: Geological Survey of Canada Bulletin, Miscellaneous Report 82.

van Heerden, I. & Bryan, M., 2007. *The storm: what went wrong and why during Hurricane Katrina–the inside story from one Louisiana scientist*, Kindle edn. New York: Viking.

Varandas da Silva, L., Piovano, E., Azevedo, D. & Aquino Neto, F., 2008. Quantitative evaluation of the sedimentary organic matter in Laguna Mar Chiquita, Argentina. *Organic Geochemistry* 39, 450–464.

Varushchenko, S.I., Varushchenko, A.N. & Klige, R.K., 1987. *Changes in the regime of the Caspian Sea and closed basins in time*. Moscow: Nauka [in Russian].

Viglizzo, E.F. & Frank, F.C., 2006. Ecological interactions, feedbacks, thresholds and collapses in the Argentine Pampas in response to climate and farming during the last century. *Quaternary International* 158, 122–126.

Villalba, R., 1994. Tree-ring and glacial evidence for the Medieval Warm Epoch and the Little Ice Age in Southern South America. *Climatic Change* 26, 183–197.

Voropaev, G.V., 1996. The problem of the Caspian Sea level forecast and its control for the purpose of management optimization. In: M.H. Glantz & F.S. Zonn, eds. *Scientific, environmental and political issues in the circum-Caspian region*. NATO Series, 2. Environment, pp. 105–117.

Voropaev, G.V., Krasnozhon, G.E. & Lahijani, H., 1998. Riverine sediments and stability of the Iranian coast of the Caspian Sea. *Water Resources* 25 (6), 747–758.

Voropaev, G.V. & Ratkovich, D.Ya., 1985. *Problem of redistribution of water resources*. Moscow: USSR Academy of Sciences [in Russian].

Weart, S., 2008. *The discovery of global warming*, 2nd edn. Cambridge, MA: Harvard University Press.

Workman, W.B., 1979. The significance of volcanism in the prehistory of subarctic northwest North America. In: P.D. Sheets & D.K. Grayson, eds. *Volcanic activity and human ecology*. New York: Academic Press, pp. 339–371.

Zárate, M., 2003. Loess of southern South America. *Quaternary Science Reviews* 22, 1987–2006.

Zonn, I.S., 1996. Assessment of the state of the Caspian Sea. In: M.H. Glantz & I.S. Zonn, eds. *Scientific, environmental and political issues in the circum-Caspian region*. NATO Series, 2. Environment, pp. 27–39.

Part IV
Geophysical Risk and Sustainability: Theory and Practice

Seismic Hazard in India – Practical Aspects and Initiatives During IYPE

R.K. Chadha

Abstract The Indian subcontinent characterizes a continent-continent collision boundary in the north viz., Himalaya, subduction zone tectonics in the east, i.e., the Indo-Burmese arc extending through Andaman and Nicobar Islands to the Sunda trench in the south and rifted/non-rifted interiors of the Indian plate i.e., the Indian Peninsular shield. All these tectonic units are sources of damaging earthquakes capable of causing loss to property and human lives. A few recent examples are the Bhuj earthquake of 2001, Jabalpur in 1997 and Latur in 1993, all occurring in the Indian shield region and claiming more than 30,000 lives, collectively. Similarly, in the Himalaya, Muzaffarabad earthquake in 2005, Chamoli in 1999 and Uttarkashi in 1991 caused heavy casualties and severe damage to property. The 2004 Sumatra earthquake in the Sunda trench ruptured a 1,200 km long fault up to the north Andaman and generated an unprecedented tsunami in the Indian ocean that claimed hundreds of thousands of lives in the south-Asian region. Thus, strategies for the assessment of seismic hazard and mitigation efforts in these regions of varied tectonics require suitable practical solutions. This paper provide glimpses of the initiatives taken in the study of seismic hazard in the country and the activities during the IYPE.

Keywords Ground deformation · Liquefaction

R.K. Chadha (✉)
National Geophysical Research Institute (CSIR),
Hyderabad – 500007, India
e-mail: rajen0555@gmail.com

Introduction

One of the ten themes of the International Year of the Planet Earth (IYPE) – *Hazards, minimizing risk, maximizing awareness* underscores the vulnerability of our planet earth to the devastating natural forces which are unleashed in the form of earthquakes, tsunami, cyclones, storm surges, hurricanes, floods, landslides, volcanoes, droughts etc. Added to these natural phenomena, the skewed development process and man made activities, globally, have further helped in altering the geosphere, the biosphere and the landscape thereby increasing the societal vulnerability to these forces. The twentieth century saw earthquake fatalities average about 100,000/decade, globally. Most of these killer earthquakes occurred in Trans-Alpide or in circum-Pacific Ring of Fire, with few in the interior of tectonic plates. The story remains the same – loss of life is generally due to low-strength masonry buildings, dwellings and inferior quality precast constructions typical in developing and underdeveloped countries. Rapid urbanization with scant respect for enforcing seismic building codes in building design is a major reason for increased risk during an earthquake. In developed countries, loss of human lives has been minimized to a large extent by the application of sound engineering practices in the construction of earthquake resistant structures. Sometimes, fatalities in these countries are due to secondary effects like fire, landslides or due to tsunami affecting the coastal regions. The 2004 tsunami in the Indian Ocean is a recent example of the vulnerability of coastal regions and islands to this natural hazard.

In India, the last two decades witnessed several damaging earthquakes in the Himalaya and in the

T. Beer (ed.), *Geophysical Hazards*, International Year of Planet Earth,
DOI 10.1007/978-90-481-3236-2_10,

peninsular shield of India viz., Uttarkashi (1991), Chamoli (1999) and Muzaffarabad (2005) in the Himalaya and Latur (1993), Jabalpur (1997) and Bhuj (2001). The islands in the Andaman and Nicobar region and the Indian coastal areas in the east, south and to a little extent south-west were severely affected by the tsunami generated by the Sumatra earthquake in 2004. These events led to two major initiatives in India. While, the Latur earthquake in 1993 changed earthquake monitoring in India from analog to digital era and led to several seismic hazard related studies, the 2004 tsunami provided an opportunity to set up the Indian Tsunami Early Warning System in the county which became operational from September 2007 onwards. This chapter reviews the practical aspects of seismic hazard at local and regional scales and the awareness on hazards brought during International Year of the Planet Earth (IYPE) in India.

Seismic Hazard in India – Practical Aspects

There are several phenomena which occur during an earthquake, for example, surface faulting, landslides, slope failure, liquefaction, lateral spreading, strong ground accelerations, site amplifications, tsunami etc. The damages could occur close to the epicenter of the earthquake or at distance even up to 300–400 km. Some of the case studies in the country are as follows:

Ground Deformation

Permanent ground deformation often occurs at the surface breaks associated with fault ruptures during an earthquake. These deformations are sometimes seen few hundreds of meters from the fault. Figure 1 shows examples of slope failure and ground cracks observed after the Uttarkashi earthquake in Himalaya in 1991.

Liquefaction

Certain type of soils, when they are saturated with water, are subjected to strong shaking due to an earthquake, they completely loose all the shear strength and flow like a liquid (Fig. 2a). Sometimes, due to strong shaking sand from subsurface will come out to the surface through fissures in the form of dykes (Fig. 2b). This phenomenon causes extreme damage by removing support to the foundations of buildings located in such areas. In India this phenomena were observed during M 8.4 Bihar-Nepal Earthquake in the Himalaya in 1934 and the recent Mw 7.7 Bhuj earthquake in 2001.

Landslides

Sloping ground or rock masses that are stable under normal loading can lose their stability during an earthquake causing effects ranging from a slow progressive

Fig. 1 Slope failure and ground cracks observed during the M 6.6 Uttarkashi earthquake in 1991

Fig. 2 (**a**) Lateral spreading (1934, Bihar–Nepal earthquake). (**b**) Liquefaction (2001 Bhuj earthquake)

Fig. 3 Landslide during the 1999 Chamoli earthquake in the Himalaya (Photo by IIT, Kanpur)

creeping of the ground to a dramatic landslide, rockfall or flow failure (Fig. 3)

Ground Shaking Amplification

Ground shaking amplification is the most important factor for earthquake hazard assessment in the far source regions, distant from the earthquake epicentre. It is often observed that earthquake damage is greater in settlements sited on soft soils than in those sited on hard soil or rock sites. This was exemplified during the 1985 Mexican earthquake in which the Mexico City comprising soft lake beds, located more than 350 km from the epicenter of the earthquake along Pacific and North American plate boundary suffered heavy damage to the buildings. The reason for such damage is due to the presence of soft soil deposits which have a strongly defined natural frequency of vibration, amplifying that part of the bedrock motion which is of similar frequency and filtering out the rest. Buildings will be affected selectively according to their own natural frequency of vibration. Such amplification will be particularly strong for distant earthquakes for which filtering of the high-frequency component of the motion has already occurred. Ground motion amplification can also occur as a result of topographical effects, but this phenomenon is not sufficiently well understood.

Resonance was observed during the Bhuj earthquake in India in 2001, where severe damage to

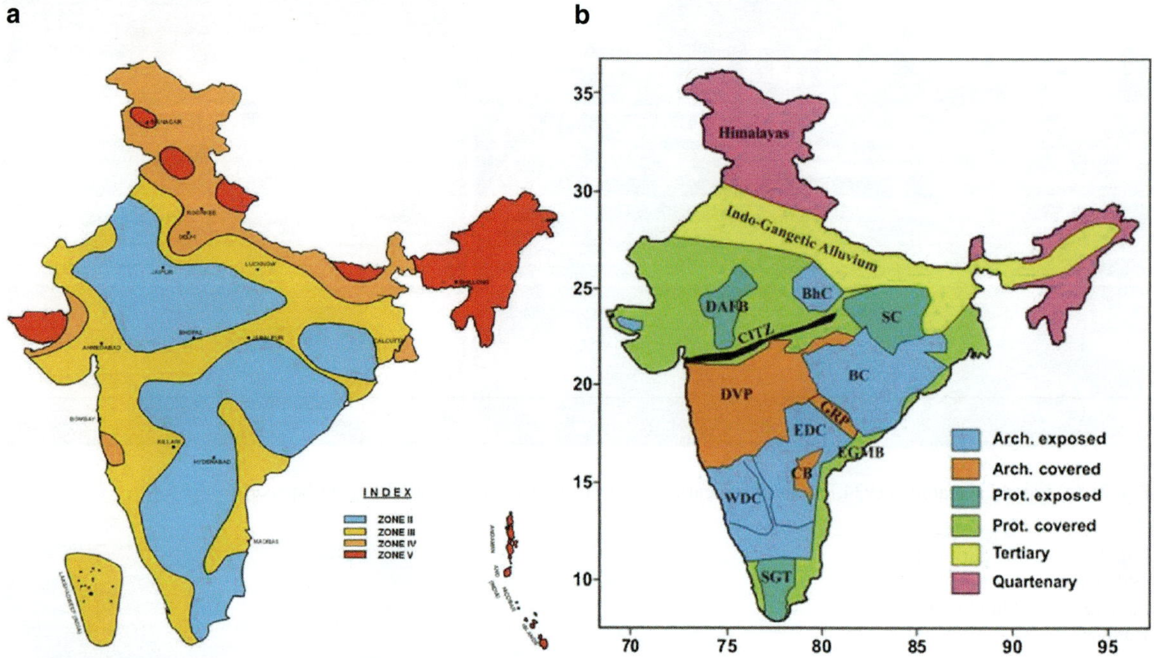

Fig. 4 (**a**) Seismic Zoning map of India. (**b**) Geological map of India showing major tectonic units

multi-storeyed buildings occurred in the city of Ahmedabad located more than 350 km from the epicenter of the earthquake in Bhuj. Figure 4a shows the Seismic Zoning map of India. Zone V is the region where earthquakes of M>7.5 have occurred and Zone II where earthquakes of M<5 occur. Zone V is mostly confined to the Himalayan belt and the entire Northeast India and Andaman and Nicobar Islands. In the western India, Bhuj region is located in Zone V. Figure 4b shows major geological units in the country. The Indo-Gangetic Alluvial Plains which parallels the Himalayan belt possesses a great seismic hazard as they comprise a thick pile of sediments varying in thickness from 500 m to more than 5 km at different locations. There are several major cities like Chandigarh, Dehradun, Lucknow, Kanpur, Allahabad, Kolkata, Gauhati and the capital New Delhi are located in the vicinity of Indo-Gangetic Plains. All these major cities are within 300–400 km of the probable earthquake sources of M 7.5 in the Himalaya and thus pose a great threat due to earthquakes.

Tsunami and Floods

Tsunamis are sequences of long-period sea waves generated by earthquakes often occurring under the sea bed. These are very common in the Pacific ocean compared to other oceans due to the subduction tectonics in the region. These waves travel long distances at high speed and gain greater heights near the shore depending on the bathymetry. Thus low lying coastal areas on the margins of the oceans are very vulnerable to this hazard. The 2004 Indian ocean tsunami generated by the M 9.3 earthquake off the coast of Sumatra was unprecedented and claimed hundreds of thousands of lives in the south Asian countries, especially Indonesia, India and Sri Lanka. Figure 5 shows the two source regions where tsunamis can be generated by great earthquakes which can affect the south Asian countries and also threaten the east coast of Africa. Flooding following earthquakes may also result from *seiches* or from the failure of reservoirs or embankments.

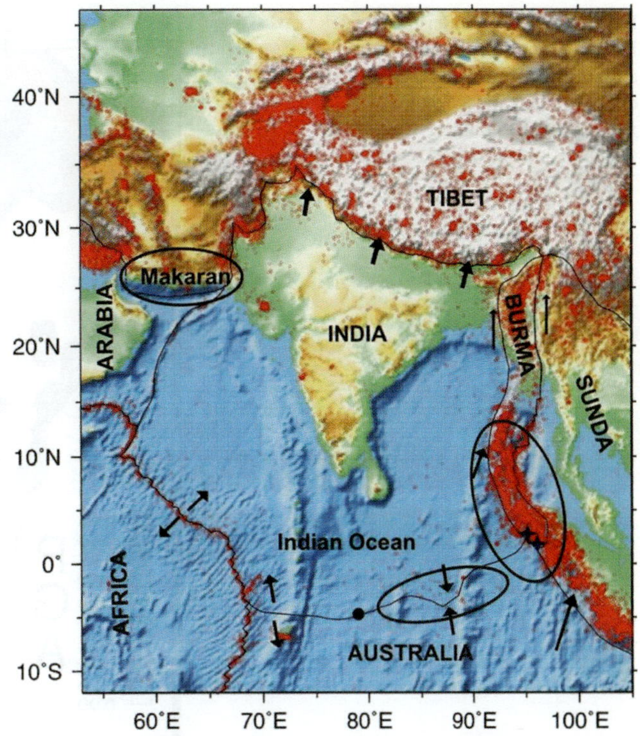

Fig. 5 Red dots show the seismicity in different tectonic zones. *Two circles* Sunda in the east and Makran in the west show tsunami sources regions in the Indian Ocean region. The *third circle* shown along the boundary of Indian Ocean and Australian plate may be a possible sources, as there are indications of the nascent boundary developing in the region

Global Initiatives

Realizing the vulnerability of mankind to all kinds of natural or human induced hazards, several global initiatives have been undertaken by world bodies, for example, the UN International Strategy for Natural Disaster Reduction (UN-ISDR) in 1994 and now the International Year of the Planet Earth (IYPE) – 2008 being organized under the auspices of International Union of Geological Sciences (IUGS) and UNESCO, to name a few. In addition, several of the GeoUnions, Commissions, Sub-Commissions of IUGG and its associations, IUGS and societies like the Natural Hazard Society (NHS) have taken several initiatives to organize meetings addressing the theme on "Hazards." A major global initiative taken up in the IYPE in 2008 is the Global Earthquake Model (GEM) Program initiated by the OECD to build an independent standard for modeling, monitoring, and communicating earthquake risk, worldwide. This Program is a public-private venture to provide authoritative, open information about seismic risk and decision tools to support mitigating actions. GEM will raise risk awareness and help post-disaster economic development, with the ultimate goal of reducing the toll of future earthquakes. The Asian Seismological Commission organized its 6th General Assembly in Thailand after the Indian Ocean Tsunami of 2004 and later the 7th General Assembly in Tsukuba, Japan devoted special sessions to Sichuan Earthquake in China in 2008 for the participants from Asia-Pacific regions.

Indian Initiatives

Seismic Hazard Assessment

Vulnerability to earthquake disaster has increased due to haphazard urbanization. More than 30,000 human lives were lost due to earthquakes in the last two decades in India. In order to minimize the loss of life, property damage and social and economic disruption caused by earthquakes, it is essential that estimates of seismic hazard be available to national decision makers and engineers for land use planning and improved building design and construction. India took a lead

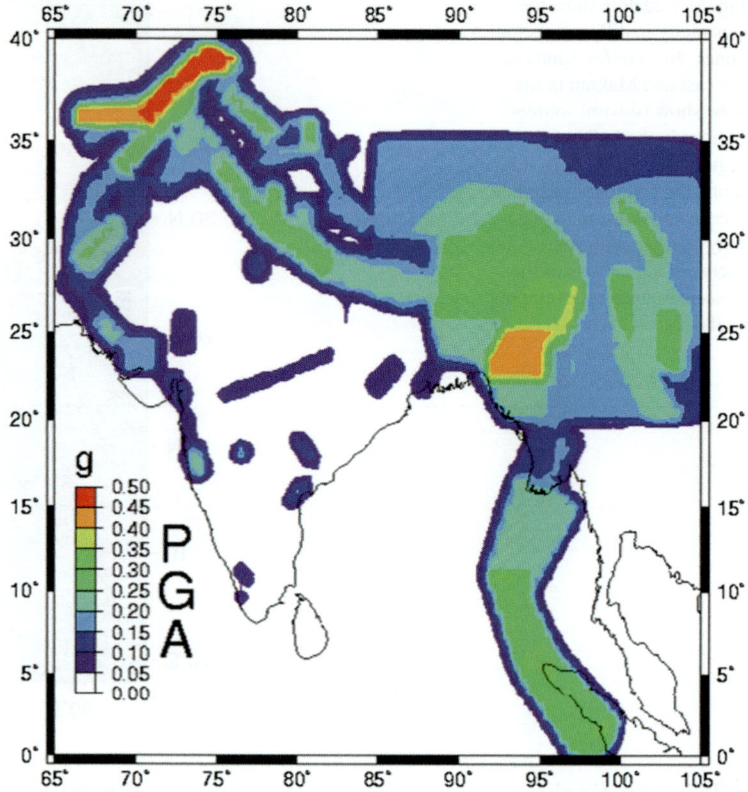

Fig. 6 Probabilistic Seismic Hazard Map of India and adjoining regions (Bhatia et al. 1997, Abstract Volume, IASPEI, Thessaloniki, Greece)

in the preparation of a Probabilistic Seismic Hazard Map of India and adjoining countries under the Global Seismic Hazard Assessment Program (GSHAP) launched in 1992 by the International Lithosphere Program (ILP) supported by ICSU (Fig. 6). This provided a great boost for seismic hazard studies in the country. Generally, such seismic hazard assessment studies can be done using either probabilistic or deterministic approach depending on the requirement.

Microzonation

In the country, there is a major thrust to use the developing technique of microzoning which promises to bring very important benefits for earthquake mitigation. In India, there are a number of major cities located in the Indo-Gangetic plains adjoining the Himalayan belt and thus vulnerable to earthquakes occurring in the Himalayas. Under the initiative of the Ministry of Earth Sciences and Geological Survey of India more than 20 cities have been taken up for microzonation in the first phase. The aim is to identify and map the variation in earthquake hazards with in a limited area, typically a city, as a result of variation in ground conditions or other characteristics. Microzonation maps can be used in conjunction with larger-scale hazard maps for urban land-use planning.

Earthquake Monitoring

For any seismic hazard assessment, a knowledge of the past earthquake occurrence in the region concerned is an essential first step, and a good catalogue is needed of all significant events of which there are records. A conscious effort was made in India, especially after the Latur earthquake in 1993 to upgrade the analog seismic stations to digital and adding additional stations in gap areas to have a better seismic monitoring to create a sound data base for earthquakes down to M 3.0. (Fig. 7a, b).

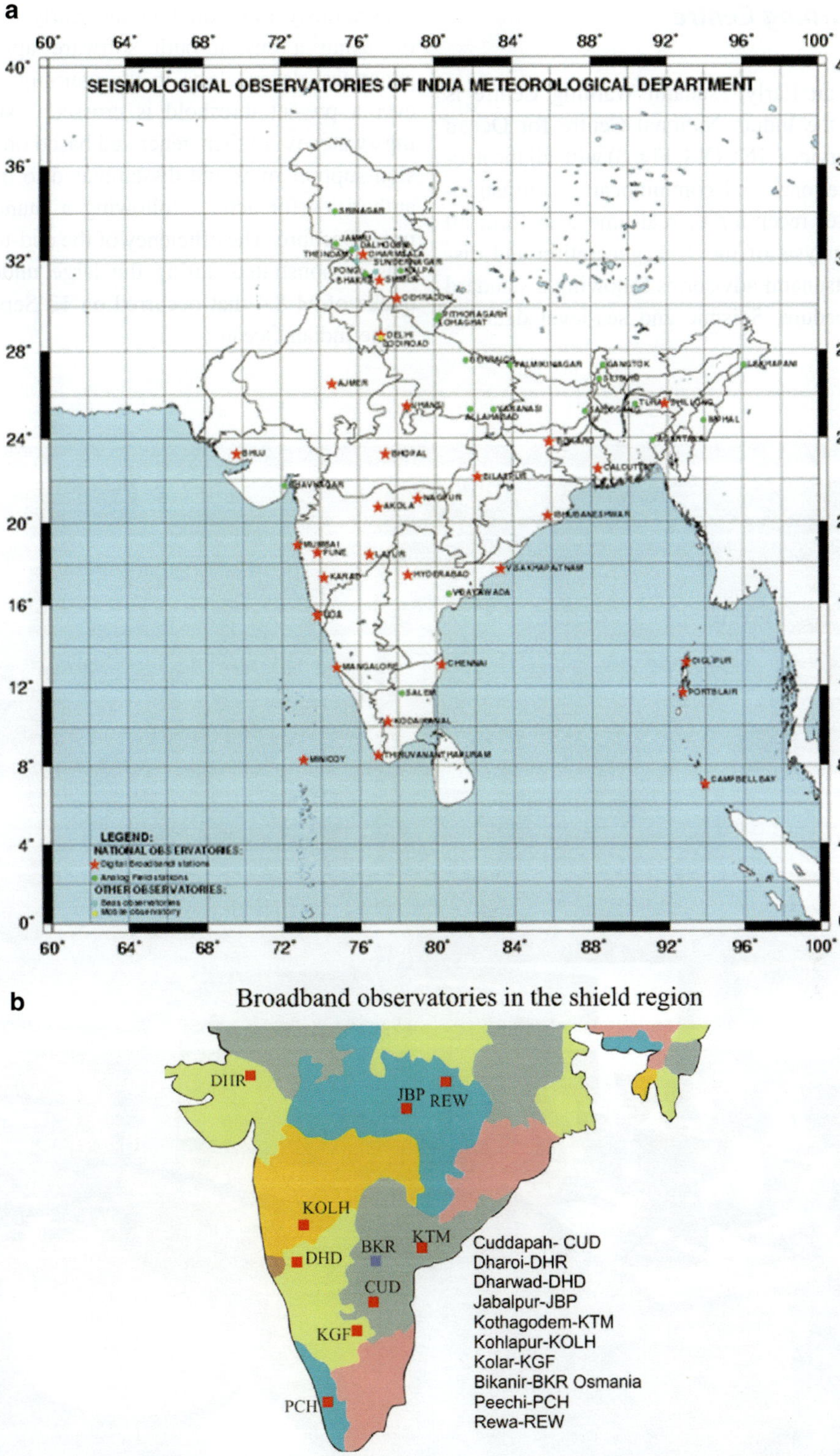

Fig. 7 (**a**) National Network of Seismological observatories operated by India Meteorological Department, New Delhi. (**b**) Ten digital broadband seismological observatories added to the National Network after the Latur earthquake in 1993

Tsunami Warning Centre

A state-of-the-art Early Tsunami Warning Centre is established at the Indian National Centre for Ocean Information Services (INCOIS, Fig. 8) with all the necessary computational and communication infrastructure that enables reception of real-time data from all the sensors, analysis of the data, generation and dissemination of tsunami advisories following a standard operating procedure. Seismic and sea-level data are continuously monitored in the Early Warning Centre using a custom-built software application that generates alarms/alerts in the warning centre whenever a pre-set threshold is crossed. Tsunami warnings/watches are then generated based on pre-set decision support rules and disseminated to the concerned authorities for action, following a Standard Operating Procedure. The efficiency of the end-to-end system was demonstrated during the large under-sea earthquake of M 8.4 that occurred on 12 September 2007 in the Indian Ocean.

Fig. 8 Early Tsunami Warning Centre at INCOIS, Hyderabad India (Source: http://www.incois.gov.in)

Hazard Mitigation Efforts

India took several concrete steps towards assessment and mitigation of seismic hazard in the country by upgrading seismic networks with digital broadband seismographs and accelographs and adding new observatories to generate high fidelity data so that a better assessment of seismic hazard can made of seismically active and stable continental regions in the country. Efforts were made to provide opportunities to earth scientists in the country to interact with international experts by way of workshops/conferences and seminars. One such Workshop by the Natural Hazard Society (NHS) and GeoRisk Commission of the IUGG was organized in India in 2004 with "Natural and Human Induced Hazards" as it theme. The Indian Geophysical Union in 2008 took up "Seismic Hazard and crustal earthquakes" as its theme for its 45th Convention. These activities increased the commitment to tackle natural hazards in the country. The Government of India supported these efforts by creating a National Disaster Management Authority (NDMA) headed by the Prime Minister of the country. The NDMA brought out disaster management guidelines on "Management of Earthquakes" (Fig. 9).

A Disaster Management Act was promulgated by the Government of India to provide for the effective management of disasters. In 2007, the IYPE in India was initiated by the President of the country (Fig. 10) during the 94th Indian Science Congress with the main theme as "Planet Earth" and a special session on Natural Hazards was organized.

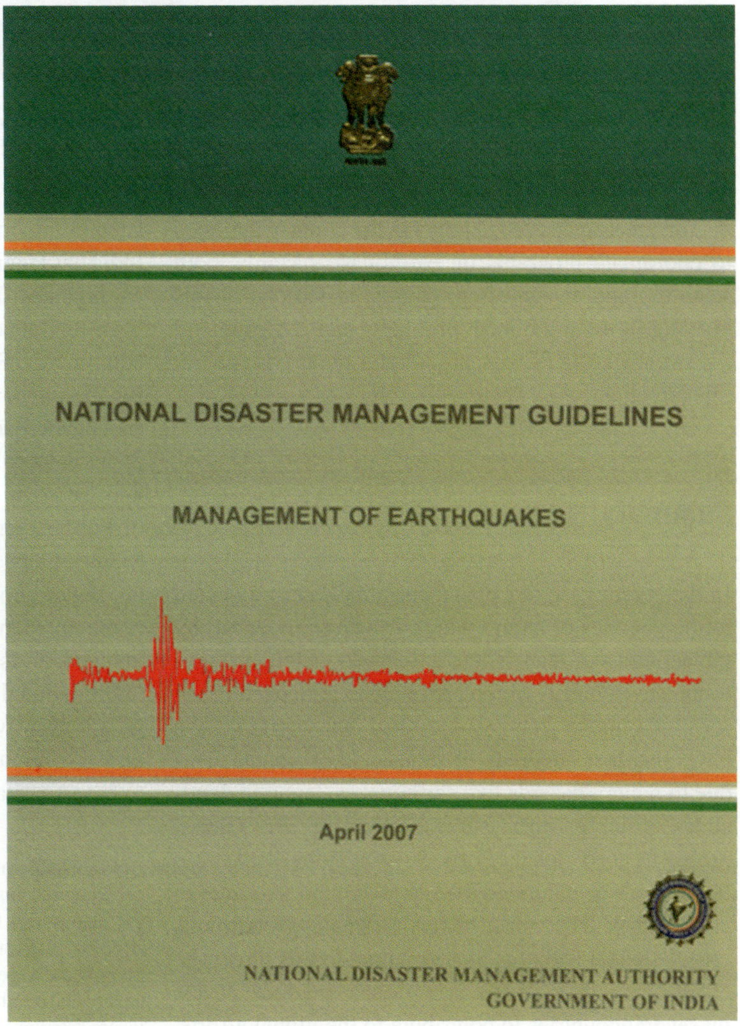

Fig. 9 Guidelines issued by National Disaster Management Authority, Government of India for Management of Earthquakes in India (2007)

Fig. 10 President of India Hon. Dr. A.P.J. Abdul Kalam releasing balloons to commemorate the Year of the Planet Earth during the 94th Indian Science Congress in Chidambaram, Tamil Nadu, India (Courtesy: Harsh Gupta)

The greatest challenge today in the country has been to coordinate programs on earthquake hazards, both scientific and awareness programs, at national levels keeping in mind the local problems and environment. An event like IYPE has provided a great stimulus in this direction.

Summary

In the recent past, our planet earth has been subjected to the vagaries of nature which is being influenced by the activities of man. There have been visible changes to the environment which are being experienced in the form of climate changes due to global warming. We have also lost hundreds of thousands of human lives due to natural hazards like earthquakes, tsunami, landslides, drought, storm surges, volcanoes etc. Global problems need global efforts. Several global efforts are being made to minimize the risk to human lives due to earthquakes. The world bodies are making enormous efforts to deal with this phenomenon. In this direction, efforts have to be made at regional and local levels by several countries to contribute to the global efforts, keeping in mind their regional and local problems. India has taken several steps to minimize the risk due to earthquakes. The earthquake monitoring systems have been upgraded and increased with the state-of-the-art equipments. Seismic hazard assessment is being done for the major cities in the country with more emphasis on Microzonation of major cities. The Government of India has shown its seriousness to tackle this problem by creating a new ministry called the "Ministry of Earth Sciences." The creation of National Disaster Management Authority both at national and state levels is yet another example to bring awareness to the common man about the possible natural and human induced disasters. Events like the UN declaration of "International Year of the Planet Earth" (IYPE), thus, go a long way to bring the global concern to save not only the human lives due to natural phenomena but also to have a sustainable development on our Planet Earth.

Acknowledgements I thank the Director, National Geophysical Research Institute (CSIR), Hyderabad, India for permitting the manuscript for publication. I thank the reviewers for constructive suggestions and finally my thanks to Dr. Tom Beer, for inviting me to contribute this article for the special publication. I acknowledge all the authors, individuals and organizations from where data have been used for the preparation of this manuscript.

Computational Geodynamics as a Component of Comprehensive Seismic Hazards Analysis

Alik Ismail-Zadeh

Abstract This paper reviews, with a few out of many examples, recent advances in computational geodynamics related to modelling of stress localization and earthquake occurrence. These studies provide a basis for a comprehensive seismic hazard analysis. Several case studies are considered: tectonic stress modelling in the southeastern Carpathians and central Apennines; dynamics of the lithospheric blocks and earthquake modelling for the Sunda arc and the Tibetan plateau; and seismic hazard assessment for the Vrancea region. Possibilities for earthquake prediction, mitigation and preparedness based on the earthquake science and computer modelling are discussed.

Keywords Computational geodynamics · Tectonic stress modelling

Introduction

Recent advances in understanding Earth dynamics and in development of computational tools permit accurate numerical modelling and forecasting that are transforming the Earth sciences. These advances have a strong impact on studies of geohazards and show significant potential to be applied to serve the sustainable development of society. The International Year of Planet Earth had chosen Hazards as one of its main scientific themes. Four key research questions were identified (Beer 2007). Among these questions are the following: *What technologies and methodologies are required to assess the vulnerability of people and places to hazards and how might these be used at a variety of spatial scales? How do geo-hazards compare relative to each other regarding current capabilities for monitoring, prediction and mitigation and what methodologies and new technologies can improve such capabilities to help civil protection at local and global scales?*

To answer such questions, natural hazard and risk assessment should be considered from a holistic point of view (from the whole to details). Particularly, a holistic comprehensive quantitative assessment of seismic hazard should be based on multidisciplinary research in (i) geodynamics (to reveal zones of tectonic stress localization), (ii) present and historical seismicity (to localize areas prone to strong events), (iii) nonlinear dynamics of the lithosphere (to analyse statistical properties of the earthquake sequences, their clustering and critical transitions), (iv) soil property (to analyse liquefaction and seismic shaking), and (v) classical hazards assessment (to determine peak ground acceleration, response spectra amplitude, and seismic intensity). This approach to seismic hazard should be accompanied by a holistic approach to earthquake prediction (e.g., Keilis-Borok and Soloviev 2003) and by a holistic approach to seismic risk (e.g., Beer and Ismail-Zadeh 2003; Cardona 2004), when a convolution of hazard, vulnerability and exposure (as functions of space and time) should be viewed also from social-psychological (e.g., resilience of community to extreme seismic events) and legal (e.g., role of law in risk reduction; see Paterson 2003) points of view.

A. Ismail-Zadeh (✉)
Geophysical Institute, University of Karlsruhe, Karlsruhe, Germany. International Institute of Earthquake Prediction Theory and Mathematical Geophysics, Russian Academy of Sciences, Moscow, Russia. Institut de Physique du Globe de Paris, France
e-mail: Alik.Ismail-Zadeh@gpi.uka.de

T. Beer (ed.), *Geophysical Hazards*, International Year of Planet Earth,
DOI 10.1007/978-90-481-3236-2_11,

The vulnerability of human civilisations to natural disasters is growing due to the proliferation of high-risk objects, clustering of populations, and destabilisation of large cities. Today a single earthquake may take up to several hundred thousand lives, and cause material damage up to several billions EURs (see Munich Re 2009) with a possible chain reaction expanding to a world-wide financial crisis and economic depression (comparable and even more severe than the 2008 financial crisis and economic recession). A large earthquake in (or close to) Tokyo might result in a world financial crisis, because many Japanese companies, which invested considerable funds in foreign enterprises, will withdraw these funds to rebuild or to restore the city infrastructure after the disaster. A large earthquake can trigger an ecological catastrophe (e.g. Chernobyl-type calamities) if it occurs in close vicinity to a nuclear power plant built in an earthquake-prone area (e.g., in Azerbaijan, Iran or elsewhere).

About a million earthquakes with magnitude >2 are registered each year; about a thousand of them are large enough to be felt; about a hundred earthquakes cause considerable damage, and once in a few decades a catastrophic event occurs. The 26 December 2004 earthquake with magnitude >9 that occurred in the Aceh-Sumatra region of the Indian Ocean generated great tsunamis, which killed more than 270,000 people and caused billions of dollars of damage (UN/ISDR Platform for the Promotion of Early Warning 2008). The occurrence of a particular earthquake is associated with dynamics of the lithosphere.

Extreme seismic events (e.g., 1755 Lisbon, 1906 San Francisco or 2004 Aceh-Sumatra earthquakes) are a manifestation of the complex behaviour of the lithosphere structured as a hierarchical system of blocks of different sizes. Driven by mantle convection these lithospheric blocks are involved in relative movement, resulting in stress localization and earthquakes. Despite the lithosphere behaving as a large non-linear system, featuring instability and deterministic chaos, some integral empirical regularities emerge, indicating a wide range of similarity, collective behaviour, and the possibility for earthquake prediction (e.g., Keilis-Borok et al. 2001). These great earthquakes, when they occur, are surprising, and society is poorly prepared to deal with them. Protecting human life and property against earthquake disasters requires an uninterrupted chain of research and civil protection tasks: from (i) understanding of the physics of earthquakes, their analysis and monitoring, through (ii) interpretation, modelling, seismic hazard assessment, and earthquake prediction, to (iii) delivery of the scientific forecasts to local authorities, public awareness, preparedness, and preventive disaster management.

The OECD Global Science Forum workshop *"Earthquake Science and its Contribution to Society"* (Potsdam, Germany, 2006) focused on the analysis of recent achievements in earthquake physics, seismic hazard and risk assessment, and earthquake prediction as well as on the role of science in increasing of awareness of governments and society on earthquakes to mitigate aftermaths of natural hazards. It was highlighted that modern super-computer facilities provide useful tools in modelling of geodynamical processes leading to earthquakes; modelling of extreme seismic events; monitoring of seismic hazard; earthquake forecasting; and modelling of seismic risks.

This paper reviews recent studies in computational geodynamics by the author and his colleagues. These studies provide a basis for a seismic hazard analysis. Section "Modelling of Lithospheric Stress" presents studies on tectonic stress localization. The studies on dynamics of the lithospheric blocks and faults as well as occurrences of large events are reviewed in section "Modelling of Earthquake Occurrence." Quantitative seismic hazard assessment is presented in section "Seismic Hazard and Risk." Moreover, possibilities for improvement of earthquake prediction are discussed in section "Earthquake Prediction" and possibilities for earthquake mitigation and preparedness based on the earthquake science and computer modelling are discussed in section "Geoscience and Preventive Disaster Management."

Modelling of Lithospheric Stress

Models of stress generation in the lithosphere are now widely used in geosciences to identify areas of stress localization and their correlation with observed seismic events (e.g., Bird and Baumgardner 1984; Ismail-Zadeh et al. 2000, 2004, 2005a, 2005b; Aoudia et al. 2007). In this section I review models of tectonic stresses in the mantle beneath the south-eastern Carpathians (the Vrancea region) and central Apennines (the Umbria-Marche region). The tectonic stress is generated as a result of heterogeneous movements of the crust and mantle. The movements can be described by a mathematical model using the equations of momentum and mass conservation with relevant boundary conditions. A discretization of the mathematical model results in the computational model, which is solved by relevant numerical methods (e.g., Naimark et al. 1998 in two-dimensional case studies; Ismail-Zadeh et al. 2001 in three-dimensional case studies).

Southeastern Carpathians

Repeated large intermediate-depth earthquakes of the southeastern (SE) Carpathians (the Vrancea region) cause destruction in Bucharest (Romania) and shake central and eastern European cities several hundred kilometres away from the hypocenters of the events. The epicentres of the mantle earthquakes in the Vrancea region are concentrated within a very small area (marked in Fig. 1). According to the historical catalogue of Vrancea events (e.g., Radu 1991), large intermediate-depth shocks with magnitudes $M_W > 6.5$ occur three to five times per century. In the twentieth century, large events at depths d of 70–180 km occurred in 1940 (moment magnitude $M_W = 7.7$, $d = 160$ km), in 1977 ($M_W = 7.5$, $d = 100$ km), in 1986 ($M_W = 7.2$, $d = 140$ km), and in 1990 ($M_W = 6.9$, $d = 80$ km) (e.g., Oncescu and Bonjer 1997).

The intermediate-depth large earthquakes gave rise to the development of a number of geodynamic models for this region. McKenzie (1972) suggested that this seismicity is associated with a relic slab sinking in the mantle and now overlain by continental crust. A seismic gap at depths of 40–70 km beneath Vrancea led to the assumption that the lithospheric slab had already detached from the continental crust (Fuchs et al. 1979). Oncescu (1984) proposed that the intermediate-depth events are generated in a zone that separates the sinking slab from the neighbouring immobile part of the lithosphere rather than in the sinking slab itself. Linzer (1996) explained the nearly vertical position of the Vrancea slab as the final rollback stage of a small fragment of oceanic lithosphere. Sperner et al. (2001) suggested a model of Miocene subduction of oceanic lithosphere beneath the Carpathian arc and subsequent soft continental collision, which transported cold and dense lithospheric material into the mantle.

Continental convergence in the SE-Carpathians ceased about 10 Ma (e.g., Jiricek 1979). At present the

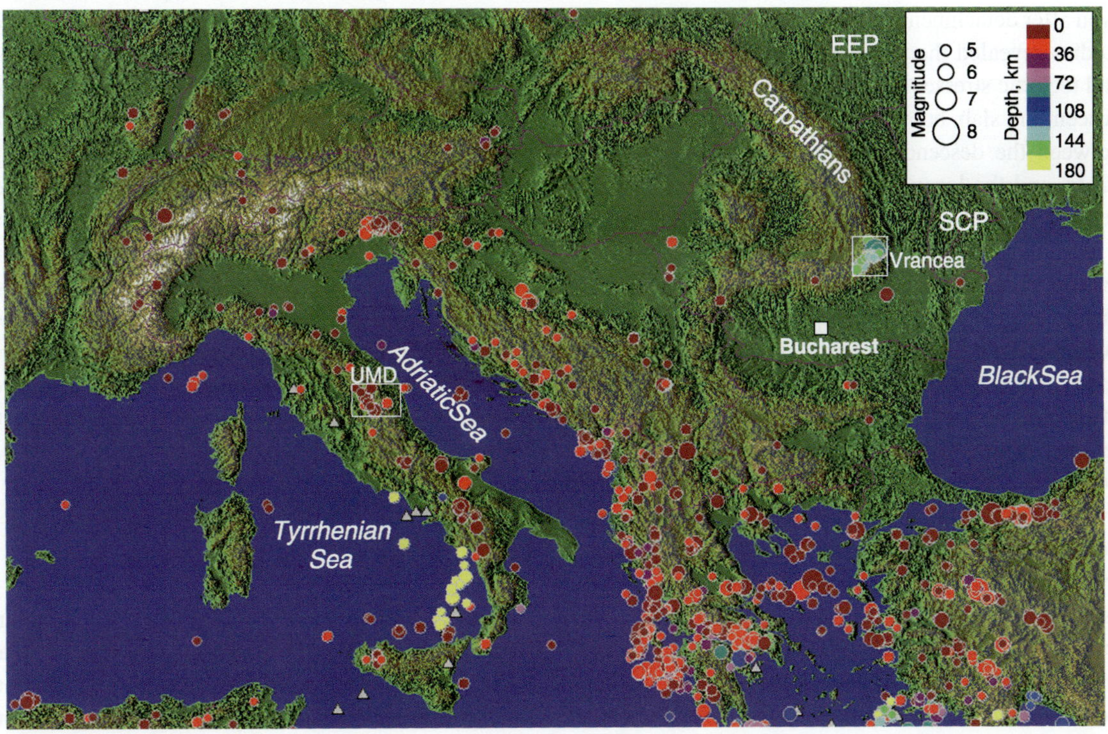

Fig. 1 Map of topography and strong earthquakes in the southern Europe

relatively cold slab beneath the Vrancea region sinks due to gravity. Hydrostatic buoyancy forces promote the sinking of the slab, but viscous and frictional forces resist the descent. The combination of these forces produces shear stresses at intermediate depths that are high enough to cause earthquakes. This was shown in two-dimensional numerical models of mantle flow and tectonic stress by Ismail-Zadeh et al. (2000). These authors recognized that the depth distribution of the annual average seismic energy released in earthquakes has a shape similar to that of the depth distribution of the modelled stress magnitude in the slab.

To evaluate the role of slab detachment in stress evolution, Ismail-Zadeh et al. (2005b) developed two-dimensional thermo-mechanical finite-element models of the post-Miocene subduction of the Vrancea slab subject to gravity forces alone. The models predicted lateral compression in the slab that were in agreement with those inferred from the stress axes of earthquakes. It was found that the maximum stress occurs in the depth range of 80 km to 200 km and the minimum stress falls into the depth range of 40 km to 80 km, which corresponds to the seismic gap. It was also shown that high tectonic stress (leading to seismic activity) is preserved in the slab for a few million years, even after detachment. The two-dimensional numerical studies revealed the principal features of mantle flow and tectonic stresses induced by a simple model of the descending slab, but they could not show a correlation between the descending high-velocity body, tectonic stress, and the locations of the Vrancea intermediate-depth earthquakes in any detail.

To analyse processes of stress generation and localization in and around the descending slab, Ismail-Zadeh et al. (2005a) developed a three-dimensional numerical model of contemporary mantle flow and stress beneath the Vrancea region. The input data of the model consisted of temperatures derived from seismic P-wave velocity anomalies and surface heat flow, crustal and uppermost mantle densities converted from P-wave velocities obtained from seismic refraction studies, geometry of the Vrancea crust and slab from tomography and refraction seismic data, and the estimated strain rate in the slab (as a result of earthquakes) to constrain the model viscosity. Ismail-Zadeh et al. (2005a) showed a correlation between the location of intermediate-depth earthquakes and the predicted localization of maximum shear stress (Fig. 2). Modelled tectonic stresses also predict large horizontal compression at depths of about 70–220 km beneath the Vrancea region, which coincides with the stress regime defined from fault-plane solutions for the intermediate-depth earthquakes. This implies that buoyancy-driven descent of the lithospheric slab beneath the Vrancea region is directly linked to intermediate-depth seismicity.

Mantle heterogeneities imaged by seismic tomography in the SE-Carpathians contain information on the present thermal state of the mantle. Ismail-Zadeh et al. (2008) developed a model of the present mantle temperature beneath the region based on *P*-wave seismic velocity anomalies (Martin et al. 2006) and combined the model with a model of crustal temperature constrained by heat flow data (Demetrescu et al. 2001). The modelled temperatures have been assimilated into the geological past using the information on the regional movement in the Early and Middle Miocene. Prominent thermal states of the lithospheric slab descending in the region have been restored from its diffuse present state. In Miocene times the slab geometry clearly shows two portions of the sinking body. The northwest-southeast oriented portion of the body is located in the vicinity of the boundary between the East European (EEP) and Scythian (SCP) platforms, and this portion of the sinking body may be a relic of a cold lithosphere that has travelled eastward. Another portion has a northeast-southwest orientation and is related to the present descending slab. Above a depth of 60 km the slab had a concave thermal shape, confirming the curvature of the Carpathian arc, and a convex surface below that depth. The slab maintained its convex shape until it split into two parts at a depth of about 220 km. Ismail-Zadeh et al. (2008) proposed that this change in the slab geometry, which is likely to be preserved until the present, can cause stress localization due to the slab bending and subsequent stress release resulting in large mantle earthquakes in the region.

Central Apennines

Central Mediterranean geology has been mainly shaped by the interplay between the Eurasian and African plates. The extremely variable structure of the lithosphere–asthenosphere system in the region is the result of its complex geodynamic history. The Cenozoic to Quaternary regional evolution has been marked

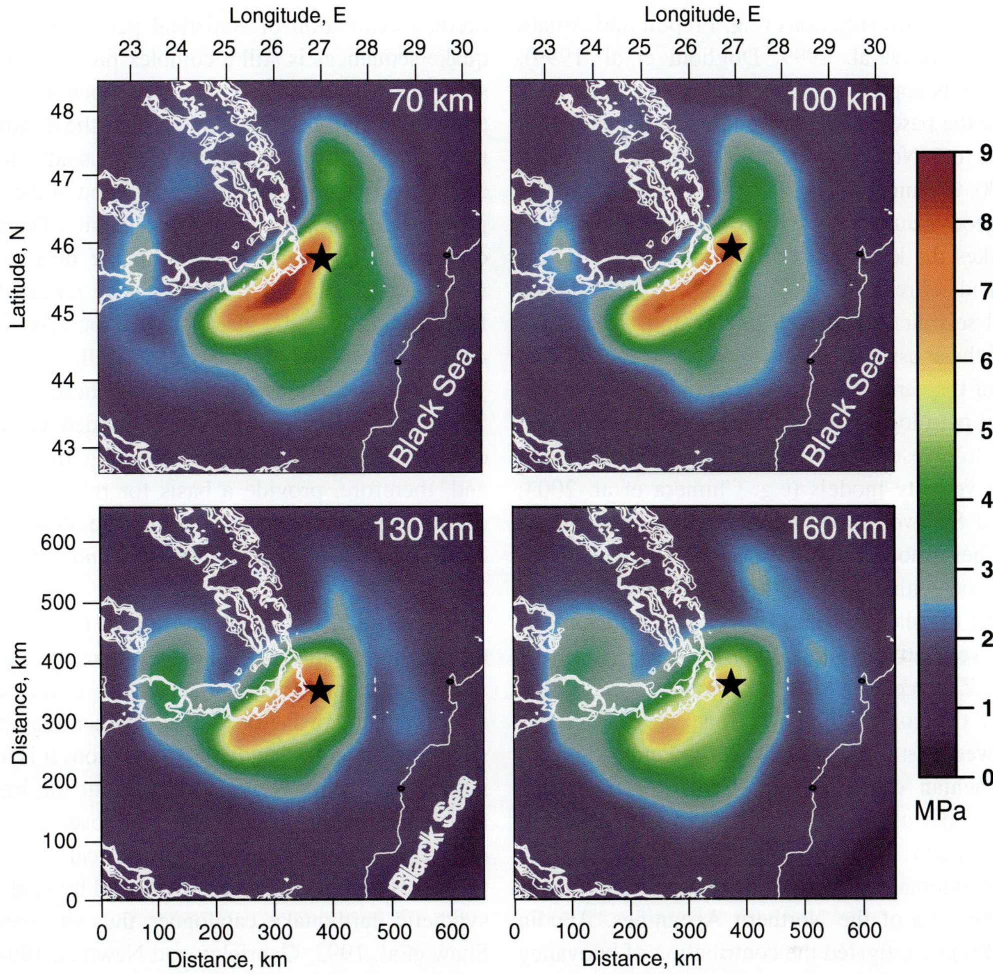

Fig. 2 Predicted maximum shear stress beneath the SE-Carpathians at different depths. Isolines present the surface topography. Star marks the location of the Vrancea intermediate-depth earthquakes (modified after Ismail-Zadeh et al. 2007b)

by the coexisting compression and tension developed at the same time between converging continental plates (e.g., Doglioni et al. 1999; Faccenna et al. 2004). However, the rate of convergence between the plates has been less significant than the east-west extension (e.g., Mantovani et al. 2002). The latter has been migrating from west to east and has been positioned behind a compression front migrating in the same direction. As a consequence, a number of extensional basins have formed behind the Apennines-Maghrebian compression front.

The eastward migration of the Apennines compression front is accompanied by a fragmentation of the Apennines lithosphere, with progressive ending of the active subduction zone from the Northern Apennines to the south. Fragmentation of the Apennines lithosphere created sectors that had an independent evolution (Locardi 1993; Sartori 2003). This may explain the variable lithosphere-asthenosphere structure in the region. The juxtaposed contraction and extension observed in the crust beneath Central Apennines has, for a long time, attracted the attention of geoscientists and is a long-standing enigmatic feature. Moreover, Selvaggi and Amato (1992) showed the sub-crustal seismicity in the Umbria-Marche geological domain (UMD) is not associated with the dipping seismic (Wadati-Benioff) zone. Several models, invoking mainly external forces, have been put forward to explain the close association of these two end-member deformation mechanisms observed by seismological and geological investigations (e.g. Frepoli and Amato 1997; Montone et al. 1999; Doglioni et al. 1999). These models appeal to interactions along plate margins or at the base of the lithosphere, or to subduction processes (e.g. Negredo et al. 1999; Wortel and Spakman 2000; Carminati et al. 2005).

The geodynamic complexity of the Central part of Italy makes the kinematics of the present day deformation and its relationships with the recent magmatism and seismicity less well understood. To unravel some of these aspects, higher resolution geophysical models of the earth structure compatible with gravity, heat-flow, petrological and geochemical data have been required to investigate the deformation. Recent Earth structure velocity models (e.g. Chimera et al. 2003; Mele and Sandvol 2003) provide the required resolution. These models exhibit clear evidence of lithospheric roots without any continuous slab. This would imply that the slab has been eroded and no engine is left to drive relative subduction processes.

Ismail-Zadeh et al. (2004) and Aoudia et al. (2007) modelled the contemporary regional tectonic stress along a west-east transect crossing the Peninsula from the Tyrrhenian coast, via the UMD, to the Adriatic coast. They used a crust/lithosphere-asthenosphere structural earth model by Chimera et al. (2003). Although external forces must have been important in the building up of the Northern Apennines, Aoudia et al. (2007) investigated the contribution of buoyancy forces with respect to the ongoing slow and complex lithospheric deformations, as revealed by very recent GPS measurements and by the unusual sub-crustal seismicity distribution. They showed that the buoyancy forces that result from the heterogeneous density distribution in the lithosphere govern the present day deformation within Central Italy and can explain regional coexisting contraction and extension at shallow depth and mantle seismicity in the UMD (Fig. 3).

Modelling of Earthquake Occurrence

Stress accumulation and its release in earthquakes are governed by non-linear hierarchical systems, which have a number of degrees of freedom and, therefore, cannot be understood by studying them piece by piece (Keilis-Borok and Soloviev 2003). Since an adequate theoretical base has yet not been well elaborated, theoretical estimation of statistical parameters of earthquake sequences is still a complex problem. Studying seismicity using the statistical and phenomenological analysis of earthquake catalogues has the disadvantage that instrumental observations cover, usually, too short a time interval compared to the duration of the tectonic processes responsible for seismic activity. The patterns of earthquake occurrence identifiable in a catalogue may be apparent and yet may not be repeated in the future. Moreover, the historical data on seismicity are usually incomplete. Numerical modelling of seismogenic processes allows overcoming these difficulties. Synthetic earthquake catalogues formed via numerical simulations may cover very long time intervals and, therefore, provide a basis for reliable estimates of the parameters of the earthquake flows (Ismail-Zadeh et al. 1999, 2007a; Soloviev and Ismail-Zadeh 2003).

It is difficult to detect the impact of a single factor on the dynamics of seismicity by analysing seismic observations, because seismicity is impacted by an assemblage of factors some of which may be more significant than that under consideration. It is also difficult (if not impossible) to single out the impact of an isolated factor by using seismic observations. This difficulty may be resolved by numerical modelling of processes that generate seismicity and by studying the synthetic earthquake catalogues thus obtained (e.g., Shaw et al. 1992; Gabrielov and Newman 1994; Allegre et al. 1995; Newman et al. 1995; Turcotte 1997; Keilis-Borok and Soloviev 2003).

Mathematical models of lithosphere dynamics are also tools for studying earthquake preparation processes and are useful in seismic hazard and earthquake prediction studies (Gabrielov and Newman 1994). An adequate model should indicate the physical basis of premonitory patterns determined empirically prior to large events. Note that available data often do not constrain the statistical significance of premonitory patterns. The model can also be used to suggest new premonitory patterns that might exist in catalogues of seismic events.

A block-and-fault dynamics (BAFD) model exploits the hierarchical block structure of the lithosphere (Alekseevskaya et al. 1977). The basic principles of the model have been developed by Gabrielov et al. (1990); the blocks of the lithosphere are separated by comparatively thin, weak, less consolidated fault zones, such as tectonic faults or lineaments. In seismotectonic

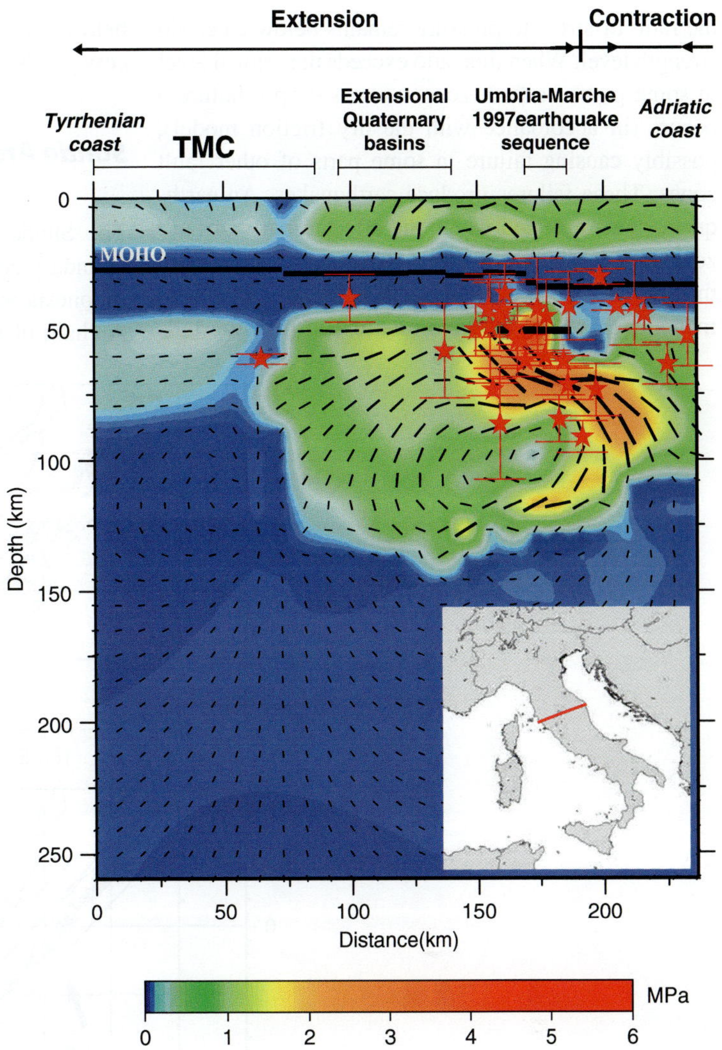

Fig. 3 Predicted maximum shear stress and compressional axes (ticks) along the profile (red line in the inset) (modified after Aoudia et al. 2007). The horizontal ticks indicate thrusting and vertical ticks indicate normal faulting. Red stars mark the hypocentres of the sub-crustal earthquakes (vertical bars indicate the depth error) recorded in the period 1965–1998 within a stripe 150 km wide along the profile. The bold black segment indicates the Moho depth (Chimera et al. 2003)

processes, major deformation and most earthquakes occur in such fault zones.

A seismic region is modelled by a system of perfectly rigid blocks divided by infinitely thin plane faults. Displacements of all blocks are supposed to be infinitely small relative to their size. The blocks interact with each other and with the underlying medium. The system of blocks moves owing to prescribed motions of boundary blocks and the underlying medium. Blocks are perfectly rigid; hence deformation takes place only in fault zones and at block bases in contact with the underlying medium. Relative block displacements take place along fault zones. Strains in the model are accumulated in fault zones. This reflects strain accumulation due to deformations of plate boundaries. Considerable simplifications are made in the model, but they are necessary to understand the dependence of earthquake flow on the main tectonic movements in a region and on its structure. This assumption is justified by the fact that the effective elastic moduli in the fault zones are significantly smaller than those within the blocks.

Lithospheric blocks interact visco-elastically with the underlying mantle. The corresponding stresses depend on the value of relative displacement. This dependence is assumed to be linearly elastic. The motion of the medium underlying different blocks may be different. Block motion is defined so that the system is in quasi-static equilibrium. The interaction of blocks along fault zones is visco-elastic (“normal state”) while the ratio of stress to pressure remains below a certain strength level. When this ratio exceeds the critical level in some part of a fault zone, a stress drop ("failure") occurs (in accordance with the dry friction model), possibly causing failure in some parts of other fault zones. These failures produce earthquakes. An earthquake starts a period where the affected parts of fault zones are in a state of creep. This state differs from the normal state by faster growth of inelastic displacements lasting until the ratio of stress to pressure falls below some level. Numerical simulation of this process yields synthetic earthquake catalogues.

Sunda Arc

The Sunda island arc marks an active convergence boundary between the Eurasian plate, which underlies Indonesia with Indian and Australian plates (Fig. 4a). A chain of volcanoes forms the topographic spine of

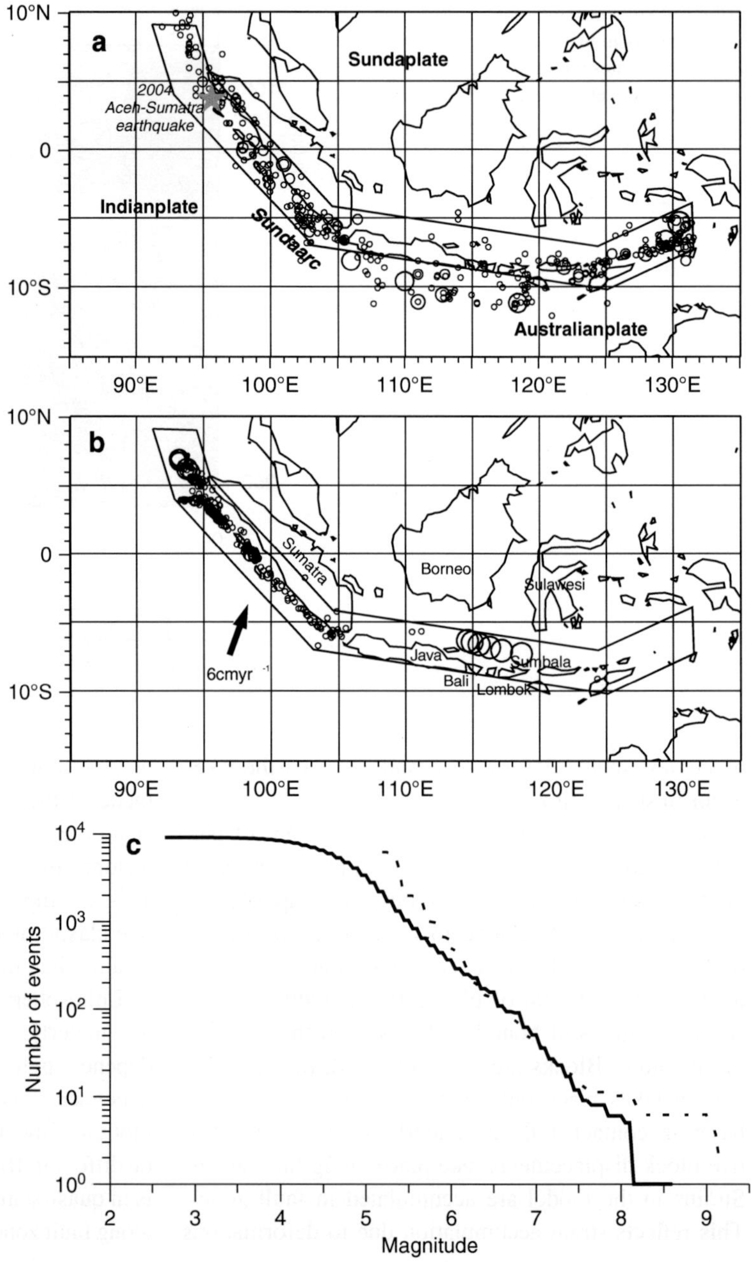

Fig. 4 BAFD model of the Sunda arc. (**a**) Observed earthquakes from 1950 to 2000 with magnitudes >6. (**b**) Modelled events with magnitudes >7. Polygonal domain represents the lithospheric block structure. (**c**) Cumulative frequency-magnitude plots for the observed and the modelled seismicity (modified after Soloviev and Ismail-Zadeh 2003)

the islands of Sumatra, Java, and Sunda. The Indian and Australian plates subduct beneath the southwestern part of the Eurasian plate along the Sunda arc. The tectonic deformation and associated stress localization in the Sunda trench caused the 2004 Aceh-Sumatra earthquakes. Seismic tomographic imaging revealed anomalies of seismic waves beneath the Sunda island arc suggesting that the lithospheric slab penetrates to the lower mantle (Widiyantoro and van der Hilst 1996).

Soloviev and Ismail-Zadeh (2003) presented a BAFD model of the Sunda arc region. The HS2-NUVEL1 global plate motion model (Gripp and Gordon 1990) is used to specify the movements. The dynamics of the lithosphere was modelled for 200 years for various parameters of the BAFD model. For the preferred case study, the regional seismicity (a) is compared with the modelled seismicity (b) in Fig. 4. This model has identified two epicentral areas prone to huge earthquakes. The first area is related to the Aceh-Sumatra region, where the magnitude 9+ mega-thrust earthquake occurred. Another area is predicted to be located between the Borneo, Sulawesi, Sumbala, Lombok, and Bali islands. Figure 4c presents the cumulative frequency-magnitude plots for the observed (solid line) and modelled (dashed line) seismicity. The slopes of the curves are close within the magnitude range from 5.5 to 8.1, and shifted to the larger magnitudes in the case of modelled events. Rundquist et al. (1998) determined that earthquakes with magnitude ≥ 6 observed in the Sunda arc region migrate from the eastern to the western part of the arc. The synthetic earthquake catalogue revealed a similar migration of the modelled earthquakes with magnitude ≥ 7.

Despite the simplicity of the BAFD model of the Sunda arc, Soloviev and Ismail-Zadeh (2003) showed that the model with movements specified by the HS2-NUVEL1 global plate motion yields synthetic seismicity having certain common features with observations. These features include the locations of larger events, the direction of migration of earthquakes, and the slope of the frequency-magnitude plot.

Tibet Plateau

Following the closure of the Mesozoic Tethys ocean, the India-Asia collision initiated the development of the Himalayan range and the Tibetan plateau and induced widespread strain in southeastern Asia and China. The Tibetan plateau is underlain by a thick crust (up to 70–80 km) as inferred from gravity anomalies and seismic profiles (Barazangi and Ni 1982; Le Pichon et al. 1992; Nelson et al. 1996). The Himalayan frontal thrust and the Longmen Shan represent abrupt and steep topographic fronts at the southern and eastern edges of the plateau (Fig. 5a).

There are three distinct views about the active deformation in Tibet that dominates the debate on the mechanics of continental deformation. One view is that the deformation is distributed throughout the continental lithosphere (e.g., Houseman and England 1996). Another view is associated with the crustal thinning and the deformation due to a channel flow within the mid-to-lower crust (e.g., Royden et al. 1997). Meanwhile there is growing evidence supporting the alternative view that a substantial part of the deformation of the continents is localized on long and relatively narrow faults and shear zones separating rigid crustal blocks (e.g., Tapponnier et al. 2001). Many of these zones cut the base of the crust (Vergnes et al. 2002; Wittlinger et al. 2004), and some extend to the base of the lithosphere (e.g., Wittlinger et al. 1998). Therefore, such deformations can be described by motions of crustal lithospheric blocks separated by faults.

Ismail-Zadeh et al. (2007a) developed a BAFD model for the Tibet-Himalayan region. They showed that the contemporary crustal dynamics and seismicity pattern in the region are determined by the north-northeastern motion of India relative to Eurasia and the movement of the lower crust overlain by the upper crustal rigid blocks. Variations in rheological properties of the fault zones and/or of the lower crust as well as in the motion of the lower crust influence the displacement rates of the crustal blocks and hence the slip rates at the faults separating the blocks. This may explain the discrepancies in the estimates of slip rates at major faults in the region based on different techniques.

Clustering of earthquakes can be considered as a consequence of the dynamics of the crustal blocks and faults in the region. The number and maximum magnitude of synthetic earthquakes change with the variations in the movements of the crustal blocks and in the rheological properties of the lower crust and the fault zones. As an example, a cluster of modelled events along the Longmen Shan fault was identified by the BAFD model (Fig. 5b). The 2008 $M = 7.9$

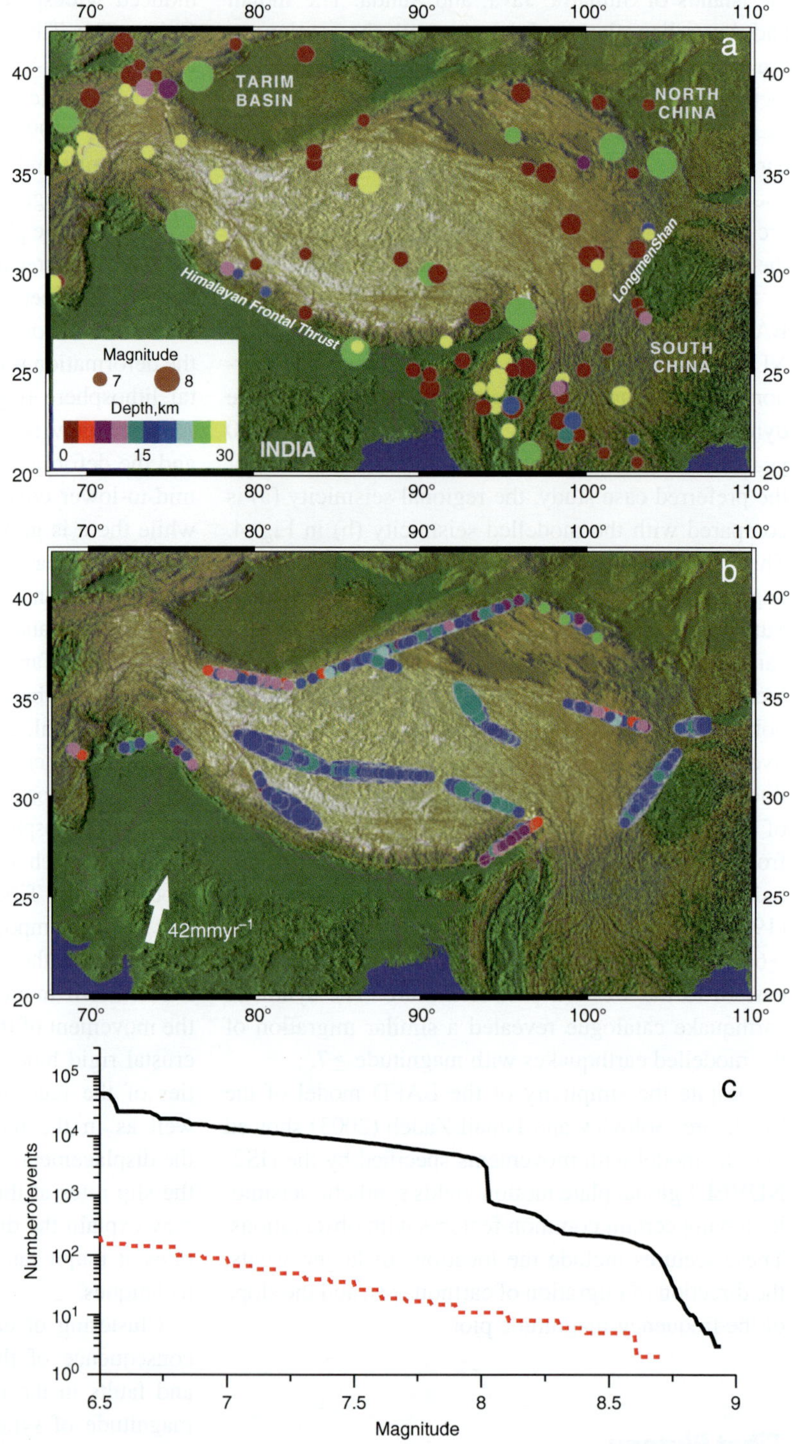

Fig. 5 BAFD model of the Tibet-Himalayan region. (**a**) Observed earthquake since 1900. (**b**) Modelled events. (**c**) Cumulative frequency-magnitude plots for the observed and the modelled seismicity. Note that the frequency-magnitude plots for synthetic events are shifted upwards due to the larger number of the events compared to the number of seismic events in the region (modified after Ismail-Zadeh et al. 2007a)

Sichuan (Wenchuan) earthquake occurred along this fault killing about 70,000 in addition to about 400,000 injured and about 20,000 missing people.

Seismic Hazard and Risk

Primary models of geodynamics, stress generation, earthquake occurrences, and strong ground motions caused by earthquakes are important inputs for seismic hazard analysis. Seismic hazard assessment in terms of engineering parameters of strong ground motion (namely, peak ground acceleration PGA, response spectra amplitude RSA, and seismic intensity) is based on information about the features of earthquake ground motion excitation (source scaling), seismic wave propagation (attenuation), and site effect in the region under consideration. Ideally, all these factors should be studied on the basis of available regional earthquake ground motion data.

Quantitative Seismic Hazard Assessment: Case Study of the SE-Carpathians

Analysis of the macroseismic and instrumental data from intermediate-depth Vrancea earthquakes reveal several peculiarities of earthquake effects (e.g., Mandrescu et al. 1988; Ivan et al. 1998; Mandrescu and Radulian 1999; Moldovan et al. 2000). They may be summarized as follows: earthquakes affect very large areas with a predominant NE-SW orientation; local and regional geological conditions can control the amplitudes of earthquake ground motion to a larger degree than magnitude or distance; and strong ground motion parameters exhibit a large variability.

To assess the seismic hazard in the SE-Carpathians, strong ground motion excitation and attenuation during the intermediate-depth Vrancea earthquakes were analysed (e.g., Oncescu et al. 1999; Radulian et al. 2000; Gusev et al. 2002). Several studies were carried out to estimate the seismic hazard in Romania using a probabilistic approach (Lungu et al. 1999; Musson 2000; Mantyniemi et al. 2003; Ardeleanu et al. 2005). The azimuth-dependent empirical attenuation models evaluated from regional strong motion data were used in some of these studies; however variations of the local site response were not taken into account.

Ismail-Zadeh et al. (2007b) performed probabilistic seismic hazard assessment (PSHA) for the region using knowledge of the soil conditions and basic features of local geology (properties of sediments down to bed-rock). Figure 6 shows results of the site-dependent PSHA for the SE-Carpathians in terms of PGA together with the PGA distribution during the large Vrancea earthquake ($M_W = 7.2$, 30 August 1986; Fig. 6a). There is a good agreement between observations and the site-dependent estimations predicted by the PSHA models both in the shape of contours and absolute values. For comparison, the PGA hazard model was also calculated without consideration of site response - for rock site conditions (Fig. 6b). Those PSHA results that do not consider geological factors have nothing in common with the distribution of ground motion amplitudes during earthquakes. Ismail-Zadeh et al. (2007b) concluded that geological factors play an important part in the distribution of earthquake ground motion parameters within the region analysed. These results can be considered as a basis for comprehensive site-dependent PSHA analysis in the region and as a basis for a new seismic code.

Seismic Risk

Problems of estimating risks of natural catastrophes are becoming highly important. In the last few decades a number of concepts of risks of natural catastrophes have been suggested and a number of international projects on safety and risk management have been conducted. Serious difficulties in decision making in these fields are concerned with strong uncertainties in data and limitations in using mathematical tools for carrying out the historical analysis and forecasting.

Seismic risk can be defined as a measure that combines, over a given time, the likelihoods and the consequences of a set of earthquake scenarios (Beer and Ismail-Zadeh 2003). The risk can be estimated as the probability of harmful consequences or expected losses (of lives and property) and damages (e.g., people injured, economic activity disrupted, environment damaged) due to an earthquake resulting from interactions between seismic hazards (*Hs*), vulnerability (*V*), and exposed values (*E*). Conventionally, seismic risk (*Rs*) is expressed quantitatively by the convolution of

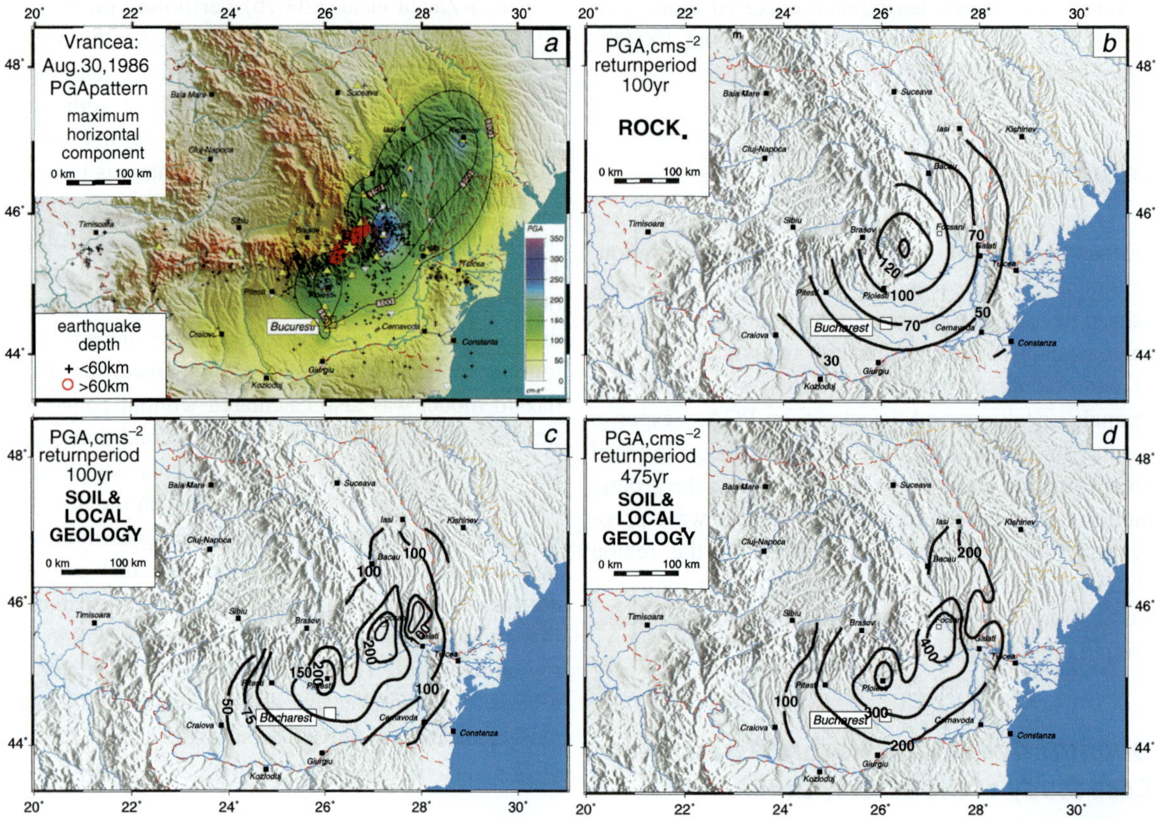

Fig. 6 Comparison of the peak-ground acceleration (maximum of two horizontal components) distribution during the $M_W = 7.2$, 30 August 1986 earthquake in Vrancea (**a**) and the PSHA results evaluated for two types of site conditions (**b**: rock and **c**, **d**: soil and local geology) and for two return periods (**c**: $T = 100$ years and **d**: $T = 475$ years). Numbers at the contours are scaled in cm s^{-2} (after Ismail-Zadeh et al. 2007b)

these three parameters: $Rs = Hs \otimes V \otimes E$ (Kantorovich et al. 1973).

An estimation of seismic risk may facilitate a proper choice in a wide variety of seismic safety measures, ranging from building codes and insurance to establishment of rescue-and-relief resources. Different representations of seismic risk require different safety measures. Most of the practical problems require estimating seismic risk for a territory as a whole, and within this territory separately for the objects of each type: areas, lifelines, sites of vulnerable constructions, etc. The choice of the territory and the objects is determined by the jurisdiction and responsibility of a decision-maker. Each specific representation of seismic risk is derived from the models of seismic hazards, the territorial distribution of population, property, and vulnerable objects, and the damage caused by an episode of strong motion.

Earthquake Prediction

Though the assessment of seismic risk is important in earthquake preparedness, it does not answer the question of *where* and *when* the next earthquake will occur. This question worries not only scientists, but also populations living in earthquake-prone regions. The catastrophic nature of earthquakes has been known for centuries due to the resulting devastation during some earthquakes. Their abruptness, along with the apparent irregularity and infrequency of large seismic events, facilitates a common perception that earthquakes are random unpredictable phenomena. For the last decade, earthquake prediction research has been widely debated: opinions on the possibilities of prediction varies from the statement that earthquake prediction is intrinsically impossible (Geller et al. 1997) to the statement that prediction is possible, but difficult

(Knopoff 1999). To predict an earthquake, someone *"must specify the expected magnitude range, the geographical area within which it will occur, and the time interval within which it will happen with sufficient precision so that the ultimate success or failure of the prediction can readily be judged. Only by careful recording and analysis of failures as well as successes can the eventual success of the total effort be evaluated and future directions charted. Moreover, scientists should also assign a confidence level to each prediction"* (Allen et al. 1976).

Four major stages can be distinguished in earthquake prediction: (i) long-term (a decadal time scale), (ii) intermediate-term (one to several years), (iii) short-term (days to weeks and months), and (iv) immediate (seconds to hours). Long-term prediction of earthquakes is essentially based on the determination of probabilities of active fault segments to rupture for the next few decades (e.g., Working group on California earthquake probabilities 1999). This kind of prediction can guide engineering and emergency planning measures to mitigate the impact of the earthquake.

An intermediate-term prediction is an update of the long-term prediction brought about by some indicators (e.g., an increase in background seismicity, clustering of events in space and time, transformation of magnitude distribution and some others). For example, an intermediate-term earthquake prediction method was designed by retrospective analysis of the dynamics of seismic activity preceding the largest events worldwide (Keilis-Borok and Kossobokov 1990). Initially tested retrospectively, the method is subject to on-going real-time experimental testing, and the results of the prediction based on the method support the idea of predictability of great earthquakes (Kossobokov 2006).

The feasibility of short-term earthquake prediction (days to weeks and months) is still controversial, and the major difficulty here is to identify short-term precursors in the background of intermediate-term alarms. One of the aims of intermediate- and short-term predictions of earthquakes is to reduce the number of false alarms and the number of failures to predict a strong event. A reduction of false alarms using, for example, the theory of optimal stopping (Feinberg and Shiryaev 2007) could improve the performance of prediction methods.

Immediate earthquake prediction is usually based on the first arrival of seismic waves and transmission of an electronic alert within a lead-time of seconds. It is used (e.g. in Japan) to shut down nuclear reactors, gas and electricity grids and to stop high-speed trains in the event of a strong earthquake.

Compared to the accuracy of weather forecasting (e.g., Kalnay 2003), the current accuracy of earthquake prediction is still too low. Our knowledge of earthquake physics and earthquake dynamics is limited to predicting strong earthquakes with a high accuracy. We do not know well (i) how earthquakes, especially large events, work; (ii) when an earthquake starts, when it stops, and what magnitude could be expected; (iii) how earthquakes cluster in terms of stress transfer; (iv) what were the initial conditions of stress state before a large event. Moreover, there is no quantitative description of earthquake physics, namely, no mathematical equations to describe non-linear dynamics of fault systems and earthquake "flow." The Navier–Stokes equations in meteorology describe atmospheric flow and hence allow weather prediction with a high accuracy for time scales ranging from one to several days.

The scientific community should use the full potential of mathematics, statistics, statistical physics, and computational modelling and the data derived from seismological (monitoring of physical parameters of earthquakes and tectonic stress, fluid migration, etc), geodetic (GPS, InSAR and other measurements of the crustal deformation), and geological (e.g., determination of the time intervals between strongest earthquakes using paleo-seismological tools) studies to improve intermediate- and short-term earthquake predictions.

Though the current accuracy of earthquake prediction is limited, any scientifically validated prediction can be useful for earthquake preparedness and disaster management, if the accuracy of the prediction (even though it is not high) is known. In this case an inexpensive low-key response to the prediction (e.g., to lower a water level in reservoirs located in the area of a predicted earthquake in order to prevent large flooding due to a possible damage of the reservoirs) would be well justified, if even a little part of the total damage due to a strong event is prevented (Keilis-Borok et al. 2004).

Geoscience and Preventive Disaster Management

Several extreme seismic events have struck the Earth and affected our society in recent times, among them: the October 2004 large Niigata earthquake (magnitude $M = 6.6$) in Japan, the December 2004 great

Aceh-Sumatra (M = 9+) earthquake and devastating tsunami in the Indian Ocean, the October 2005 large earthquake in Pakistan (M = 7.6), and the most recent May 2008 Sichuan (M = 7.9) earthquake in China. Such natural events are rare, but not unexpected. The earthquake and tsunami in the Indian Ocean and the earthquakes in Pakistan and in China resulted in large humanitarian disasters because of weak preventive disaster management. Fortunately, the Niigata earthquake did not lead to extreme loss of human life, but became one of the costliest natural disasters of the twenty-first century (economic losses were estimated to be US$ 28 billions; Munich Re 2005).

Scientists know about historical devastating earthquakes and tsunamis, such as those, which occurred in the Indian Ocean region or earthquakes in Himalayas, Tibet, and the Japanese islands. Time is an important variable in natural disaster management, especially when it concerns extreme events. An extreme seismic event, in general, cannot be predicted in full details. So far, seismology can put confidence limits on uncertainty, although the limits are very wide in terms of the time, place and magnitude of an anticipated earthquake, which gives insufficient information for disaster management. Nevertheless hazard preparedness is vital for society. The less often natural events occur (and the large extreme events are rare by definition), the more often disaster managers postpone preparedness for the events.

Ismail-Zadeh and Takeuchi (2007) wrote: "The tendency to reduce the funding for preventive disaster management of natural catastrophes rarely follows the rules of responsible stewardship for future generations neither in developing countries nor in highly developed economies." The investment to avoid losses tends not to be easily accepted in political decision making as compared with that to gain positive benefits. It is because the benefit of preventing losses, however long lasting it is, is not easily visible while the positive benefit is obvious and can easily be agreed by people. A large investment is made, when a big disaster due to an earthquake occurs, and the investment decreases till the next large earthquake (Fig. 7). If about 5–10% of the funds, necessary for recovery and rehabilitation after a disaster, would be spent to mitigate an anticipated earthquake, it could in effect save lives, constructions, and other resources. The reaction of media and the societal attention to disasters follow the same cycle. Following V. Klemes and R. Tannehill (Tannehill 1947), we can call this cycle the "seismo-illogical cycle" meaning that a large investment is made when a big earthquake disaster occurs and the investment decreases till the next disaster occurs. If a seismic cycle is marked by an *increase* of tectonic stresses toward an earthquake (stress drop), the "seismo-illogical cycle" is characterised by a *decrease* of funding toward the next large seismic event.

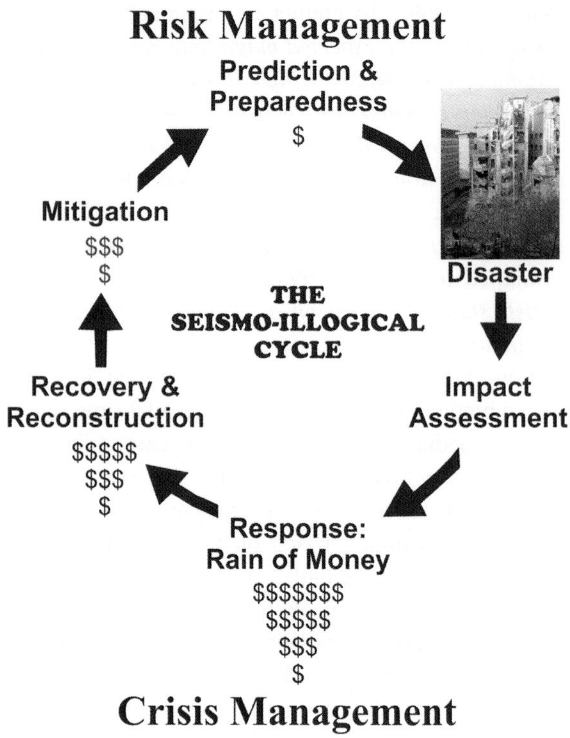

Fig. 7 Diagram representing the seismo-illogical cycle in seismic risk and earthquake disaster management

How to convince disaster-management authorities paying forward, in advance of rare extreme events, to mitigate or even to prevent in some cases inflicting humanitarian disasters? What is the role of geoscientists in preventive disaster management of catastrophic events? Ismail-Zadeh and Takeuchi (2007) believe that science should provide the "brains" for preventive disaster management.

One of the roles of seismologists is to assist disaster managers in terms of timely delivery of reliable forecasts on potentially large earthquakes and on their possible aftermath. These forecasts should be based on physics of the seismogenic processes, statistical analysis of data, and computer models of stress generation and release in earthquakes. At present the role of science is limited by inaccurate (in a deterministic sense) predictions of earthquakes. But this fact should not push scientists to give up the predictions of extreme seismic events. They can enhance the study of accurate temporal and spatial prediction of such events, that is, prediction of the two important parameters for preventive disaster management: *where* and *when*. "The frequency and scope of loss of major natural catastrophes will continue to increase dramatically throughout the world. Unless drastic measures are taken soon to prevent it, this trend will be intensified considerably by the ever more evident warming of the atmosphere, the resultant increase in sea level, and the intensification of storm and precipitation processes. In its own interest, the insurance industry must assume a major role in implementing preventive measures in order to ensure that it can provide cover for natural hazards over the long term" (Berz 2004).

After the 2004 Aceh-Sumatra catastrophic event, the Commission on Geophysical Risk and Sustainability of the International Union of Geodesy and Geophysics recommended to develop multidisciplinary and multinational research programs and networks on geophysical hazards and risks in the affected countries in order to integrate diverse data streams, to improve understanding of the natural phenomena associated with the disasters, to enhance predictive modelling capability, and to generate and to disseminate timely and accurate information needed by decision makers and the public (IUGG GeoRisk Commission Statement 2005). Scientists must act today and implement state-of-the-art measures to protect society from rare but recurrent extreme natural catastrophes. Otherwise we will witness again and again the tragic aftermaths of disasters that could have been avoided.

Conclusion

Advances in understanding of natural hazards and modern computational tools make considerable contribution to reduction of earthquake disasters, but their danger keeps escalating. This review illustrates that reversal of that trend does require a deep integration of knowledge on earthquake physics and analysis of seismic observation with theoretical development and numerical analysis. And computational geodynamics plays an essential role in identification of the places of large seismic events and in seismic hazard analysis.

References

Alekseevskaya MA, Gabrielov AM, Gvishiani AD et al. (1977) Formal morphostructural zoning of mountain territories. J Geophys 43: 227–233

Allegre CJ, Le Mouël JL, Chau HD et al. (1995) Scaling organization of fracture tectonics (SOFT) and earthquake mechanism. Phys Earth Planet Inter 92: 215–233

Allen CR, Edwards W, Hall WJ et al. (1976) Predicting earthquakes: a scientific and technical evaluation – with implications for society. Panel report on earthquake prediction, Committee on Seismology, National Research Council. U.S. National Academy of Sciences, Washington, DC

Aoudia A, Ismail-Zadeh AT, Romanelli F (2007) Buoyancy-driven deformation and contemporary tectonic stress in the lithosphere beneath central Italy. Terra Nova 19: 490–495

Ardeleanu L, Leydecker G, Bonjer KP et al. (2005) Probabilistic seismic hazard map for Romania as a basis for a new building code. Nat Haz Earth Sys Sci 5: 679–684

Barazangi M, Ni J (1982) Velocities and propagation characteristics of Pn and Sn beneath the Himalayan arc and Tibetan plateau: possible evidence for underthrusting of Indian continental lithosphere beneath Tibet. Geology 10: 179–185

Beer T (2007) The natural hazards theme of the International Year of Planet Earth, Nat Haz 42(3): 469–480

Beer T, Ismail-Zadeh A, (eds) (2003) Risk science and sustainability. Kluwer Academic Publishers, Dordrecht

Berz G (2004) Natural disasters and climate change: concerns and possible counter-measures from the viewpoint of an international reinsurer. Report, Munich Reinsurance Co., Munich, 15 p

Bird P, Baumgardner J (1984) Fault friction, regional stress, and crust-mantle coupling in southern California from finite element models. J Geophys Res 89: 1932–1944

Cardona O (2004) The need for rethinking the concepts of vulnerability and risk from a holistic perspective: a necessary review and criticism for effective risk management. In: Bankoff G, Frerks G, Hilhorst D (eds) Mapping vulnerability: disasters, development and people. Earthscan Publishers, London

Carminati E, Negredo AM, Valera JL et al. (2005) Subduction-related intermediate-depth and deep seismicity in Italy: insights from thermal and rheological modelling. Phys Earth Planet Inter 149: 65–79

Chimera G, Aoudia A, Sarao A et al. (2003) Active tectonics in central Italy: constraints from surface wave tomography and source moment tensor inversion. Phys Earth Planet Inter 138: 241–262

Demetrescu C, Nielsen SB, Ene M et al. (2001) Lithosphere thermal structure and evolution of the Transylvanian depression – insight from new geothermal measurements and modelling results. Phys Earth Planet Inter 126: 249–267

Doglioni C, Harabaglia P, Merlini S et al. (1999) Orogens and slabs vs. their direction of subduction. Earth Sci Rev 45: 167–208

Faccenna C, Piromallo C, Crespo-Blanc A et al. (2004) Lateral slab deformation and the origin of the western Mediterranean arcs. Tectonics. DOI: 10.1029/2002TC001488

Feinberg EA, Shiryaev AN (2007) Quickest detection of drift change for Brownian motion in generalized Bayesian and minimax settings. Stat Decis 24: 445–470

Frepoli A, Amato A (1997) Contemporaneous extension and compression in the northern Apennines from earthquake fault-plane solutions. Geophys J Int 129: 368–388

Fuchs K, Bonjer K, Bock G et al. (1979) The Romanian earthquake of March 4, 1977. II. Aftershocks and migration of seismic activity. Tectonophysics 53: 225–247

Gabrielov AM, Levshina TA, Rotwain IM (1990) Block model of earthquake sequence. Phys Earth Planet Inter 61: 18–28

Gabrielov A, Newman WI (1994) Seismicity modelling and earthquake prediction: a review. In: Newman WI, Gabrielov A, Turcotte DL (eds) Nonlinear dynamics and predictability of geophysical phenomena. Geophysical Monograph 83, IUGG vol. 18. American Geophysical Union, Washington DC

Geller RJ, Jackson DD, Kagan YY et al. (1997) Earthquakes cannot be predicted. Science 275: 1616–1617

Gripp AE, Gordon RG (1990) Current plate velocities relative to the hotspots incorporating the NUVEL-1 global plate motion model. Geophys Res Lett 17(8): 1109–1112

Gusev A, Radulian M, Rizescu M et al. (2002) Source scaling of intermediate-depth Vrancea earthquakes. Geophys J Int 151: 879–889

Houseman G, England P (1996) A lithospheric-thickening model for the Indo–Asian collision. In: Yin A, Harrison TM (eds) The tectonic evolution of Asia. Cambridge University Press, New York

Ismail-Zadeh AT, Aoudia A, Panza GF (2004) Tectonic stress in the central Apennines due to buoyancy of the lithosphere. Doklady Earth Sci/Trans Russ Acad Sci 395(3): 369–372

Ismail-Zadeh AT, Keilis-Borok VI, Soloviev AA (1999) Numerical modelling of earthquake flows in the southeastern Carpathians (Vrancea): effect of a sinking slab. Phys Earth Planet Inter 111: 267–274

Ismail-Zadeh AT, Korotkii AI, Naimark BM et al. (2001) Numerical modelling of three-dimensional viscous flow under gravity and thermal effects. Comp Math Math Phys 41(9): 1331–1345

Ismail-Zadeh AT, Le Mouël JL, Soloviev A et al. (2007a) Numerical modelling of crustal block-and-fault dynamics, earthquakes and slip rates in the Tibet–Himalayan region. Earth Planet Sci Lett 258: 465–485

Ismail-Zadeh AT, Müller B, Schubert G (2005a) Three-dimensional modelling of present-day tectonic stress beneath the earthquake-prone southeastern Carpathians based on integrated analysis of seismic, heat flow, and gravity observations. Phys Earth Planet Inter 149: 81–98

Ismail-Zadeh AT, Müller B, Wenzel F (2005b) Modelling of descending slab evolution beneath the SE-Carpathians: implications for seismicity. In: Wenzel F (ed) Perspectives in modern seismology, LNES, vol. 105. Springer-Verlag, Heidelberg

Ismail-Zadeh AT, Panza GF, Naimark BM (2000) Stress in the descending relic slab beneath the Vrancea region, Romania. Pure Appl Geophys 157: 111–130

Ismail-Zadeh A, Schubert G, Tsepelev I et al. (2008) Thermal evolution and geometry of the descending lithosphere beneath the SE-Carpathians: an insight from the past. Earth Planet Sci Lett 273: 68–79

Ismail-Zadeh AT, Sokolov V, Bonier K (2007b) Geodynamics, seismicity and seismic hazard of the south-eastern Carpathians. Nat Haz 42: 493–514

Ismail-Zadeh AT, Takeuchi K (2007) Preventive disaster management of extreme natural events. Nat Haz 42: 459–467

IUGG GeoRisk Commission (2005) Statement on the greatest earthquake and tsunami of the early XXI century and the need for urgent action to reduce natural disasters in the Indian Ocean region and elsewhere. International Union of Geodesy and Geophysics, Boulder, Colorado (http://www.iugg-georisk.org/c_statement.html) Accessed 7 January 2009

Ivan IA, Enescu BD, Pantea A (1998) Input for seismic hazard assessment using Vrancea source region. Rom J Phys 43: 619–636

Jiricek R (1979) Tectonic development of the Carpathian arc in the Oligocene and Neogene. In: Mahel M (ed) Tectonic profiles through the western Carpathians. State Geological Institute of Dionyz Stur, Romania

Kalnay E (2003) Atmospheric modeling, data assimilation and predictability. Cambridge University Press, Cambridge

Kantorovich L, Keilis-Borok VI, Molchan G (1973) Seismic risk and principles of seismic zoning. In: Keilis-Borok VI (ed) Computational and statistical methods for interpretation of seismic data. Nauka, Moscow

Keilis-Borok V, Davis C, Molchan G et al. (2004) Earthquake prediction and disaster preparednes: interactive algorithms. EOS/Trans. AGU 85(47), Fall Meeting Suppl., Abstract S22B-02.

Keilis-Borok VI, Ismail-Zadeh AT, Kossobokov VG (2001) Non-linear dynamics of the lithosphere and intermediate-term earthquake prediction. Tectonophysics 338 (3–4): 247–259

Keilis-Borok VI, Kossobokov VG (1990) Premonitory activation of earthquake flow: algorithm M8. Phys Earth Planet Inter 61: 73–83

Keilis-Borok VI, Soloviev AA (2003) Nonlinear dynamics of the lithosphere and earthquake prediction. Springer, Heidelberg

Knopoff L (1999) Earthquake prediction is difficult but not impossible. Nature Debates. http://www.nature.com/nature/debates/earthquake. Accessed 10 March 2009

Kossobokov V (2006) Quantitative earthquake prediction on global and regional scales. In: Ismail-Zadeh AT (ed) Recent geodynamics, georisk and sustainable development in the Black Sea to Caspian Sea region. American Institute of Physics Conference Proceedings, vol. 825. Melville, NY

Le Pichon X, Fournier M, Jolivet L (1992) Kinematics, topography, shortening, and extrusion in the India–Eurasia collision. Tectonics 11(6): 1085–1098

Linzer HG (1996) Kinematics of retreating subduction along the Carpathian arc, Romania. Geology 24: 167–170

Locardi E (1993) Dynamics of deep structures in the Tyrrhenian–Apennines area and its relation to neotectonics. Il Quaternario 6: 59–66

Lungu D, Cornea T, Nedelcu C (1999) Hazard assessment and site dependent response for Vrancea earthquakes. In: Wenzel et al. (eds) Vrancea earthquakes: tectonics, hazard and risk mitigation. Kluwer Academic Publishers, Dordrecht

Mandrescu N, Anghel M, Smalbergher V (1988) The Vrancea intermediate-depth earthquakes and the peculiarities of the seismic intensity distribution over the Romanian territory. St Cerc Geofiz Geogr Geofizica 26: 51–57

Mandrescu N, Radulian M (1999) Macroseismic field of the Romanian intermediate-depth earthquakes. In: Wenzel et al. (eds) Vrancea earthquakes: tectonics, hazard and risk mitigation. Kluwer Academic Publishers, Dordrecht

Mantovani E, Albarello D, Babbucci D et al. (2002) Trench-arc-backarc systems in the Mediterranean area: examples of extrusion tectonics. J Virt Expl 8: 131–147
Mantyniemi P, Marza VI, Kijko A et al. (2003) A new probabilistic seismic hazard analysis for the Vrancea (Romania) seismogenic zone. Nat Haz 29: 371–385
Martin M, Wenzel F, the CALIXTO working group (2006) High-resolution teleseismic body wave tomography beneath SE-Romania – II. Imaging of a slab detachment scenario. Geophys J Int 164: 579–595
McKenzie DP (1972) Active tectonics of the Mediterranean region. Geophys J R Astr Soc 30: 109–185
Mele G, Sandvol E (2003) Deep crustal roots beneath the northern Apennines inferred from teleseismic receiver functions. Earth Planet Sci Lett 211: 69–78
Moldovan IA, Enescu BD, Ionescu C (2000) Predicting peak ground horizontal acceleration for Vrancea large earthquakes using attenuation relations for moderate shocks. Rom J Phys 45: 785–800
Montone P, Amato A, Pondrelli S (1999) Active stress map of Italy. J Geophys Res 104: 25595–25610
Munich Re (2005) Annual review of natural catastrophes, topics Geo. Münchener Rückversicherungs-Gesellschaft, Munich, 56 p
Munich Re (2009) Annual reports on natural catastrophes; http://www.munichre.com/en/publications/default.aspx?category=17. Accessed 22 March 2009
Musson RMW (2000) Generalised seismic hazard maps for the Pannonian basin using probabilistic methods. Pure Appl Geophys 157: 147–169
Naimark BM, Ismail-Zadeh AT, Jacoby WR (1998) Numerical approach to problems of gravitational instability of geostructures with advected material boundaries. Geophys J Int 134: 473–483
Negredo AM, Barba S, Carminati E et al. (1999) Contribution of numeric dynamic modelling to the understanding of the seismotectonic regime of the northern Apennines. Tectonophysics 315: 15–30
Nelson KD, Zhao W, Brown LD et al. (1996) Partially molten middle crust beneath southern Tibet: synthesis of project INDEPTH results. Science 274: 1684–1688
Newman WI, Turcotte DL, Gabrielov AM (1995) Log-periodic behaviour of a hierarchical failure model with application to precursory seismic activation. Phys Rev E 52: 4827–4835
Oncescu MC (1984) Deep structure of the Vrancea region, Romania, inferred from simultaneous inversion for hypocenters and 3-D velocity structure. Ann Geophys 2: 23–28
Oncescu MC, Bonjer KP (1997) A note on the depth recurrence and strain release of large Vrancea earthquakes. Tectonophysics 272: 291–302
Oncescu MC, Bonjer KP, Rizescu M (1999) Weak and strong ground motion of intermediate depth earthquakes from the Vrancea region. In: Wenzel F et al. (eds) Vrancea earthquakes: tectonics, hazard and risk mitigation. Kluwer Academic Publishers, Dordrecht
Paterson J (2003) Science for risk reduction and sustainable development: the role of law. In: Beer T, Ismail-Zadeh A (eds) Risk science and sustainability. Kluwer Academic Publishers, Dordrecht
Radu C (1991) Strong earthquakes occurred on the Romanian territory in the period 1901–1990 (in Romanian). Vitralii 3: 12–13
Radulian M, Vaccari F, Mandrescu N et al. (2000) Seismic hazard of Romania: deterministic approach. Pure Appl Geophys 157: 221–247
Royden LH, Burchfiel BC, King RW et al. (1997) Surface deformation and lower crustal flow in eastern Tibet. Science 276: 788–790
Rundquist DV, Vladova GL, Rozhkova VV (1998) Regularities of migration of the seismic activity along island-arcs. Doklady Earth Sci/Transactions Russ Acad Sci 360(2): 263–266
Sartori R (2003) The Tyrrhenian backarc basin and subduction of the Ionian lithosphere. Episodes 26: 217–221
Selvaggi G, Amato A (1992) Subcrustal earthquakes in the northern Apennines (Italy): evidence for a still active subduction? Geophys Res Lett 19: 2127–2130
Shaw BE, Carlson JM, Langer JS (1992) Patterns of seismic activity preceding large earthquakes. J Geophys Res 97: 479–487
Soloviev AA, Ismail-Zadeh AT (2003) Models of dynamics of block-and-fault systems. In: Keilis-Borok VI, Soloviev AA (eds) Nonlinear dynamics of the lithosphere and earthquake prediction. Springer, Heidelberg
Sperner B, Lorenz F, Bonjer K et al. (2001) Slab break-off – abrupt cut or gradual detachment? New insights from the Vrancea region (SE Carpathians, Romania). Terra Nova 13: 172–179
Tannehill R (1947) Drought: its causes and effects. Princeton University Press, Princeton
Tapponnier P, Zhiqin X, Meyer B et al. (2001) Asymmetric, stepwise rise and growth of the Tibet Plateau. Science 294: 1671–1677
Turcotte DL (1997) Fractals and chaos in geology and geophysics, 2nd edn. Cambridge University Press, Cambridge
UN/ISDR Platform for the Promotion of Early Warning (2008) Evaluation and strengthening of early warning systems in countries affected by the 26 December 2004 tsunami. Final Report of the UN Flash Appeal Project. UN/ISDR, Bonn, Germany
Vergnes J, Wittlinger G, Hui Q et al. (2002) Seismic evidence for stepwise thickening of the crust across the NE Tibetan Plateau. Earth Planet Sci Lett 203: 25–33
Widiyantoro S, van der Hilst R (1996) Structure and evolution of lithospheric slab beneath the Sunda arc, Indonesia. Science 271: 1566–1570
Wittlinger G, Tapponnier P, Poupinet G (1998) Tomographic evidence for localized lithospheric shear along the Altyn Tagh fault. Science 282: 74–76
Wittlinger G, Vergnes J, Tapponnier P (2004) Teleseismic imaging of subducting lithosphere and Moho offsets beneath western Tibet. Earth Planet Sci Lett 221: 117–130
Working Group on California Earthquake Probabilities (1999) Earthquake probabilities in the San Francisco Bay region: 2000–2030 – a summary of findings. U.S. Geological Survey Open-file Report 99–517
Wortel MJR, Spakman W (2000) Subduction and slab detachment in the Mediterranean–Carpathian region. Science 290: 1910–1917

Hazards in the Coastal Zones Related to Groundwater–Seawater Interaction Processes

Y. A. Kontar, Yu.R. Ozorovich, and A.T. Salokhiddinnov

Abstract Hazards related to groundwater–seawater (GW–SW) interactions in the coastal zone have been underestimated (Kontar 2008). This paper considers two case studies: one in Central Asia (Aral Sea region) and one in the Indian Ocean (December 2004 tsunami) that are important examples whose method of treatment provide insight into future feasibility studies of hazards in the coastal zone related to GW–SW interaction processes. The Aral Sea region is known as an ecological disaster zone. To provide reasonable living conditions for the coastal zone population, it is necessary to drastically improve the quality of the water used for human needs by developing a source of safe and sustainable groundwater input to the Aral Sea region. In the Indian Ocean tsunami waves, which affected thousands of kilometers of coastal zone in SE Asia, caused an ecological disaster by the large inflow of salt seawater into coastal aquifers. The tsunami has created an accelerating process of salt-water intrusion and fresh-water contamination in affected regions that now require drastic remediation measures. Analytical approaches have been developed for analysis of coastal water balance and temporal evolution of water basins and coastal aquifers after hazardous events.

Keywords Hazards, Coastal zone · Groundwater–seawater interactions · Salt-water intrusion · Tsunami · Submarine groundwater discharge

Y.A. Kontar (✉)
Institute of Natural Resource Sustainability, University of Illinois at Urbana-Champaign, Champaign, Illinois 61820, USA
e-mail: kontar@isgs.illinois.edu

Introduction

Hydrological and geophysical risk analysis related to groundwater–seawater interactions in the coastal zone have been underestimated either due to economic limitations or underestimation of their significance. This is particularly true for tsunami-created salt-water intrusion into coastal aquifers, even though most tsunami hazard assessments have in the past relied on scenarios or deterministic type models (Geist and Parsons 2006). Furthermore increasing mineralization of potable water because of intensive water diversions and also the abundance of highly toxic pollutants (mainly pesticides) in water, air and food contribute to the deterioration of the coastal population's health (Glantz and Zonn 2005).

This paper considers two case studies that are important examples whose method of treatment provide insight into future feasibility studies of hazards in the coastal zone related to GW–SW interaction processes: the Aral Sea, and the Indian Ocean tsunami.

The territory of the Aral Sea Region in Central Asia has been described as an ecological disaster coastal zone (Zavialov 2005). It is now obvious that, in order to provide reasonable living conditions to the coastal population, it is necessary to drastically improve the quality of their water. Due to intensive pollution by industrial wastes and by drainage waters from irrigated fields, the Syr Darya and Amu Darya rivers can no longer be considered as a source of safe and sustainable water supply. The population's water supply must be achieved through a more comprehensive use of fresh and even sub-saline groundwater resources from coastal aquifers.

T. Beer (ed.), *Geophysical Hazards*, International Year of Planet Earth,
DOI 10.1007/978-90-481-3236-2_12, © Springer Science+Business Media B.V. 2010

The December 2004 tsunami in the Indian Ocean caused an ecological disaster affecting thousands of kilometers of coastal zone in SE Asia (Ismail-Zadeh et al. 2005; Lay et al. 2005). Many coastal wetlands were affected in the short term by the large inflow of salt seawater and littoral sediment deposited during the tsunami, and in the longer-term by changes in their hydrogeology caused by changes to coastlines and damage to sea-defenses. Many water quality and associated problems were generated by the tsunami. The tsunami created an accelerating process of salt-water intrusion and fresh-water contamination in affected regions that now requires drastic remediation measures. According to the International Commission on groundwater–seawater Interaction (CGSI) these measures have to be economically feasible, environmentally sound and socially acceptable.

Case Study One – Aral Sea

Analytical Approaches for Analysis of the Aral Sea Region Water Balance Under Hazardous Events

Changed ambient conditions in the Aral Sea Drainage Basin (ASDB) in Central Asia have led to drastically decreased river discharges into the Aral Sea. This decrease has led to ongoing dessication of the Aral Sea and to particularly adverse environmental effects, in terms of both the affected number of people and degree of environmental degradation in the ASDB (Shibuo et al. 2005; 2006). Because of intensive water diversions from the Amu Darya and Syr Darya rivers for irrigation of newly developed lands in Central Asia, the Aral Sea has shrunk to one-third of its surface area and divided into two parts – the Large and the Small Aral Seas (Fig. 1), fed separately by the Amu Darya and Syr Darya rivers, respectively (Bortnik and Chistyaeva 1990; Glantz and Zonn 2005; Kontar et al. 2000a, b, 2003; Salokhiddinnov and Khakimov 2004; Zavialov 2005).

A distributed basin-scale hydrological balance modeling approach has been used to estimate the relative influence of agricultural irrigation and climate change on observed decreases of river discharges in the ASDB (Shibuo, et al. 2005). Results show that water losses through evapotranspiration increased as a result of higher temperatures in the basin after 1950. However, these increases in evapotranspiration loss due to rising temperatures alone are smaller than the water gains caused by increased precipitation in the ASDB over the same time period. Climatic changes have therefore not contributed to the observed drying of the rivers in the basin, at least not so far. By contrast, the evapotranspiration loss increases as a result of the expanded agricultural irrigation in the area can fully explain the decreased river discharges and the present water scarcity in the ASDB. It was further shown (Shibuo et al. 2006), that the largest increase (1.85°C) in seasonal average temperature in the basin has occurred in the winter, whereas the smallest increase (0.69°C) has occurred in the summer. This result is consistent with a surface temperature cooling effect of intense irrigation in the summer, which should have increased since the 1950s due to the evapotranspiration increase implied by the major irrigation expansion in the ASDB.

The dramatic shrinkage of the Aral Sea that has occurred since 1960 has also affected groundwater resources in the region, and this effect needs to be clarified and quantified because groundwater is essential for meeting freshwater demands in the area (Jarsjö and Destouni 2004; Jarsjö et al. 2006).

We have been working on compilation, generalization and analysis of available hydrological and hydrogeological materials, including data from boreholes and active water collections and sampling. Modeling of groundwater flow into the Aral Sea provides an assessment of the groundwater role in the water and salt balance of the Aral Sea. Also we have been working on quantative assessment of regional groundwater inflow through elucidation of the main regularities and peculiarities of groundwater–seawater interactions, groundwater flow formation and its distribution in different natural-climatic, hydrogeological and anthropogenic conditions, to estimate and predict the groundwater flow and its spread in the Aral Sea region. Estimates have been conducted to predict groundwater spreading as a main part of its replenishable resources under the conditions of developing desertification in the Aral Sea region and assessment of rational groundwater use perspectives under minimum adverse effect of groundwater intake on the environment in the Aral Sea basin.

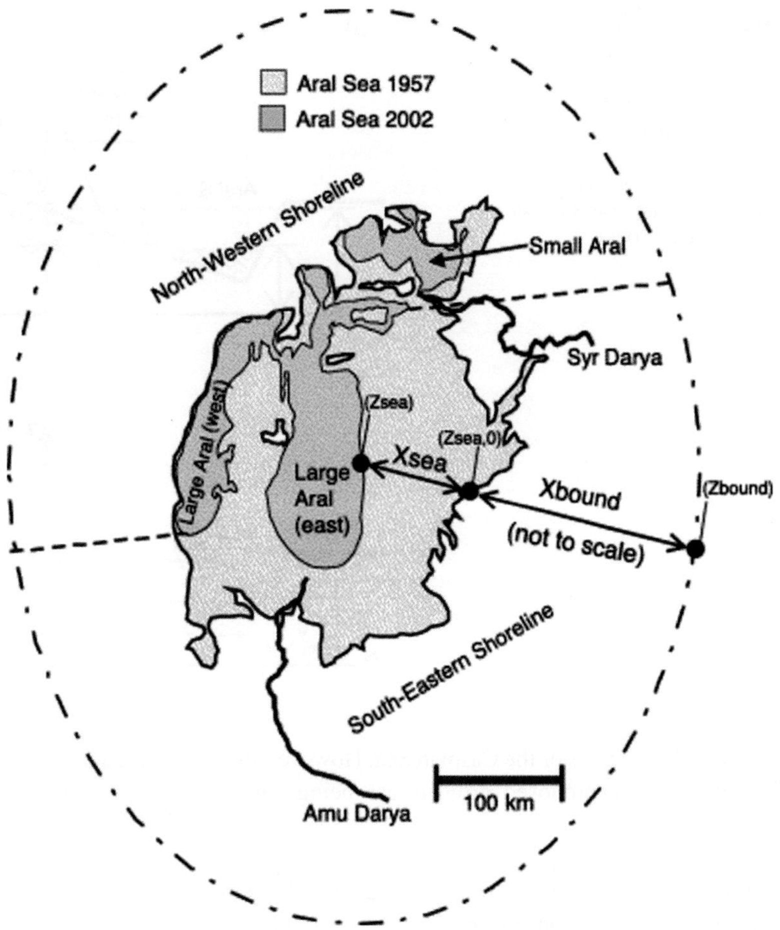

Fig. 1 Schematic view of the Aral Sea according to a satellite image (DAAC GSFC). Illustration of the distances: X_{sea} between the original (pre-1960) shoreline at elevation $z_{sea,0}$ and the shoreline of 2002 at elevation z_{sea}; and X_{bound} between the original shoreline and an up gradient location at which the groundwater table is no longer influenced by the lowering of the sea surface (dashed-dotted line; ground elevation z_{bound}). The dashed borderlines illustrate the division between the southeastern shoreline and the northwestern shoreline. The basic features of the shoreline were obtained from a NOAA satellite image of 25th September 2002 (Jarsjö and Destouni 2004)

Assessment of Impacts of Desertification on the Groundwater Discharge into the Aral Sea

The current runoff delivered by the Syr Darya seems to be enough to keep the Small Aral Sea in a quasi-stationary state. However, the Large Aral Sea level is still declining and soon may reach a critical state, after which it would be impossible to take any reasonable action for preserving it as an ecologically viable system. The Large Aral Sea surface level has dropped by more than 23 m relative to the value of 1961 (Zavialov 2005).

This depends not only on the water volume delivered by the Amu Darya to the sea, but also on the peculiarities of the Aral Sea's relief. Because the Large Sea, in its turn, has already divided into two, nearly-isolated, parts – eastern and western, water comes to the western part only through the eastern. Hereafter for the sake of convenience they will be called the East and West Seas, respectively. Further water level decline will lead to complete isolation of the West Sea from the East Sea, as a result the former, which now has more water than the latter, may eventually lose almost all of this water due to evaporation.

The lowering of the Large Aral Sea level by more than 23 m since 1961 influences the slope of the groundwater table and the groundwater flow conditions in the ASDB, as shown in Fig. 2. The horizontal sea shore retreat X_{sea} is then one of the parameters that determines and limits the magnitude of this influence (Jarsjö and Destouni 2004).

To resolve different aspects (hydrogeological, economic, ecological, political, etc.) of the Aral Sea crisis, at least, 300 projects have been proposed (Micklin 2004; Mirabdullaev et al. 2004), including ones that suggest restoring the Sea at the expense of Russian

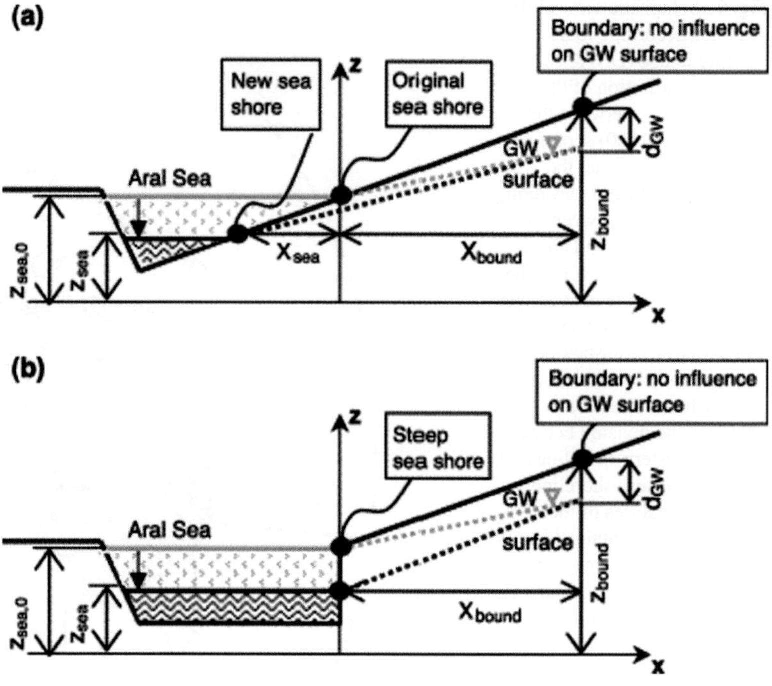

Fig. 2 Effects on regional groundwater surface slope of the changed hydrologic conditions caused by the on-going Large Aral Sea level lowering, for (**a**) shallow shore bathymetry, and (**b**) steep shore bathymetry (Jarsjö and Destouni 2004)

north-flowing rivers or the Caspian Sea. However, only a few of these proposed projects are being carried out. These minimal responses to the Aral Sea situation explicitly or implicitly assume that the Aral Sea can not be preserved. Activities now are geared toward diminishing the possible consequences of its shrinkage.

Meanwhile, unless the Small Aral Sea is preserved at a reasonable level, preferably close to the current level, the effects of these attempts will apparently be minimal.

This case study presents a water-level equation-based approach and its application to the analysis of the water balance of the Large Aral Sea and its temporal changes due to natural processes and anthropogenic impacts. The role of the East and West Seas in the water balance of the Large Sea is also discussed.

Development of Models for Quantative Assessment of the Groundwater Role in Water and Salt Balance of the Aral Sea

The Large Aral Sea now is very close to being divided into two independent seas, which will be facilitated by annual fluctuation of the Amu Darya water flux together with the effect caused by the difference between water level elevation and falling rates. Changes of water level, h, of any basin can be described by the following equation

$$\frac{dh}{dt} = \frac{a}{S(h)} - b, \qquad (1)$$

Where a is the resulting water discharge rate from all sources (rivers, ground water, etc) except precipitation; b is the evaporation rate from the current basin surface area, S, reduced by precipitation to this surface. The first term on the right side of Eq. (1) describes the elevation rate of the water level, and the second one – its rate of fall; in some time periods; of course, precipitation may dominate evaporation, then $b < 0$. All parameters in Eq. (1) are time dependent. The parameters a and b are subject to variations with respect to the seasons and time of day. These variations are complicated due to the irregularity of precipitation, as well as due to anthropogenic impacts. The surface area of the basin implicitly depends on time through its dependence on the water level, and this dependence may be also complicated.

Thus, the differential Eq. (1), in spite of its simplicity at first sight, does not have a closed form solution with respect to h. Of course, such solutions can be found for some specially chosen model systems or this equation can be integrated numerically by a step by step approach for a given time dependence of a and b and $S(h)$. However, in this work we will be interested only in approximate analytical solutions of Eq. (1), which are applicable for any system and, therefore, are much more valuable then specific solutions.

Note that an equation similar to Eq. (1) can be written for water volume changes too. Such an equation is often used in approaches to water balance estimations, but its solutions, to our best knowledge, have not been obtained. Water volume change is itself a derivable quantity that can be estimated using water level, surface changes, and hypsometry.

In contrast, the left side of Eq. (1) is a directly and reliably measurable quantity using modern altimetry techniques. Therefore, this equation and its solutions enable one to extract information on the state and dynamics of a basin, which is not possible in the case of an equation for water volume changes. In particular, actual values of the parameters a and b at a given time also can be recovered by inverse approaches, mapping theoretical curves or solutions of Eq. (1) onto actual water level changes. In most cases, experimental estimates with acceptable accuracy are not possible for every water budget constituent. Furthermore, such estimates may not always derive from the same time period (or this period may be too long) to use them safely for the period of interest.

In order to obtain solutions of Eq. (1), we introduce the following approximations. First, the period of time under consideration can be divided into a number of time intervals (days, months, seasons, years) and values of the variables a and b can be considered to be equal to their mean values in the corresponding intervals. Such an approximation, the accuracy of which can be systematically improved by decreasing of the length of the time intervals, is reasonable given the uncertainty in experimental values of the water budget constituents and their possible irregular temporal changes. Because of this, numerical integration, mentioned above, will not have any advantage in accuracy.

Second, the real ASDB can be replaced by a convenient mathematical equivalent, elementary volume, such as a cone, a paraboloid of revolution, etc, that allows one to obtain simple solutions of Eq. (1), as well as water volume changes using only water level changes. Clearly, a real basin can not be replaced by a single elementary volume for the entire range of possible water level changes. Therefore, the basin must be divided into parts, each of which can be defined by volume with appropriate parameters (Fig. 3).

For an elementary cone that we denote as Cup-2, the surface area-water level relation is

$$S = kh^2, \qquad (2)$$

Where $k = \pi/\text{tg}^2\,\alpha$, α is the angle between the base and directrice of the cone. For a hyperbolic paraboloid of revolution, Cup-1, this relation reads

$$S = kh_c h, \qquad (3)$$

where $k = 2\pi$, and h_c^{-1} is the curvature of the corresponding parabola. For the sake of completeness we must include in this family, at least, a volume resembling a swimming pool with $S = \text{const}$, Cup-0.

These three volumes (in the sequence of Cup-0, Cup-1, Cup-2) mathematically correspond to the first three terms of a polynomial series in h approximating S. If one retains two terms in this series, one gets (3) by choosing an appropriate reference point for h; this point marks the bottom of the volume described by (3). A three-term series can be reduced to (2) in a particular case. The general case and series with more terms lead to solutions of ever increasing complexity.

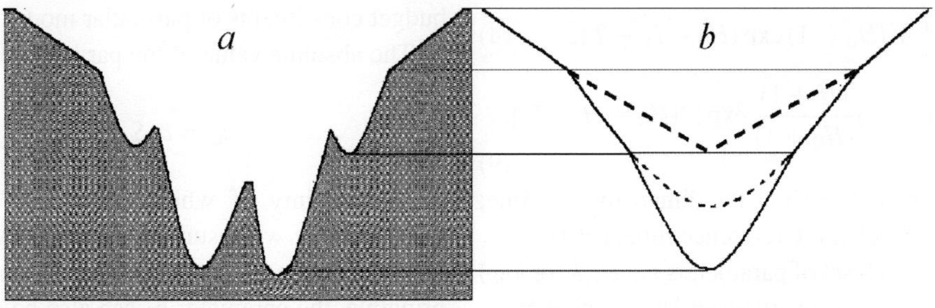

Fig. 3 A realistic seabed relief (**a**) and its representation (**b**) by a set of elementary volumes. Here, it is assumed that the two bottom depressions in (**a**) can share water in any water level

We will not be interested in such solutions, because there is no need to obtain a single solution for the whole range of possible water level changes in a particular basin, which would be rather inconvenient to use and non-transferable from a given basin to others. Instead, we will divide a given basin into limited intervals based on its depth, where simple models like Cup-1 and Cup-2 will be satisfactory. As one can see below in the case of the Large Aral Sea, these intervals are few and not too narrow. In fact, they will be appropriate during long time periods. It is probable that other seas and lakes also need elementary volumes with appropriate parameters (k and kh_c) for adequate mathematical description.

Eq. (1) can be rewritten as follows

$$\frac{1}{b_0}\frac{dh}{dt} = \frac{a_0}{b_0 S} - 1 \quad (4)$$

or, accounting for the surface area-water level relations (2) and (3), as

$$\tau\frac{dH}{dt} = H^{-n} - 1. \quad (5)$$

Here H is the dimensionless water level

$$H = \frac{1}{b_0\tau}h, \quad (6)$$

Where τ is the time constant of the water basin

$$\tau_{\text{Cup}-1} = \frac{a_0}{kh_c b_0^2} \quad \text{and} \quad \tau_{\text{Cup}-2} = \sqrt{\frac{a_0}{kb_0^3}} \quad (7)$$

for models Cup-1 ($n = 1$) and Cup-2 ($n = 2$), respectively; a_0 and b_0 are the mean values of parameters a and b for the given time interval.

Solutions of Eq. (5) for Cup-1 and Cup-2 are

$$H = 1 + (H_0 - 1)\exp(H_0 - H - T), \quad (8)$$

$$H = 1 + (H_0 - 1)\frac{(H+1)}{(H_0+1)}\exp[2(H_0 - H - T)], \quad (9)$$

respectively, where $T = t/\tau$ is the dimensionless time; H_0 is the water level at the reference time, $t = 0$.

Note that for each set of parameters a_0, b_0, k (or kh_c) there is a particular stationary water level, where $H_0 = 1$, that is $a_0 = b_0 S$. If these parameters have changed due to climatic and anthropogenic factors, then $H_0 \neq 1$, and the water level will tend to a new stationary state corresponding to the new values of the parameters.

It should be stressed that the Eqs. (8) and (9) do not explicitly depend on the parameters a_0, b_0, k (or kh_c). Figure 4 presents typical universal curves for a number of values of H_0, which are applicable for any water basin. As seen from this figure, the time for reaching the stationary state depends on H_0, τ and the type of volume and approximately equals to 5τ and 3τ for Cup-1 and Cup-2, respectively.

Solution of Eq. (1) for Cup-0 ($S = S_0 = \text{const}$) is trivial:

$$h = h_0 + (H_0 - 1)b_0 t; \quad H_0 = a_0/b_0 S_0. \quad (10)$$

Here, if $H_0 \neq 1$, water level persistently rises ($H_0 > 1$) or falls unless all water volume in the basin disappears ($H_0 < 1$).

Note that the models Cup-1 and Cup-2 (see Fig. 4), unlike Cup-0, are asymmetric with respect to water rise and fall. This means that water level moves from one state to another and back in different time intervals; a water level rise takes more time than a fall of the same extent.

Water volume change is

$$\Delta V(t) = \frac{kh_c}{2}[h^2(t) - h^2(0)] \text{ for Cup} - 1, \quad (11)$$

$$\Delta V(t) = \frac{k}{3}[h^3(t) - h^3(0)] \text{ for Cup} - 2. \quad (12)$$

Water deficit or excess for a given state of the basin can be obtained using known values of its surface and rate of water level variation as follows

$$\Delta a_0 = S(dh/dt + b_0) - Sb_0 = S(dh/dt), \quad (13)$$

which does not assume particular values of the water budget constituents or particular model of the basin.

The absolute value of the parameter a_0 is clearly

$$a_0 = b_0 S + \Delta a_0, \quad (14)$$

the uncertainty of which is defined by that of b_0 (Everywhere, we assume that water level and surface changes can be obtained with the desired accuracy). In principle, the parameter b_0 can also be recovered from water level changes. This needs some modification of the approach, using more detailed and precise data on

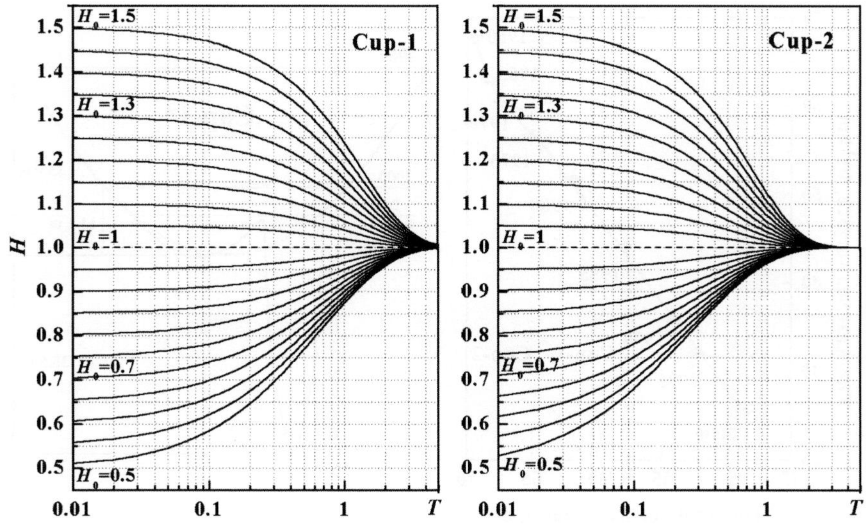

Fig. 4 Typical universal curves for predicting of water level temporal evolution for the water basin models of Cup-1 and Cup-2

water level variations, which will be presented in a further paper.

Based on the current state and dynamics of the basin, the expected stationary water level, h_{st}, can be found from the following equation

$$h_{st} = h\left(1 + \frac{1}{b_0}\frac{dh}{dt}\right)^{1/n}. \quad (15)$$

The basin's time constant is related to h_{st}, $\tau = h_{st}/b_0$, and amounts to many years for realistic lakes and seas, because their level is much higher than the height of the water column evaporated during one year. However, from the fact that the evaporation rate varies not only during year, but also during the same day, one can conclude that water level never takes a truly stationary value, changing always around some value averaged over a long time period. Hereafter, we mean such a quasi-stationary state, when we talk about a stationary state.

Assessment, Analysis and Prediction of the Aral Sea State

Figure 5 depicts the dependence of the Aral Sea surface on its water level as measured with respect to the deepest point of the Sea, which is 16 m below the absolute level defined with respect to the Baltic Sea. Two time periods, before 1975 and after 1975, can be distinguished. In the first period, the Aral Sea follows the model (Cup-0)+(Cup-2):

$$S = 56.7 + 0.578(h - 65)^2, \quad (16)$$

but in the second period – the model Cup-1:

$$S = 1.956(h - 35.52), \quad (17)$$

where S is measured in thousands of square kilometres.

The water level of the Large Sea now is 13.1 m above the bottom of the volume described by Eq. (17). This level does not coincide with the absolute water level, 48.61 m. There are, at least, another two volumes of the sea corresponding to its eastern and western parts, respectively.

The estimated total water volume of the Aral Sea is

$$V = 1.956 \int_0^{h-35.52} h\,dh = 0.978(h - 35.52)^2, \text{ km}^3. \quad (18)$$

Here the formula (18) is used for the period before 1975, too (Cup-1* thus describes volumes for two periods); the calculations using formula (18) give practically the same results as those using formula (16) and (17) separately. The results for 1960 and 2000 are given in Table 1.

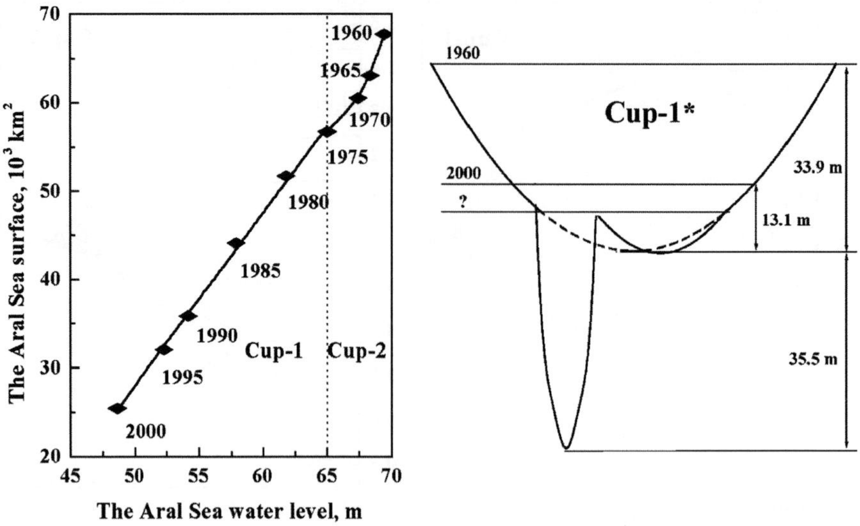

Fig. 5 The Aral Sea surface versus its water level (*on the left*) with respect to the deepest seabed point and the Aral Sea model (*on the right*) based on elementary volumes (Cup-1* stands for Cup-1 and Cup-2; see text). Two bottom volumes correspond to the western (deep volume *on the left*) and eastern (shallow volume *on the right*) parts of the Large Aral Sea (Fig. 1).

Table 1 Total water volume in the Aral Sea (km^3)

Year	Calculated water volume, using formula (18)	Known volume from published data (Zavialov 2005)	Difference
1960	1115.84	1108	+7.84
2000*	147.08	152.8	–5.72

* Without the small Aral Sea

Figure 6 shows water level variations in the Aral Sea from 1912 to the present, which can be approximated by the formula

$$h = 47.61 + \frac{21.44}{1+\eta}, \mathrm{m} \qquad (19)$$

where $\eta = \exp[(t–1984.81)/6.61]$, and t stands for year.

The sharp water level decline after 1960 is known to relate to the beginning of new land development and intense diversion of water resources from the Amu Darya and Syr Darya.

Substituting formulae (17) and (19) in formula (13) for water deficit, we find Δa_0 is

$$\Delta a_0 = -\left[76.70 + \frac{136.02}{1+\eta}\right]\frac{\eta}{(1+\eta)^2}, \ \mathrm{km}^3/\mathrm{year}.$$

According to formula (20), water deficit for 1985 was about 36 km^3/year, but it was only 7.6 km^3/year for 2000 (Fig. 7, dashed line).

Figure 7 also depicts actual water deficit (solid line) calculated by the formula (13) using actual water level variations which are smoothed by the formula (19) that reflect annual fluctuations of water budget constituents, primarily inflow from the Amu Darya and Syr Darya.

From observations over long time periods, the mean values of evaporation rate and precipitation for the Aral Sea have been estimated as about 1 m/year and 0.13 m/year, respectively, i.e. $b_0 = 0.87$ m/year. The surfaces of the West and East Aral Seas now are about 6×10^3 and 15.8×10^3 km^2 respectively. This means that in order to preserve either of these seas in their current state the parameter a must be 5.2 and 13.4 km^3/year, respectively. Preservation of both will require 18.6 km^3/year. Our estimates for 1991–2000 using formula (14) indicate that a ranges from 6 to 18 km^3/year, including climatic fluctuations. Therefore, under the current water diversion regime both seas cannot be preserved at the present level.

Let us consider the role of fluctuations of Amu Darya flow on the Aral Sea's behaviour. For the sake of simplicity suppose that the water level is

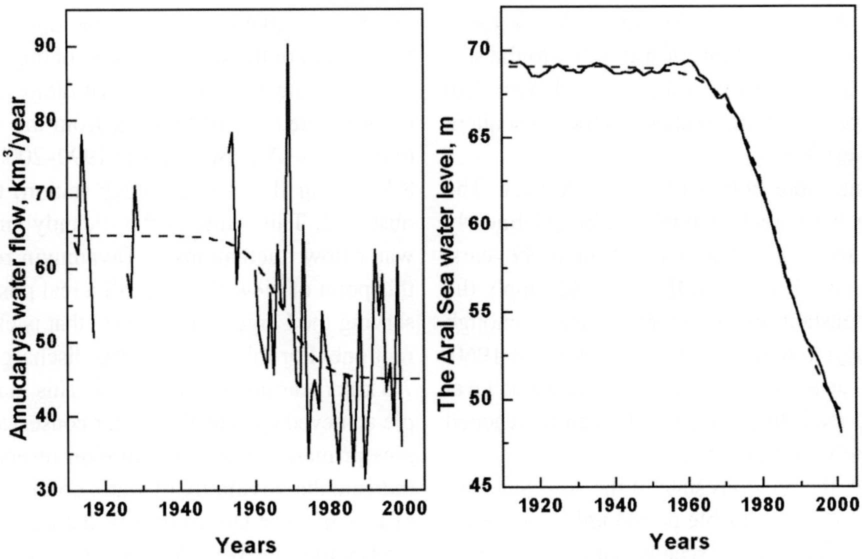

Fig. 6 (*Left*) Amu Darya water flow (*solid line*) at the Tuyamuyun reservoir and its trend (*dashed line*); (*Right*) Aral Sea water level change (*solid line*) and its trend (*dashed line*)

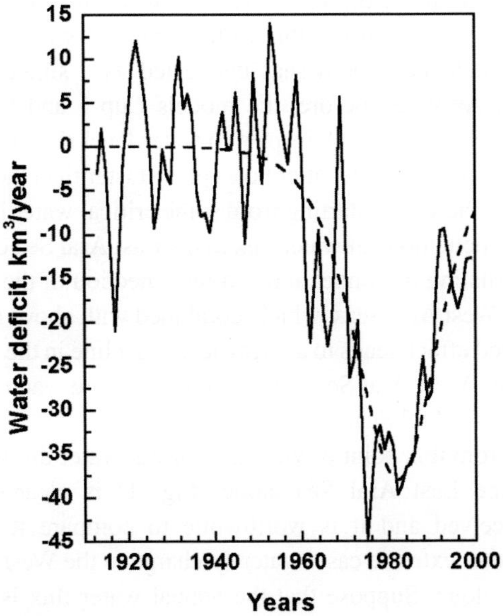

Fig. 7 Water deficit for the Aral Sea (*solid line*) and its trend (*dashed line*)

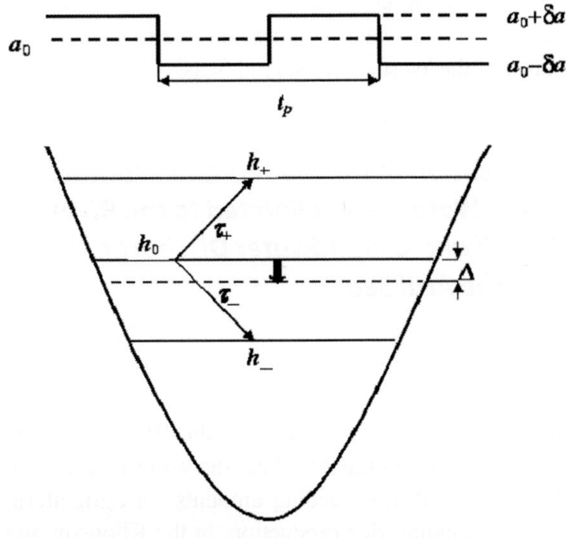

Fig. 8 Schematic description of effect of difference in water rise and fall rates (see text)

quasi-stationary, b_0 is constant, but a changes stepwise around the mean value a_0 with an amplitude of δa, i.e. it takes the value $a_0+\delta a$ in the first half-period $(0, t_p/2)$ and $a_0-\delta a$ in the second half-period $(t_p/2, t_p)$. If t_p were $t_p>5\tau_\pm$ (here we suppose model Cup-1), where$\tau_\pm = (a_0 \pm \delta a)/(b_0^2 k h_c)$, the water level could travel between two extreme values of h_+ and h_- ($h_\pm = (a_0 \pm \delta a)/(b_0 k h_c)$) (Fig. 8).

However, in reality $t_p \ll 5\tau_\pm$ (see Fig. 8 and the last paragraph of section II), hence, the water level never reaches these extreme values, but it always comes closer to h_- than to h_+ , because $\tau_- < \tau_+$. Consequently, the mean value of water level (Fig. 8, dashed line) is always lower than the arithmetic mean value

$h_0 = (h_+ + h_-)/2$ and this deviation, Δ, increases with increasing of fluctuations of water flux to the sea. This may accelerate division of the East and West Aral Seas into independent water bodies and the consequent death of the West Sea.

There is one more comment in this respect. The water level in the Cup-1* model would fall linearly from the level for 1960 down to its bottom in 39 years, if $a = 0$ during all this period. If one could supply the basin with a constant amount of water that is enough ($a = b_0 k h_c h_{1960}$) to reach the stationary state of 1960, it would take, at least, 200 years, though water level values that are close to those for 1960 can be reached during noticeably shorter times.

In any event, the previous state of the Aral Sea cannot be restored in a reasonable period unless it is supplied with water inflow that exceeds that for 1960. This is clearly impossible without water diversion from the northern rivers. Nevertheless, it is urgently important to discuss possible ways for keeping the sea in a state, close to the current one, as well as ways for preventing tremendous water volume loss due to the unequal status of the West and East Aral Seas.

Application of the Models and Estimate of the Water Fluxes Related to the River Renoff and Groundwater Discharge into the Aral Sea

It should be noted that the annual water diversion from the Amu Darya was almost stabilized above Tuyamuyin reservoir in the middle of the 1980s. Between this reservoir and the Aral Sea, the water of the Amu Darya is used in exceeding amounts for agricultural needs, including rice production, in the Khorezm and Karakalpakstan regions of the Republic of Uzbekistan. Therefore, the current water deficit of the sea is almost entirely caused by water diversion in these regions.

Formula (19) describes correctly the tendency of water level changes only before 1990. According to it, however, the stabilization of the Large Aral Sea is expected at the level of 47.61 m and only after 2020, but the water level of 48.11 m, expected in 2009, already had been reached in 2001. This deviation of water level prediction from the actual water level, noted above, perhaps indicates the beginning of another noticeable increase in water diversion after 1990, most likely due to restoration of a system of lakes south of the southern shore of the Large Aral Sea.

According to data from hydrological gauging stations situated 50–100 km far from the southern shore of the Large Aral Sea, during 1990–2000 no more than 8 km^3/year fluxes (Samanbay station) to the sea was observed. This value is significantly smaller than the water flow fluctuations at Tuyamuyin reservoir. From this point of view, there is still a real possibility of preserving the Large Sea at a level that is close to the current one. For this an additional discharge not less than 7.6 km^3/year must be provided. This may be in principle achieved by effective water conservation technologies primarily below Tuyamuyin reservoir. Estimates indicate the possibility of water conservation in excess of 10 km^3/year. Unfortunately, there is also a great possibility that the West Aral Sea will be separated completely from the East Aral Sea before any real actions on water conservation start.

The situation will became worse due to the annual fluctuations of the Amu Darya flow, even though the mean value of water flux to the Aral Sea will be sufficient to keep both seas connected as a single system. As stated before, the models Cup-1 and Cup-2 are asymmetric with respect to water level rise and fall (Figs. 4 and 8); the latter takes place faster than the former. Therefore, starting from some critical water level, the fluctuations of water flux to the East Aral Sea cause alternating disconnection and reconnection of the East and West Aral Seas, which, combined with above mentioned effect, leads to an irrevocable decline in the level of the West Aral Sea, which now has more water than the East Aral Sea.

From this point of view the current water discharge to the East Aral Sea alone (Fig. 1) is clearly ill-conceived and it is worthwhile to compare it with another extreme case: water discharge to the West Aral Sea alone. Suppose that the annual water flux is sufficient to compensate evaporation only from a surface area that is equal to that of the East Aral Sea. In the case of current water discharge no more water then comes to the West Aral Sea, resulting in eventual evaporation of more than half the water volume of the Large Aral Sea. In the alternative water discharge case, the water of the West Aral Sea will be preserved by only about 40% of the total water flux; the East Aral Sea will receive the remainder and will lose no more half of its water. Thus, more than 25% water of the Aral Sea can be preserved.

According to our estimates and data from Samanbay station, water flux to the Large Sea lies between two values, 5.2 and 13.4 km^3/year, corresponding to preservation of either the West or East Aral Seas. Consequently, is not possible to preserve the East Aral Sea at the current water diversion level from the Amu Darya and the East Aral Sea level will be significantly lowered in any case. The current water flux is sufficient not only to preserve the West Aral Sea, but also to make it circulate, enabling one to manage its salinity. In this case the East Aral Sea will act as a salt sink and this is perhaps the best role it can play in the future.

In summary for this case study, we have developed a convenient analytical approach for analysis of the water balance of the ASDB and its temporal changes, which is based on the water level equation and its solutions obtained by replacing the real basin by a set of elementary volumes with specified parameters. This approach has a number of advantages compared to the usual approaches of water balance analysis. In particular, it enables one to avoid use of unreliable data.

The approach has been applied to the analysis of the current state of the Aral Sea and its possible behaviour under different scenarios. The numerical results obtained and comparative analysis of the role of the East and West Aral Seas in the Large Aral Sea water balance show that the water discharge to the East Aral Sea alone will accelerate death of the Aral Sea at the current water diversion level below Tuyamuyin reservoir. The redirection of the Amu Darya water flux from the East Aral Sea to West Aral Sea allows one to prevent more than 25% of the loss of current water volume of the Large Aral Sea, which can not be achieved by other means.

Preservation of the Large Aral Sea at a level that is close to the current one requires an additional discharge of not less than 7.6 km^3/year, which may be in principle achieved by effective water conservation technologies, first of all, below Tuyamuyin reservoir. However, the Large Aral Sea is now very close to being divided into two independent seas, which will be facilitated by the annual fluctuation of the Amu Darya water flux together with difference between water level rise and fall rates. Therefore, the preservation of the Aral Sea as an ecological system with reasonable water volume can not achieved without providing independent water inflow (not through the eastern part) to the western part of the Large Aral Sea. How this can be achieved technically is another issue that must be discussed separately, but it is not acceptable to lose more than half of the current water volume of the Large Aral Sea just because of ill-advised water discharge to the East Aral Sea alone.

Case Study Two – Indian Ocean

Analytical Approaches for Analysis of Coastal Aquafers After Hazardous Events

The 9.0 magnitude earthquake of 26 December 2004 that occurred off the west coast of Northern Sumatra in Indonesia was the fourth largest earthquake in the world since 1900 (Gusiakov 2009). This earthquake generated a tsunami that was among the deadliest natural disasters in modern history. The devastating tsunami waves caused a terrible humanitarian disaster affecting thousands of kilometers of the coastal belt in SE Asia (Ismail-Zadeh et al. 2005; Gupta 2005; Gupta and Sharma 2006; Lay et al. 2005; Shaw 2006). Many coastal wetlands were affected by the large inflow of salty seawater and littoral sediments that were deposited during the tsunami, with longer-term effects that include changes in the local hydrogeology caused by changes to coastlines and damage to sea-defences. Serious problems pertaining to salinity changes encountered in coastal south India during this tsunami (Fig. 9) and the consequent loss of fertility of agricultural land have been reported in requests for remedial measures to revitalize economic growth (Havidan et al. 2006).

The tsunami mainly affected the states of Tamil Nadu, Kerala and Andhra Pradesh and the Union Territory of Pondicherry, all in south India, as well as the Andaman and Nicobar Islands of India in the Bay of Bengal (Nirupama 2009; Tsunami Report 2005). Many coastal water quality and associated problems generated by the tsunami are related to past and on-going contamination of terrestrial groundwaters, because those groundwaters are now seeping out along the shorelines affected by the tsunami. For example, chronic inputs of fertilizer and sewage on the land surface over several decades have resulted in increased groundwater nitrogen concentrations which, because of slow yet persistent discharge along the coast, eventually result in coastal marine eutrophication

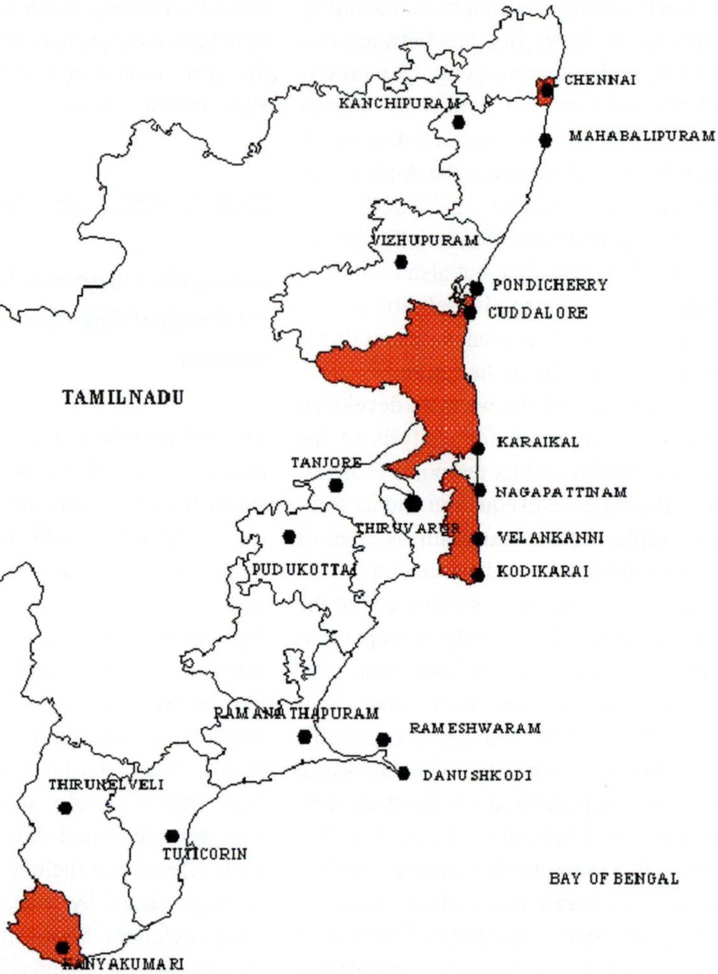

Fig. 9 Coastal areas most affected by the December 2004 tsunami in Tamil Nadu, India

(Alagarswami 1993; Sonak 2000; Ramesh 2001). Such inputs may thus contribute to the increased occurrences of coastal hypoxia, detrimental algal blooms, and associated ecosystem consequences. Tsunamis, in addition to increasing the magnitude of salt-water intrusion, can significantly increase these nutrient pollution problems by sweeping land based nutrients into coastal waters.

Research Concept

The coastal zone of the Indian Ocean is an intensively populated region where the tsunami waves tragically altered land forms and land use practice, causing perturbations and impacts on the "river-sediment-soil-groundwater" system. Saltwater intrusion caused severe modification to the hydraulic regime of coastal aquifers, the position of the salt-fresh water interface, the chemical composition of surface and subsurface waters, and adversely affected the fauna and flora. Studies on the impact of tsunami-wave-induced changes on coastal zone soils and sediments are being focused on interfaces and transition zones characterized by steep gradients. Shifts in these boundary lines and transition zones are sensitive indicators for ongoing processes and trends. Specifically, we will investigate the tsunami-related shifts of the subsurface salt-/freshwater interface, phase switches between dissolved and free gas (e.g., methane), relocation of the oxic-anoxic boundary in soils and sediments, and changes of the spatial distribution of key organisms sensitive to changes in salinity, nutrients, oxygen, and methane concentrations.

The scientific and technological objectives can best be covered by well coordinated interaction of data mining (Zektser 2000), expert knowledge (Kontar et al. 2003; Lobkovsky et al. 2003), well focused field work (Bokuniewicz et al. 2004; Burnett et al. 2002, 2003; Kontar and Ozorovich 2006; Taniguchi et al. 2006), and modeling (Kontar 2007, 2008). Data acquisition in the tsunami-affected transition zone between land and sea requires innovation, so cost and time efficient techniques are being applied by the partners. This includes shallow seismic, geoelectric and radar systems for characterization of the subsurface aquifer system (Kontar and Ozorovich 2006). Investigations will not be limited to the availability of basic parameters such as salinity, hydraulic heads or major chemical components, and will include a strong expertise in trace gases, isotopes, and radio-nuclides suitable for dating of coastal groundwater, detection of subsurface freshwater discharge, and to derive kinetics of remineralisation processes (Valiela et al. 2002). Data mining and field work will be compiled for spatial modeling by Geo-Information System (GIS). There is some data on the seawater intrusion in the coastal areas over the past decade (coastal Tamil Nadu region) on the subsurface groundwater discharge in selected location of tsunami affected regions (see Fig. 9). This data will supplement these studies. The project will concentrate on this concept because there is sufficient data and experience.

Plan of Implementation

This research concept combines intensive exploitation and compilation of existing data sources complemented by landside and marine field research in India, geo-statistical analysis, and numeric modeling to achieve an integrated view on the impact of perturbations on groundwater quality and ecology, shifts of the salt/fresh water interface in the areas affected by tsunami. For several tsunami affected key regions land and marine soils and sediments, hydrology, biogeochemistry of groundwater and coastal waters, and ecological indices will be addressed from existing data sources and by targeted field research, respectively. Target zones, considered as representative for tsunami affected areas, were identified as: (1) Karstic coasts off North Sentinel, Nicobar, (2) soft, glacial sediments characteristic for delta system of Adyar River at the confluence with the Bay of Bengal and (3) the coastal lagoon system in Pichavaram Mangrove, Tamil Nadu. Karstic environments at Nicobar have been experiencing strongly increasing coastal urbanization, and, thus, undergo predominantly anthropogenic perturbations. The delta system of the Adyar River coast is of predominantly of glacial origin with diffusive subterranean and submarine flow. Its coastal lagoon system in Pichavaram Mangrove, Tamil Nadu is situated within a humid temperate climate within the target regions that are affected by the tsunami, sea level rise, and an increasing number of storm events.

Historical Data/Data Mining and Compilation

Within the project's initial phase existing data and published pre-investigations will be used to gather information needed to characterize the coastal soil-water continuum at the selected tsunami-effected regions. A database management system (DBMS) will be coupled to GIS and topographic and bathymetric charts, remote sensing data, thematic maps of the coastal zone (land and sea) and attribute data like biological and geochemical measurements that will be made available for modeling purposes. The DBMS will be used as a base to proceed from with field work. Format and structure of data will be harmonized with all partners in order to have compatible data sets for integrative modeling.

Coastal Aquafer Contamination Assessment

The three-dimensional study of salinization of waters and soils along the east and west coasts of India, with particular reference to the coasts of Tamil Nadu and Car–Nicobar Islands caused by the tsunami of 26 December 2004, is proposed to be studied by geophysical (resistivity and self-potential), geochemical and hydrological methods. In some cases, the tsunami-caused salinization may have been superimposed on anthropogenic salinization caused by excessive exploitation of groundwater. Within this project the localization and quantification of submarine groundwater discharge, the efflux of dissolved constituents and trace gases from sub-seafloor aquifers, and the shifts in the subsurface freshwater/seawater interface towards the shore will be investigated by geochemical analyses of dissolved and particulate components (including stable and radio-isotopes, trace gases, and xenobiotics). The importance of sub-seafloor pathways will be estimated by studying interfacial transport and reaction processes with techniques especially appropriate for the individual regions. These activities including natural tracer studies (^{222}Rn, ^{226}Ra), physical parameter measurements, major ion and nutrient measuments, and sediment pore water sampling. They are all planned to be performed hand-in-hand with bathymetric and shallow seismic surveys. Special emphasis is focused on the interfacial exchange of carbon, nitrogen, and phosphorous compounds. Their processes will be observed throughout all research lines. For consideration of shifts in the landside freshwater/saltwater interface advanced radar systems and geo electric techniques will be applied along with seafloor photography. The modeling efforts of this project will be partitioned into three branches: (1) Modeling of the water cycle in coastal areas considering groundwater renewal and submarine groundwater discharge (2) shifts of the freshwater-saltwater interface and (3) transport reaction modeling.

Contaminated Groundwater Seepage and Tracing Sewage

Sewage is a generic term for the fecal waste from animals although it is usually applied to human derived materials. This highlights one of the principal problems with assessing sewage inputs – is it human or derived from other, predominantly agricultural, sources? Notwithstanding this issue, the term sewage used here will predominantly apply to human wastes. Unlike most contamination, sewage is not a single compound, element or even class of compounds. Rather it is primarily a mixture of organic and inorganic components along with intact biological entities (bacteria and viruses); together, this makes a very complex mixture. This mixture changes from region to region, diurnally and with distance from source due to partitioning between solid and solution phases. Some of the more water-soluble components will be moved with the liquid phase while others are principally associated with the solid or particulate phase and may only move small distances. Therefore, different tracers may exist for each environment and at a range of distances from potential sources. Chemicals that are intrinsic to sewage such as the stanols and sterols associated with human faecal matter, additives like detergents, microbiological communities present in the wastewaters and effects caused by sewage to communities will be examined.

Agricultural Solutions for use of Coastal Contaminated Aquaferers

A systematic study pertaining to salinity problems encountered in coastal south India during tsunami, and the consequent loss of fertility of agricultural land including remedial measures to revitalize economic growth in the region will be conducted. The study suggests that the tsunami affected area will be unproductive due to (a) deposition of littoral sediments containing heavy minerals on the surface; and (b) contamination of soil and ground water aquifer due to sea water intrusion. To revive the fertility of the land, the following measures are suggested: (1) Removal of littoral deposits - the littoral deposits containing heavy minerals should process through weathering and change into clay minerals or soil. Moisture, temperature climatic condition, salinity stressed vegetation and time factor are main components that affect weathering (artificial weathering). Although the scraping of the inert littoral sediments from the agricultural field is a permanent solution, it is not economically viable due to the large size and extent of the affected area. (2) Desalinization of soil and groundwater - the surface water and groundwater of the tsunami- effected area are affected by seawater intrusion into the aquifer. To prevent salt-water contamination, over pumping and unnecessary mining of groundwater should cease. Earth-filled barriers may be constructed along the coast to limit the salt-water intrusion. Further, earthen bunds should be provided in flat areas so that rain water gets trapped and can percolate through the soil.

Social Impacts of Tsunami and the Feasibility of Remedial Actions

The specter of groundwater contamination looms over industrialized, suburban and rural areas (Zektser

2000). The sources of groundwater contamination in tsunami affected countries are many and the contaminants numerous. The disposal of domestic wastewater is accomplished in many areas through the use of septic tanks and drain canals. The tsunami accelerated these processes of contamination in affected regions, and they now require drastic remediation measures. These measures have to be economically feasible, environmentally sound and socially acceptable. Interventions to secure livelihood through the innovative solutions will consider the involvement of genders as a part of the social analysis. The main consideration of this project is to make sure the proposed remedial actions are appropriate for local institutional, environmental, social and economic conditions. The further analyses will be carried out through field surveys and standard environmental/economic/econometric analyses.

Acknowledgement We are grateful to Professor Tom Beer for valuable scientific ideas, tremendous help and co-operation. Also we are grateful to Dr. Slava Gusiakov for his useful comments which helped us to improve the manuscript. This work has been performed thanks to the support of the Swiss National Foundation SCOPES Project, contract No. 7 IP 65663 to develop an information network to assess the water-related environmental problems of the Aral Sea region and the European Commission grants ICA2-CT-2000-10023, INTAS-ARAL-00-1003, and INTAS-ARAL-00-1014.

References

Alagarswami, K., 1993. *Sustainable Management and Development of Coastal Aquaculture*. MSSRF, Chennai, India.

Bokuniewicz, H., E. Kontar, M. Rodrigues and D.A. Klein, 2004. Submarine Groundwater Discharge (SGD) Patterns through a Fractured Rock: a Case Study in the Ubatuba Coastal Area, Brazil. *AAS Revista, Asociacion Argentina de Sedimentologia*, 11, 1, 9–16.

Bortnik, V.N. and S.P. Chistyaeva, 1990. *Hydrometeorology and Hydrochemistry of the USSR Seas, The Aral Sea*, vol. VII. Gidrometeoizdat, Leningrad, p. 196. (in Russian)

Burnett, W.C., J. Chanton, J. Christoff, E.A. Kontar, S. Krupa, M. Lambert, W. Moore, D. O'Rourke, R. Paulsen, C. Smith, L. Smith and M. Taniguchi, 2002. Assessing Methodologies for Measuring Groundwater Discharge to the Ocean. *EOS*, 83, 11, 117, 122–123.

Burnett, W.C., J.P. Chanton and E.A. Kontar, 2003. Submarine Groundwater Discharge. *Biogeochemistry*, vol 66. Kluwer Academic Publishers. The Netherlands, p. 202.

Geist, E.L. and T. Parsons, 2006. Probabilistic Analysis of Tsunami Hazards. *Natural Hazards*, 37, 3, 277–314.

Glantz, M.H. and I.S. Zonn, 2005. *The Aral Sea: Water, Climate and Environmental Change in Central Asia*. World Meteorological Organization (MWO), Geneva, p. 97.

Gupta, H., 2005. Mega-tsunami of the 26th December 2004: Indian Initiative for Early Warning System and Mitigation of Oceanogenic Hazards. *Episodes*, 28, 1, 2–5.

Gupta, M. and A. Sharma, 2006. Compounded Loss: the Post Tsunami Recovery Experience of Indian Island Communities. *Disaster Prevention and Management*, 15, 67–79.

Gusiakov, V.K., 2009. Tsunami History – Recorded. In: *The Sea, Tsunamis* (Ed by Robinson, A., and E. Bernard). Harvard University Press, Cambridge, vol. 15, pp. 23–53.

Havidan, R., T. Wachtendorf and J. Kendra, 2006. A Snapshot of the 2004 Indian Ocean Tsunami: Societal Impacts and Consequences. *Disaster Prevention and Management*, 15, 163–177.

Ismail-Zadeh, A., T. Beer, P. Dunbar, V. Gusiakov, A. Jayawardena, E. Kontar, V. Kossobokov, U. Shamir, R. Singh, K. Takeuchi, G. Tetzlaff and Z. Wu, 2005. Urgent Action to Reduce the Effects of Natural Disasters in the Indian Ocean Region and Elsewhere. European Geosciences Union, Geophysical Research Abstracts, vol. 7, 04824, SRef-ID: 1607-7962/gra/EGU05-A-04824.

Jarsjö, J., I. Alekseeva, C. Schrum and G. Destouni, 2006. Simulation of Groundwater–Seawater Interactions in the Aral Sea Basin by a Coupled Water Balance Model. In: *From Uncertainty to Decision Making, ModelCare 2005* (Ed by Bierkens, M.F.P., K. Kovar, and J.C. Gehrels). International Association of Hydrological Sciences (IAHS) Red Book Series, Paper no. IAHS 304-30-145.

Jarsjö, J. and G. Destouni, 2004. Groundwater Discharge into the Aral Sea After 1960. *Journal of Marine Systems*, 47, 109–120.

Kontar, Y., 2007. Groundwater–Seawater Interactions in Tsunami Affected Areas: Solutions and Applications. In: *A New Focus on Groundwater–Seawater Interactions*. IAHS Press, Wallingford, vol. 312, pp. 19–27.

Kontar, Y., 2008. Hazards in the Coastal Zones. *Symposium PEH-01: Hazards: Minimizing Risk, Maximizing Awareness*. In: *IYPE Hazards Megasymposium CD-ROM*, Convener T. Beer. *The 33rd International Geological Congress*, Oslo, Norway, 6–14 August 2008.

Kontar, E.A. and Yu.R. Ozorovich, 2006. Geo-Electromagnetic Survey of Fresh/Salt Water Interface in the Coastal Southeastern Sicily. *Continental Shelf Research*, 26, 7, 843–851.

Kontar, E.A., A. Salokhiddinnov, N. Takhirov, Y. Azhigaliev and R. Bigarinov, 2003 Assessment of Groundwater–Seawater Interactions in the Aral Sea Basin and Pollution Control. *Proceedings of the XXIII General Assembly of the International Union of Geodesy and Geophysics* (IUGG2003), Sapporo, Japan, 30 June–11 July 2003, A139.

Kontar, E.A., A.Yu. Tkachev, I.S. Zektser, L.I. Al'piner, Yu.L. Obyedkov, A.V. Shapovalov, R.K. Ikramov, G.V. Stulina, 2000a. Groundwater Flow and Ecological Stability of the Environment in the Basin of Aral Sea. *XXV General Assembly of European Geophysical Society (EGS), Millennium Conference on Earth, Planetary & Solar Systems Sciences*, Nice, France, 25–29 April 2000.

Kontar, E.A., I.S. Zektser, L.I. Elpiner, R.K. Ikramov and G.V. Stulina, 2000b. Groundwater Flow and Ecological Stability of the Environment in the Basins of Central Asia Lakes. *Proceedings of the Hydro 2000, 3rd International Hydrology and Water Resources Symposium*, 20–23 November 2000, Perth, Western Australia, vol. 2, pp. 621–627.

Lay, T., H. Kanamori, C. Ammon, M. Nettles, S. Ward, R. Aster, S. Beck, S. Bilek, M. Brudzinski, R. Butler, H. DeShon, G. Ekström, K. Satake and S. Sipkin, 2005. The Great Sumatra-Andaman Earthquake of December 26, 2004. *Science*, 308, 1127–1133, DOI:10.1126/science.1112250, 2005

Lobkovsky, L.I., E.A. Kontar, I.A. Garagash and Yu.R. Ozorovich, 2003. Monitors and Methods for Investigation of Submarine Landslides, Seawater Intrusion and Contaminated Groundwater Discharge as Coastal Hazards. In: *NATO "Risk Science and Sustainability: Science for Reduction of Risk and Sustainable Development of Society"* (Ed by Beer, T. and A. Ismail-Zadeh). Kluwer Academic Publishers. The Netherlands. pp. 191–207.

Micklin, P.P., 2004 The Aral Sea crisis. In: *Dying and Dead Seas, NATO ARW/ASI Series* (Ed by Nihoul, J., P. Zavialov, P. Micklin). Kluwer Publishing, Dordrecht, pp. 99–123.

Mirabdullaev, H., I.M. Joldasova, Z.A. Mustafaeva, S. Kazakhbaev, S.A. Lyubimova and B.A. Tashmukhamedov, 2004. Succession of Ecosystems of the Aral Sea during its Transition from Oligohaline to Polyhaline Waterbody. *Journal of Marine Systems*, 47, 101–108.

Nirupama, N., 2009. Socio-Economic Implications Based on Interviews with Fishermen Following the Indian Ocean Tsunami. *Natural Hazards*, 48, 1–9.

Ramesh, R., 2001. Effect of Land-Use Change on Groundwater Quality in a Coastal Habitat of South India. In: *Impact of Human Activity on Groundwater Dynamics* (Ed by Gehrels, H., J. Peters, E. Hoehn, K. Jensen, C. Leibundgut, J. Griffoen, B. Webb, and W.-J. Zaadnoordijk) Proc. Symp. at Maastricht, The Netherland, July 2001. IAHS Publication no. 269, pp. 161–166.

Salokhiddinnov, A.T. and Z.M. Khakimov, 2004. Ways the Aral Sea Behaves. *Journal of Marine Systems*, 47, 127–136.

Shaw, R., 2006. Indian Ocean Tsunami and Aftermath: Need for Environment-Disaster Synergy in the Reconstruction Process. *Disaster Prevention and Management*, 15, 5–20.

Shibuo, Y., J. Jarsjö and G. Destouni, 2005. Modeling Groundwater–Seawater Interactions in the Aral Sea Region. In: *Groundwater and saline intrusion* (Ed by Araguás, L., E. Custodio, M. Manzano). Insitututo Geologico y Minero de España, Serie: Hidrogeología y Aguas Subterráneas No 15, Madrid, pp. 163–171.

Shibuo, Y., J. Jarsjö and G. Destouni, 2006. Bathymetry-Topography Effects on Saltwater-Fresh Groundwater Interactions Around the Shrinking Aral Sea. *Water Resources Research*, 42, W11410, DOI:10.1029/2005WR004207.

Sonak, S., 2000. Aquaculture in India. In: *Proc. On Land Use, Land Cover Changes and Modeling in Coastal Areas* (Ed by Ramachandran, S.). Anna University, Chennai, India, pp. 20–32.

Taniguchi, M., W. Burnett, H. Duevalova, E. Kontar, P. Povinec and W. Moore, 2006. Submarine Groundwater Gischarge Measured by Seepage Meters in Sicilian Coastal Waters. *Continental Shelf Research*, 26, 7, 835–842.

Tsunami Report, 2005. Government of India. http://www.dod.nic.in/tsunami3.pdf.

Valiela, I., J.L. Bowen and K.D. Kroeger, 2002. Assessment of Models for Estimation of Land-Derived Nitrogen Loads to Shallow Estuaries. *Applied Geochemistry*, 17, 935–953.

Zavialov, P.O., 2005. *Physical Oceanography of the Dying Aral Sea*. Springer-Verlag/Praxis, Chichester, p. 158.

Zektser, I.S., 2000. *Groundwater and the Environment: Applications for the Global Community*. Lewis Publishers, Boca Raton, p. 175.

Part V
GeoHazards and Risks – Observation and Assessment

Mega Tsunami of the World Oceans: Chevron Dune Formation, Micro-Ejecta, and Rapid Climate Change as the Evidence of Recent Oceanic Bolide Impacts

Viacheslav Gusiakov, Dallas H. Abbott, Edward A. Bryant, W. Bruce Masse, and Dee Breger

Abstract This paper deals with the physical and environmental effects resulting from oceanic impacts by sizable comets, and the rates and risks associated with such cosmic impacts. Specifically, we investigate two sets of probable oceanic impact events that occurred within the last 5,000 years, one in the Indian Ocean about 2800 BC, and the other in the Gulf of Carpentaria (Australia) about AD 536. If validated, they would be the most energetic natural catastrophes occurring during the middle-to-late Holocene with large-scale environmental and historical human effects and consequences. The physical evidence for these two impacts consists of several sets of data: (1) remarkable depositional traces of coastal flooding in dunes (chevron dunes) found in southern Madagascar and along the coast of the Gulf of Carpentaria, (2) the presence of crater candidates (29-km Burckle crater about 1,500 km southeast of Madagascar which dates to within the last 6,000 years and 18-km Kanmare and 12-km Tabban craters with an estimated age of AD 572±86 in the southeast corner of the Gulf of Carpentaria), and (3) the presence of quench textured magnetite spherules and nearly pure carbon spherules, teardrop-shaped tektites with trails of ablation, and vitreous material found by cutting-edge laboratory analytical techniques in the upper-most layer of core samples close to the crater candidates.

Although some propose a wind-blown origin for V-shaped chevron dunes that are widely distributed around the coastlines of the Indian Ocean and in the Gulf of Carpentaria, we have evidence in favor of their mega tsunami formation. In southern Madagascar we have documented evidence for tsunami wave run-up reaching 205 m above sea-level and penetrating up to 45 km inland along the strike of the chevron axis. Subtly the orientation of the dunes is not aligned to the prevailing wind direction, but to the path of refracted mega-tsunami originating from Burckle impact crater.

The results of our study show that substantive oceanic comet impacts not only have occurred more recently than modeled by astrophysicists, but also that they have profoundly affected Earth's natural systems, climate, and human societies. If validated, they could potentially lead to a major paradigm shift in environmental science by recognizing the role of oceanic impacts in major climate downturns during the middle-to-late Holocene that have been well documented already by different techniques (tree-ring anomalies, ice-, lake- and peat bog-cores).

Keywords Oceanic impact · Comets · Tectonic tsunamis

Introduction

Tsunamis belong to the class of long-period oceanic waves generated by underwater earthquakes, submarine or subaerial landslides or volcanic eruptions. They are among the most dangerous and complex natural phenomena, being responsible for great loss of life and extensive destruction of property in many coastal areas. The tsunami phenomenon includes three overlapping but quite distinct physical stages: the generation of a wave by any external force that disturbs a water column, the propagation of that wave at high

V. Gusiakov (✉)
Tsunami Laboratory, ICMMG SD RAS, Novosibirsk 630090, Russia
e-mail: gvk@sscc.ru

T. Beer (ed.), *Geophysical Hazards*, International Year of Planet Earth,
DOI 10.1007/978-90-481-3236-2_13, © Springer Science+Business Media B.V. 2010

speed in the open ocean, and finally, the propagation of the tsunami wave through shallow coastal water and the inundation of dry land by run-up. Most tsunamis occur in the Pacific, but they are known in all other areas of the world including the Atlantic and the Indian Oceans, the Mediterranean and many marginal seas. Tsunami-like phenomena can occur even in lakes, large man-made water reservoirs and large rivers.

Destruction from tsunamis results from three main factors: inundation by salt water, impact dynamics, and erosion. Considerable damage is also caused by floating debris that enhances the destructive force of flooding. Flotation and drag forces can destroy frame buildings, overturn railroad cars and move large ships far inland. Ships in harbors and port infrastructure can be damaged by the strong current and surge generated by even a weak tsunami.

An average height for a tsunami, generated by an earthquake with magnitude of 7.5–8.0 (the range where most tsunamigenic earthquakes occur), is between 3 and 10 m along 100–300 km of the coastline closest to the epicenter. This height is still within the range of the largest possible storm surges for many coastal locations. However, tsunami have a longer wavelength and can penetrate inland to much greater distances reaching in many places several hundreds of meters and sometimes several kilometers. The current velocity in a tsunami flood can exceed 10 m/s, being the most important factor in producing great damage and the loss of human lives.

Tsunamis caused by bolide impact undoubtedly share many characteristics in common with those described above, however they also possess a suite of unique features that reflect the size of the impactor, its extraterrestrial composition, the enormous temperatures and pressures that accompany high velocity impact, and the fact that bolide impacts can occur ubiquitously in the world's oceans. Our purpose in this paper is threefold. First, we will set the stage for our discussion of impact generated tsunami events, including climatic implications, by placing these types of events within the context of the historical record of tsunamis caused by seismic, landslide, and volcanic geophysical events. Second, rather than focusing on recent general simulation and modeling exercises that further our understanding of the nature of oceanic impact tsunami propagation and the potential size of resultant coastal run-ups (e.g. Hills and Mader 1997; Gisler et al. 2003), we will draw attention to existing Holocene coastal landforms and sedimentary signatures that appear to be the likely product of tsunamis generated by bolide impacts. Finally, we will focus upon two middle-to-late Holocene oceanic impact events, one that we hypothesize may have occurred in the abyssal Indian Ocean around 4,800 years ago, and the other in the shallow waters of Australia's Gulf of Carpentaria around 1,500 years ago, in order to show the potential of impact generated tsunamis to affect coastlines.

Regular Tectonic Tsunamis – Analysis of Available Historical Data

The Global Tsunami DataBase (GTDB) Project is a world-wide catalog and database on tsunamis and tsunami-like events that covers the period from 2000 BC to the present (Gusiakov 2003). It currently contains nearly 2,250 historical events with 1,206 of these located in the Pacific, 263 in the Atlantic, 125 in the Indian Ocean, and 545 from the Mediterranean Sea. The geographical distribution and intensity of the tsunamis are shown in Fig. 1. Most of the tsunamis were generated along subduction zones and the major plate boundaries in the Pacific, Atlantic and Indian Oceans and in the Mediterranean region. Very few historical events occurred in the deep ocean or central parts of the marginal seas, except several cases of small tsunamis that originated along the middle-ocean ridges and some major transform faults.

Most oceanic tsunamis (up to 75% of all historical cases) reported in historical catalogs are generated by shallow-focus earthquakes capable of transferring sufficient energy to the overlying water column to generate significant waves at the shore. The rest are due to landslide (7%), volcanic (5%), meteorological (3%) events and to water waves from explosions (less than 1%). Up to 10% of all the reported coastal run-ups still have an unidentified source (Gusiakov 2009).

All destructive tsunamis can be divided into two categories: *local* (or *regional*) and *trans-oceanic*. For local tsunamis, the destructive effect is confined to the nearest coast (from 100 to 500 km) located within one hour of propagation. In all tsunamigenic regions of the world oceans, most damage and casualties come from local tsunamis. Far less frequent but potentially much more hazardous are trans-oceanic tsunamis capable of widespread destruction. Formally, this category

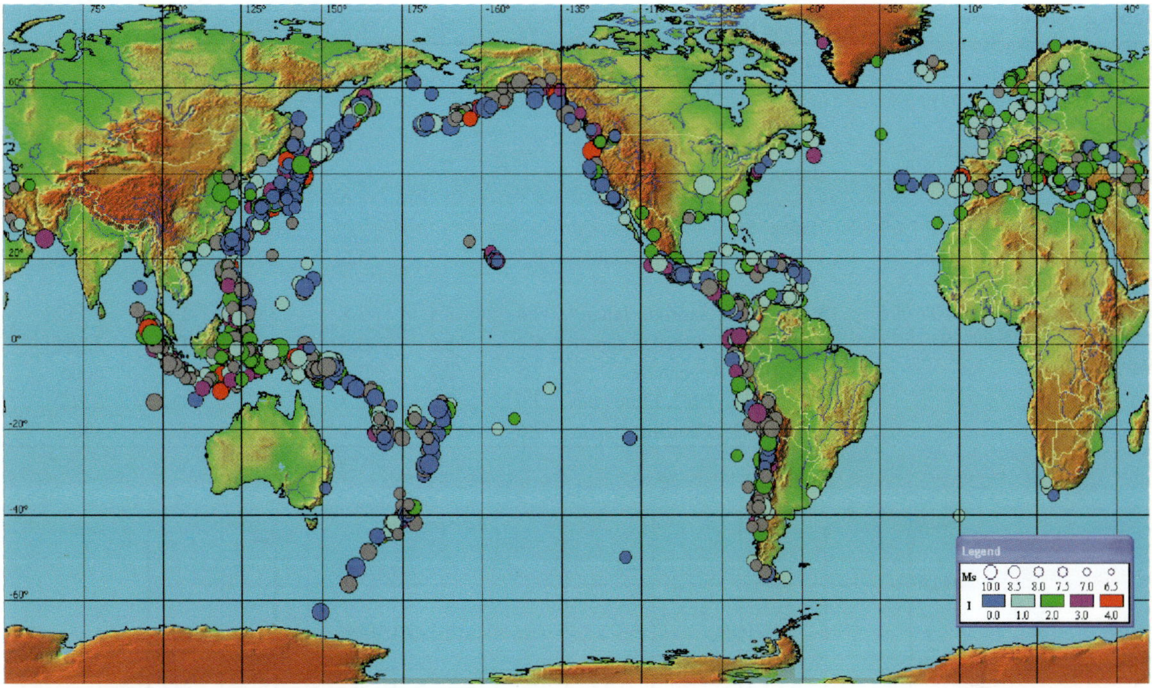

Fig. 1 Visualization of the global historical tsunami catalog. A total of 1965 tsunamigenic events with identified sources are shown for the period from 2000 BC to present time. Size of *circles* is proportional to event magnitude (for seismically induced tsunamis); *color* represents tsunami intensity on the Soloviev-Imamura scale

includes the events that have run-ups higher than 5 m at a distance of more than 5,000 km from the source. Historically, all trans-oceanic tsunamis have originated in the Pacific with only two cases occurring in other regions (the 1755 Lisbon tsunami in the Atlantic and the 2004 Sumatra tsunami in the Indian Ocean). The basic parameters of the 11 known trans-oceanic tsunamis are listed in Table 1. By total energy, they are the largest documented tsunamigenic events, but they are not the largest ones in terms of maximum run-up observed at the coast and number of fatalities. Ten of the largest regional tsunamigenic events together with their basic source parameters are listed in Table 2. All of them were highly destructive locally, and some had high rates of mortality, but they did not generate an ocean-wide tsunami.

One of the most important aspects of seismogenic tsunamis is their potential maximum run-up height in the near-field. Available historical data, summarized in Tables 1 and 2 and presented graphically in Fig. 2,

Table 1 List of historical trans-oceanic tsunamis (see text for definition). *M*–magnitude (macroseismic, M_S or M_W), *I*–tsunami intensity on the Soloviev-Imamura scale, H_{maxNF}–maximum reported run-up in the near field in m, H_{maxFF}–maximum reported run-up in the far field (more than 5,000 km) in m, FAT–number of reported fatalities due to tsunami

Date and place	M_S	I	H_{maxNF}, m	H_{maxFF}, m	FAT
1 November 1755, Lisbon	8.5	4.0	30.0	7.0	30,000
7 November 1837, Chile	8.5	3.0	8.0	6.0	many
13 August 1868, Chile	9.1	3.5	15.0	5.5	612
15 June 1896, Sanriku	7.4	3.8	38.2	5.5	27,122
3 February 1923, Kamchatka	8.3	3.5	8.0	6.1	3
1 April 1946, Aleutians	7.9	4.0	42.2	20.0	165
4 November 1952, Kamchatka	9.0	4.0	18.0	9.1	>10,000
9 March 1957, Aleutians	9.1	3.5	22.8	16.1	none
22 May 1960, Chile	9.5	4.0	15.2	10.7	1,260
28 March 1964, Alaska	9.2	4.5	68.0	4.9	221
26 December 2004, Sumatra	9.3	4.5	50.9	9.6	229,866

Table 2 List of some of the largest regional seismogenic tsunamis in the historical catalogs. M_S–surface wave magnitude, M_W–moment-magnitude, I–tsunami intensity on the Soloviev-Imamura scale, H_{max}–maximum reported run-up in m, CAU-cause of tsunami (T-tectonic, L–landslide), FAT–number of reported fatalities due to tsunami

Date and Place	M_S	M_W	I	H_{max}, m	CAU	FAT
9 July 1586, Lima, Peru	8.5	-	3.5	26.0	T	many
31 January 1605, Shikoku, Japan	8.0	-	3.5	30.0	T	many
2 December 1611, Sanriku, Japan	8.1	-	4.0	25.0	T	4,783
28 October 1707, Nankaido, Japan	8.1	-	4.0	25.7	T	30,000
23 December 1854, Nankaido, Japan	8.3	-	3.0	28.0	T	5,000
15 June 1896, Sanriku, Japan	7.4	8.5	3.8	38.5	T	27,122
2 March 1933, Sanriku, Japan	8.3	8.6	3.5	29.3	T	3,064
9 July 1956, Aegean Sea	7.5	7.7	3.0	30.0	TL	none
12 December 1992, Flores Sea	7.6	7.7	2.7	26.2	TL	2,200
12 July, 1993, Okushiri, Japan	7.6	7.7	3.1	31.7	T	198

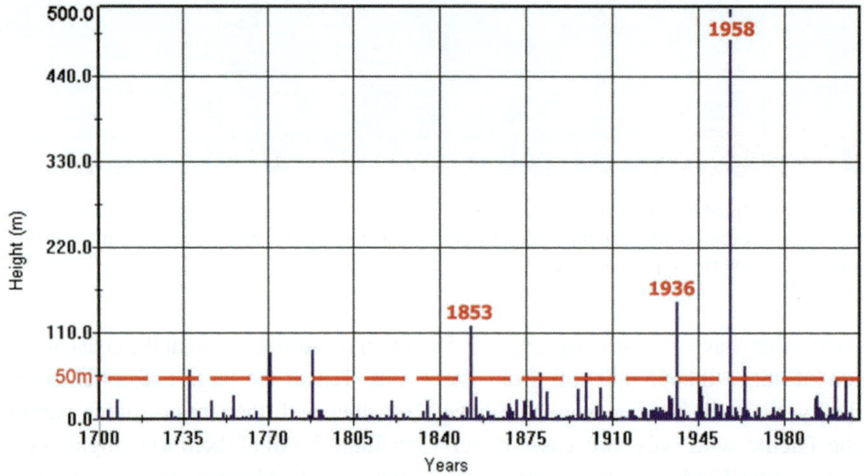

Fig. 2 Maximum run-up heights observed at the nearest coast for historical tsunamis during the last 300 years

suggests that this value cannot exceed 35–50 m even for the largest possible submarine earthquakes. Maximum run-up values of 60–70 m reported for the 1771 Ishigaki, 1788 Sanah-Kodyak, and 1737 Kamchatka tsunamis are not very reliable. They are based on single anecdotal reports, which are not always confirmed by recent geological investigation (see, e.g. Pinegina and Bourgeois 2001), and may have been produced by a locally generated landslide accompanying the main tectonic tsunami. As demonstrated for the 1964 Alaska earthquake, all the major run-ups that exceeded 25–30 m along the Alaskan coast were generated by slides from the fronts of the local deltas. These waves arrived at the coast almost immediately after the earthquake and were followed by the main seismically-induced tsunami that typically had a height of 12–18 m (Lander 1996)

Slide-generated tsunamis result from rock and ice falling into the water, and sudden submarine landslides or slumps. They can produce extremely high water splash along a coastline, up to 50–70 m, with the highest historical record being 525 m in Lituya Bay, Alaska in 1958. In general, the impact is not widely extended along the coast. The energy of a landslide-generated tsunami rapidly dissipates as waves travel away from the source. However, in some cases, if the landslide covers a large depth range or occurs over a long duration of slide movement, it can focus the tsunami energy into a beam narrower than the equivalent seismic source (Iwasaki 1997).

Based on analysis of available historical data, we can conclude the following: (1) tsunami run-ups generated by the largest possible seismogenic and volcanic events rarely exceed 45–50 m at the nearest coast and 15–20 m run-up in the far-field, and (2) landslide-generated tsunamis can be highly destructive locally, but never flood any extended part of the coast.

It is therefore of considerable interest that the world's coastlines contain prominent erosional and depositional features of catastrophic water currents and waves of much higher magnitude over wide areas.

Erosional and Depositional Traces of Large-Scale Water Impact on the Coast

The most common signature of tsunami is the deposition of landward tapering sandy units up to 50-cm-thick sandwiched between finer material and peats on flat coastal plains. While similar lenses can be deposited by individual surging waves during tropical cyclones, such units are rarely longer than 10–20 m and do not form continuous deposits behind modern beaches. Tsunami sand units form part of a coherent landward thinning splay of fining sediment extending up to 10 km or more inland. The thickness of laminae decreases landward while that of an individual unit decreases upwards, implying waning energy conditions. All these characteristics match transport of sediment-rich flows by tsunami across marsh surfaces. Anomalous sand layers can have an erosional basal contact and incorporate rip-up clasts of muddier sediment.

The emphasis on sand units as a signature of tsunami ignores other depositional phenomena – namely the transport and imbrication of boulders. The clearest evidence of this is the movement of large boulders onshore (Bryant 2008). For example, the Sea of Japan tsunami of 26 May 1983 produced a tsunami over 14 m high. A large block of concrete weighing over 1,000 tons was moved 150 m from the beach over dunes 7 m high. Boulders transported by tsunami have also been found in paleo-settings. For example, on the reefs of Rangiroa, Tuamoto Archipelago in the Southeast Pacific, individual coral blocks measuring up to 750 m^3 have been linked to tsunami rather than to storms. On Hateruma and Ishigaki Islands in the South Ryukyu Islands, coralline blocks measuring 100 m^3 have been emplaced 30 m above present sea level, 2.5 km from the nearest beach (Kawana and Nakata 1994). These boulders have been dated and indicate that tsunamis with a local source, have washed over the islands seven times in the last 4,500 years. Two of the largest events occurred 2,000 years ago and during the great tsunami of 24 April 1771. In the Leeward Islands of Netherlands Antilles in the Caribbean, boulders weighing up to 280 tons have been moved 100 m by repetitive tsunamis most likely occurring 500, 1,500 and 3,500 years ago (Scheffers 2004). In fact, detailed cataloging of anomalous boulders from the literature indicates that they are the prevalent signatures of tsunami events on most coasts (Kelletat 2008).

Australian Imbricated Boulder Fields, Cavitation, and Vortex Structures

Along the East Coast of Australia, anomalous boulders are incompatible with the storm wave regime (Bryant 2008). For example, exposed coastal rock platforms along this coast display little movement of boulders up to 1–2 m in diameter, despite the presence of 7- to 10-m-high storm waves. Boulders are also found in completely sheltered locations along the coast. At Bass Point, which extends 2 km seaward from the coast, a boulder beach faces the mainland coast rather than the open sea. Similarly at Haycock Point, rounded boulders, some with volumes of 30 m^3 and weighting 75 tons, have been piled into a jumbled mass at the base of a ramp that begins 7 m above a vertical rock face on the sheltered side of the headland. Perhaps the most dramatic deposits are those containing piles of imbricated boulders (Fig. 3). These piles take many forms, but include boulders up to 105 m^3 in volume and weighing as much as 285 tons. The boulders lie en echelon one against the other like fallen dominoes, often in parallel lines. At Jervis Bay, New South Wales, blocks weighing almost 100 tons have clearly been moved in suspension and deposited in this fashion above the limits of storm waves on top of cliffs 33 m above present sea level. The longest train of imbricated boulders exists at Tuross Head where 2-to 3-m-diameter boulders stand as sentinels one against the other, over a distance of 200 m at an angle to the coast (Young et al. 1996).

Fig. 3 Massive imbricated boulders at height of up to 40 m on the southeastern coast of Australia

Tsunamis can also sculpture bedrock in a fashion analogous to the s-forms produced by high velocity catastrophic floods or surges from beneath icecaps in sub-glacial environments (Kor et al. 1991). S-forms include features such as muschelbrüche, sichelwannen, V-shaped grooves, cavettos, and flutes. They have been linked to paleo-floods in Canada, the northwestern United States, Scandinavia, Britain, the Alps, and the Northern Territory, Australia. Tsunami flows at velocities above 10 m/s have the hydrodynamic potential over rocky headlands to generate cavitation or small vortices capable of producing sculptured forms. Cavitation is a product of high-velocity flow as great as 10 m/s in water depths as shallow as 2 m deep (Baker 1981). At these velocities, small, low pressure, air bubbles appear in the flow. These bubbles are unstable and immediately collapse, generating impact forces up to thirty thousand times greater than normal atmospheric pressure. Cavitation bubble collapse is highly corrosive. Cavitation features generated by mega tsunami are widespread along the New South Wales coast of eastern Australia and consist of impact marks, drill holes, and sinuous grooves (Bryant 2008). The spatial organization of s-forms on headlands, often above the limits of storm waves, is a signature of tsunami in the absence of any other definable process (Bryant and Young 1996). Impact marks appear as pits or radiating star-shaped grooves on vertical faces facing the flow. It would be simple to suggest that such features represent the impact mark of a rock hurled at high velocity against a vertical rock face; however, such marks have also been found in sheltered positions or tucked into undercuts where such a process is unlikely (Fig. 4f). Drill holes are found over a range of locations on tsunami-swept headlands (Fig. 4e). Their distinguishing characteristic is a pit several centimeters in diameter bored into resistant bedrock. While it would be easy to attribute these features to marine borers, they often occur profusely above the limit of high tide.

The term flute describes long linear forms that develop under unidirectional, high-velocity flow in the coastal environment. These are noticeable for their protrusion above, rather than their cutting below, bedrock surfaces (Fig. 4d). In a few instances, flutes taper downstream and are similar in shape to rock drumlins and rattails described for catastrophic flow in subglacial environments (Kor et al. 1991). In all cases, the steeper end faces the tsunami wave, while the spine is aligned parallel to the direction of tsunami flow. Flutes span a range of sizes, increasing in length to 30–50 m as slope decreases. However, their relief rarely exceeds 1–2 m.

On flat surfaces, longitudinal vortices give way to vertical ones that can form potholes (Fig. 4b, c, e).

Fig. 4 Different forms of sculptured bedrock along the southeastern Australian coast. They are believed to be a result of cavitation and/or impact of high velocity water currents

Potholes are usually attributed to mechanical abrasion under normal ocean wave action; however many exhibit features of high velocity flow. They form at different scales from forms up to 70 m in diameter to smaller features with dimensions of 4–5 m. The potholes tend to develop as flat-floored, steep-walled rectangular depressions, usually within the zone of greatest turbulence. While bedrock jointing may control this shape, a pothole's origin as a bedrock-sculptured feature is unmistakable where the inner walls are inevitably undercut or imprinted with cavettos. In places where vortices have eroded the connecting walls between potholes, a chaotic landscape of jutting bedrock with a relief of 1–2 m can be produced (Fig. 4b, e). This morphology – termed hummocky topography – forms where flow is unconstrained and turbulence is greatest. These areas occur where high-velocity water flow has changed direction suddenly, usually at the base of steep slopes or on the seaward crest of headlands. In the latter case, they can be situated well above sea level and the effects of wind-generated waves (Fig. 4b, e).

Large-scale features can usually be found sculptured or eroded on rock promontories, which protrude seaward onto the continental shelf (Bryant 2008). Such features require extreme run-up velocities that can only be produced by the higher or longer waves (mega tsunami) generated by large submarine landslides or asteroid impacts in the ocean. One of the most common features of high-velocity overwashing is the stripping of joint blocks from the fronts of cliffs or platforms forming inclined surfaces or ramps (Fig. 4d). In many cases, this stripping is aided by the detachment of flow from surfaces, a process that generates enormous lift forces that can pluck joint-controlled rock slabs from the underlying bedrock. Where standing waves have formed, bedrock plucking can remove two or three layers of bedrock from a restricted area, leaving a shallow, closed depression on the ramp surface devoid of rubble and unconnected to the open ocean. Ramps are obviously controlled structurally and have an unusual juxtaposition beginning in cliffs up to 30 m above sea level and sloping down flow, often into a cliff. If these high velocities are channelized, erosion can produce linear canyon features 2–7 m deep and pool-and cascade features incised into resistant bedrock on the lee side of steep headlands (Fig. 4c, e).

Perhaps the most impressive features are whirlpools formed in bedrock on the sides of headlands. Whirlpools and smaller potholes are commonly formed under catastrophic flow in the channeled scabland of Washington State. In coastal environments, whirlpools often contain a central plug of rock and show evidence of smaller vortices around their rim.

Fig. 5 Vortex structures along the Australian SE coast. They are believed to be carved into resistant coastal rocks by water currents with velocity V > 10 m/s and depth D > 5 m. Cavitation is also involved in the erosional process

Whirlpools can reach 50–70 m in diameter with central plugs protruding 2–3 m vertically upwards from the floor of the pit at the quiescent centre of the vortex (Fig. 5b, e). One of the best examples occurs on the south side of Atcheson Rock south of Bass Point, New South Wales (Fig. 5e). Here a large vortex, spinning in a counterclockwise direction, produced smaller vortices rotating around its edge on the up flow side of a headland. The overall whirlpool is 10 m wide and 8–9 m high. The central plug stands 5 m high and is surrounded by four 3-m-diameter potholes, one of which bores another 3 m below the floor of the pit into resistant basalt. The counterclockwise rotation of the overall vortex produces downward-eroded helical spirals that undercut the sides of the pit, forming spiral benches. Circular or sickle-shaped holes were drilled, by cavitation, horizontally into the sides of the pothole and into the wall of the plug. Under exceptional circumstances, the whirlpool can be completely eroded, leaving only the plug behind (background of Fig. 5c).

Chevron Dunes

Many coastlines of the world exhibit sets of large V-shaped chevron-like dunes – symmetrical sand dunes that are similar in their lancet-form, showing strong parallelism, often at different angles to the shoreline (Bryant 2001; Kelletat and Scheffers 2003). Some typical examples of chevron dunes from different parts of the world are shown in Fig. 6. The term "chevron" was first used to describe wind-blown dunes by Maxwell and Haynes (1989) in south-western Egypt and northern Sudan, where the chevrons consist of sinuous, parallel, blade-shaped deposits of sand 10–30 cm high and 0.13–1.2 km in length which actively migrate over darker coarse sands. Hearty et al. (1998) and Kindler and Strasser (2000) applied this term to coastal dunes. They used this term in the Bahamas to describe nested parabolic shaped dunes formed during the Last Interglacial age and attributed them to megastorm deposits. Bryant (2001) recognized that certain coastal chevrons could be formed by megatsunamis because their limits are far above those of any storm surges reported in the literature. Only dominant winds or high-energy water currents can be responsible for the formation of chevrons. In this paper, we do not limit the term "chevron" to any one process but use it in the broadest sense to refer to any sedimentary feature within the coastal landscape with this distinctive V-shaped morphology.

Since Google Earth satellite imagery become available in early 2005, a wide distribution of chevron-type geomorphologic features along many coastlines of the world (Fig. 7) was quickly identified (Scheffers et al. 2008). Actually, they can be found at almost

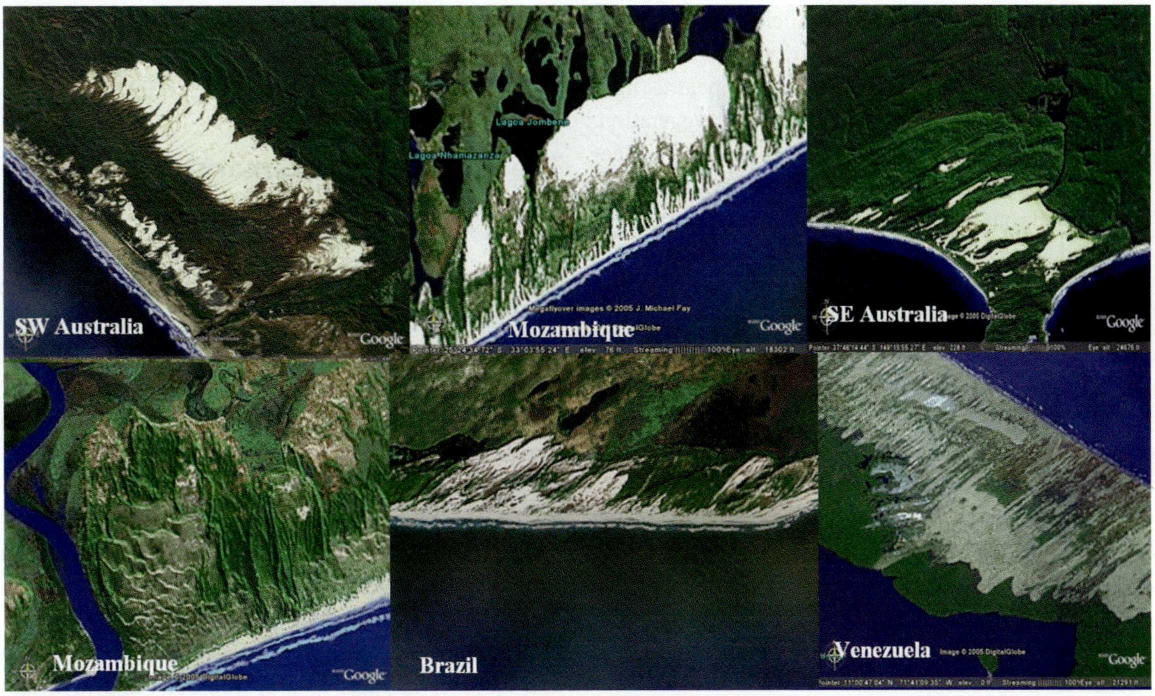

Fig. 6 Examples of chevron dunes from different coastlines of the world. They reach heights of more than 100 m with inland penetration of up to 10 km and are believed to be a result of impact of destructive oceanic waves with possible later modification made by wind

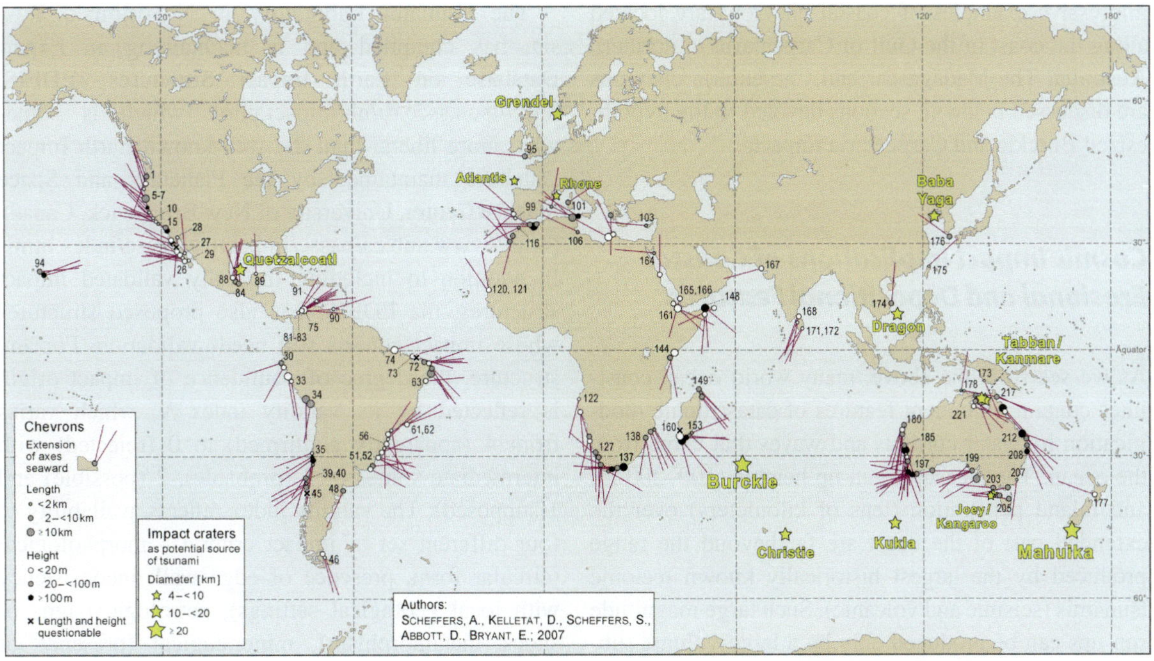

Fig. 7 Location map and orientation of chevron dunes identified on satellite images (Scheffers et al., 2008)

any undisturbed coast having sufficient sand supply from beaches and shallow water areas. On aerial photographs and satellite images chevrons appear as light colored areas with fresh sand and sparse vegetation. But in many places, they are densely covered by vegetation indicating an older age. An important feature of coastal chevrons is that they consist of unsorted marine sand and quite often contain fist-sized cobbles, pieces of coral rock and mud clasts. In some cases, they are not aligned to the dominant wind direction.

Obviously, some chevrons, or parts of them, have undergone a partial remobilization of sand by wind with formation of superimposed dunes of different shapes: parabolic, irregular or barchan types. The total length of chevrons, i.e. their landward extension, clearly depends upon coastal topography. Where the coast is flat, larger chevrons may occur. Coastal lowlands with wide beaches are the best source areas for long chevrons, but they also exist along cliffed coasts, where there is no visible source of available sand (Scheffers et al. 2008).

The most spectacular chevrons on the planet are found at the southernmost tip of Madagascar where they reach heights of more than 200 m above present sea level with inland penetration of up to 45 km. Other impressive examples of coastal chevrons are located along the coast of the Gulf of Carpentaria in northern Australia. The Madagascar and Carpentaria chevrons are discussed below in sections devoted to the hypothesized Burckle and Carpentaria impacts.

Cosmic Impact Implications of Coastal Erosional and Depositional Features

As we see from the above, many world ocean coastlines contain prominent features of catastrophic modification by water currents and waves that came from the ocean. The measured run-up heights (100–200 m) and inland penetration (tens of kilometers) over the extended part of the coast are far beyond the range produced by the largest historically known tectonic tsunamis (seismic and volcanic). Such large magnitude run-ups can be produced only by a large-volume submarine landslide or by oceanic impacts. In the latter case, the impacts may also be responsible for major environmental downturns during the Holocene that have been indicated by environmental reconstructions using tree-ring records, lake-bottom sediments analysis and ice-core data.

The Search for Quaternary and Holocene Period Cosmic Impacts

Our team's search for recent Earth impact events involves different tools depending on the likely target area (oceanic versus terrestrial), relative size of the impactor, estimated age of the event (Holocene vs. Quaternary period), and whether the impactor was an asteroid or comet. For oceanic impacts, we use satellite altimetry along with the search for coastal tsunami sedimentary signatures and the presence of impact indicators in deep sea and coastal sediment cores, such as ejecta and layers of high magnetic susceptibility. For middle and late Holocene impacts, we have discovered a treasure trove of information on the nature and dates of impacts in oral traditions and mythologies (e.g. Masse 2007; Masse and Masse 2007; Bryant et al. 2007). And recently, we have begun to realize that larger impacts may be signaled by substantive changes in the Earth's paleoclimate.

The Tsunami Laboratory in Novosibirsk, Russia, has compiled and is maintaining an Expert Database on Earth Impact Structures (EDEIS, http://tsun.sscc.ru/nh/impact.php), which is somewhat more liberal than the well-known Earth Impact Database maintained by the Planetary and Space Science Centre, University of New Brunswick, Canada (http://www.unb.ca/passc/ImpactDatabase/index.html). In addition to including the fully validated impact structures, the EDEIS lists also proposed structures whose impact genesis still needs validation. For any structure, the degree of confidence of impact origin is reflected by its validity index V, which varies from 4 (apparently confirmed) to 0 (rejected) with intermediate values of 3 (probable), 2 (possible) and 1(supposed). The validity index reflects availability of four different set of impact criteria – morphological (circular form, presence of edge wall, inconsistency with local geological settings), petrologic (signs of shock metamorphism), mineralogical (presence of high pressure minerals, PDF structures, etc.) and chemical (presence of extraterrestrial elements and materials). Currently, the database contains 905 structures with 206 of them having V= 4, 200 with

Fig. 8 Map of the Earth impact structures, divided by the age of formation (905 structures for the whole Earth history). Source of data – Expert Database on Earth Impact Structures (EDEIS, http://tsun.sscc.ru/nh/impact.php)

V=3, 391 with V=2 and 39 with V=1 (Fig. 8). Sixty-nine records in the database have validity index of 0, because the proposed impact origin was disproven by additional study.

Using this index, we have identified 115 Quaternary period structures (38 with V=4, 38 with V=3, 34 with V=2 and 5 with V=1) containing a subset of 57 Holocene period structures (19 with V=4, 19 with V=3, 15 with V=2 and 4 with V=1). Not all, including the V=4 and 3 categories, will necessarily eventually be validated. For example, it could be demonstrated by future studies that mineral specimens thought to be extraterrestrial can actually be reproduced by terrestrial means, and that circular features on the seabed currently defined using satellite sea surface altimetry may prove otherwise. However, this index helps to address some of the significant information hidden or missing from standard lists of validated impact structures (e.g., Table 3) based on current statistical estimates of the rates, risks, and effects of cosmic impact.

Table 3 demonstrates clearly the problem of our current reliance on completely validated impacts by which to judge the rates and risks of cosmic impact. In particular, recent so-called "globally catastrophic" impacts ($\geq 10^6$ megatons [Mt]), modeled to occur on average once every million years (Bobrowsky and Rickman 2007; Morrison et al. 2002; Toon et al. 1997), and even more frequent regionally catastrophic impacts (ca. 10^3–10^5 Mt), are greatly underrepresented in such lists (Masse 2007; Masse et al. 2007). Our assumption is that the majority of such impacts would have occurred in the ocean, where only a handful of crater structures have been thus far identified, and none during the past 15 million years. Table 3 does not include known airbursts (such as the 1908 Tunguska event) or tektite fields for which a crater has not yet been definitively established. Table 3 also clearly demonstrates how smaller craters become increasingly more difficult to identify back through time.

Therefore, standard lists of validated impact structures are missing several geologically recent key impact events that could alter the manner in which scientists and planners currently model and estimate cosmic impact risks and effects. These include hypothesized impact events profoundly affecting human populations, some of which may also have been a trigger for climate change (Fig. 9). Some are previously

Table 3 Validated impact structures during the past 15 million years. Asterisks (*) highlight probable regionally catastrophic impacts while the plus (+) signify a globally catastrophic impact (>10^6 Mt). (Adapted with minor date changes from the University of New Brunswick Earth Impact Database, December 2008: http://www.unb.ca/passc/ImpactDatabase)

Impact structure name	Location of impact structure (terrestrial = T; oceanic = O)	Diameter in km of largest crater (and number of known associated craters)	Estimated date of impact ca. Years before present (ad 2009)
Sikhote alin	Russia (T)	0.027(122)	52
Wabar	Saudi Arabia (T)	0.116(3)	305
Sobolev	Russia (T)	0.053(1)	<1,000
Haviland	United States (T)	0.015(1)	<1,000
Kaalijärv	Estonia (T)	0.110(9)	2,400–2,800?
Campo del cielo	Argentina (T)	0.050(20)	4,200–4,700
Henbury	Australia (T)	0.157(11)	<4,700
Rio cuarto	Argentina (T)	4.500(11)	4,500–5,000
Macha	Russia (T)	0.300(1)	<7,000
Ilumetsa	Estonia (T)	0.080(3)	7,400–7,700
Morasko	Poland (T)	0.100(8)	<10,000
Tenoumer	Mauritania (T)	1.900(1)	21,400
Barringer	United States (T)	1.186(1)	49,000
Odessa	United States (T)	0.168(7)	<50,000
Lonar	India (T)	1.830(1)	52,000
Boxhole	Australia (T)	0.170(1)	54,000
Amguid	Algeria (T)	0.450(1)	<100,000
Tswaing	South africa (T)	1.130(1)	220,000
Kalkkop	South africa (T)	0.640(1)	<250,000
Dalgaranga	Australia (T)	0.024(1)	270,000
Wolfe creek	Australia (T)	0.080(1)	<300,000
*Zhamanshin	Kazakhstan (T)	14.000(1)	900,000
Veevers	Australia (T)	0.080(1)	<1,000,000
Monturaqui	Chile (T)	0.460(1)	<1,000,000
*Bosumtwi	Ghana (T)	10.500(1)	1,070,000
New quebec	Canada (T)	3.440(1)	1,400,000
Telemzane	Algeria (T)	1.750(1)	<3,000,000
Aouelloul	Mauritania (T)	0.390(1)	3,000,000
*El'gygytgyn	Russia (T)	18.000(1)	3,500,000
Roter kamm	Namibia (T)	2.500(1)	3,700,000
+Kara-kul	Tajikistan (T)	52.000(1)	<5,000,000
*Karla	Russia (T)	10.000(1)	5,000,000
Bigach	Kazachstan (T)	8.000(1)	5,000,000
Steinheim	Germany (T)	3.800(1)	15,000,000
*Ries	Germany (T)	24.000(1)	15,100,000

known and have been the subject of much study, such as the large Australasian strewn tektite field event (ca. 800,000 years BP) and Eltanin (ca. 2,615 million years BP), for which craters have not yet been identified (Masse 2007). Others, more controversial, include Burckle crater ("Flood Comet" event … see Abbott et al. 2005; Masse 1998, 2007) that may be associated with worldwide stories of a Great Flood (Noah's Flood) and the boundary change from middle to late Holocene around 4,800 years BP; the Tabban and Kanmare structures that may be associated with the AD 536–545 "years without a summer" climatic event (Baillie 2007); the Rio Cuarto impact in Argentina (Schultz and Lianza 1992) that may be associated with human population replacement around 6,000 to 3,000 years BP (Masse and Masse 2007; Barrientos and Masse 2009); the Chiemgau crater field in southern Germany that may relate to cultural changes in the 1st millennium BC (http://www.chiemgau-impact.com; Masse 2007); Mahuika crater just south of New Zealand that may be related to the beginning of the Little Ice Age at around AD 1450 (Bryant et al. 2007); and faunal extinctions and major climatic changes and/or climate downturns during the Younger Dryas stadial event that may have been caused by an impact/airburst over the Laurentide ice sheet at around 12,900 years BP (Firestone et al. 2007). The fact that Burckle, Chiemgau, Mahuika, Tabban – Kanmare , and the Younger Dryas Event are all suggested as comet impacts is interesting – and controversial given that comets are thought to make up less than 10% of the impact risk (Bobrowsky and Rickman 2007).

Current standards for verifying the ejecta from impact craters on land are largely based on three techniques: analysis of thin sections, verification of iridium anomalies, and verification of shocked quartz. These techniques are either unsuitable or more difficult to apply in the case of abyssal impact craters. Conventional thin sections have the best quality and polish when they are made from rock fragments a centimeter or more in diameter. However, moderate sized impactors that strike an unconsolidated substrate composed of unlithified sediments do not produce centimeter sized rock fragments. Instead, we find glassy fragments that rarely exceed 500 micrometers (μm) in diameter.

Our studies of oceanic cores indicate that glass and mineral fragments are usually very small in impacts produced by moderate-sized impactors in unconsolidated unlithified sediments. The glass and mineral fragments usually range between 63 and 250 μm in diameter. These fragments can be made into thin sections but they are more altered by seawater and are more difficult to polish and to analyze by microprobe techniques. So far, we have been unable to achieve an excellent polish on fragments of silica rich minerals or glass in this micrometer size range. The quickest method of verifying shocked minerals is electron backscatter diffraction. This technique requires an excellent polish on the mineral grains or the results are unsatisfactory. The second technique, that of verification by iridium anomalies, depends on a low background level of iridium in the layers above and below the projected impact layer. Because deep sea sediments

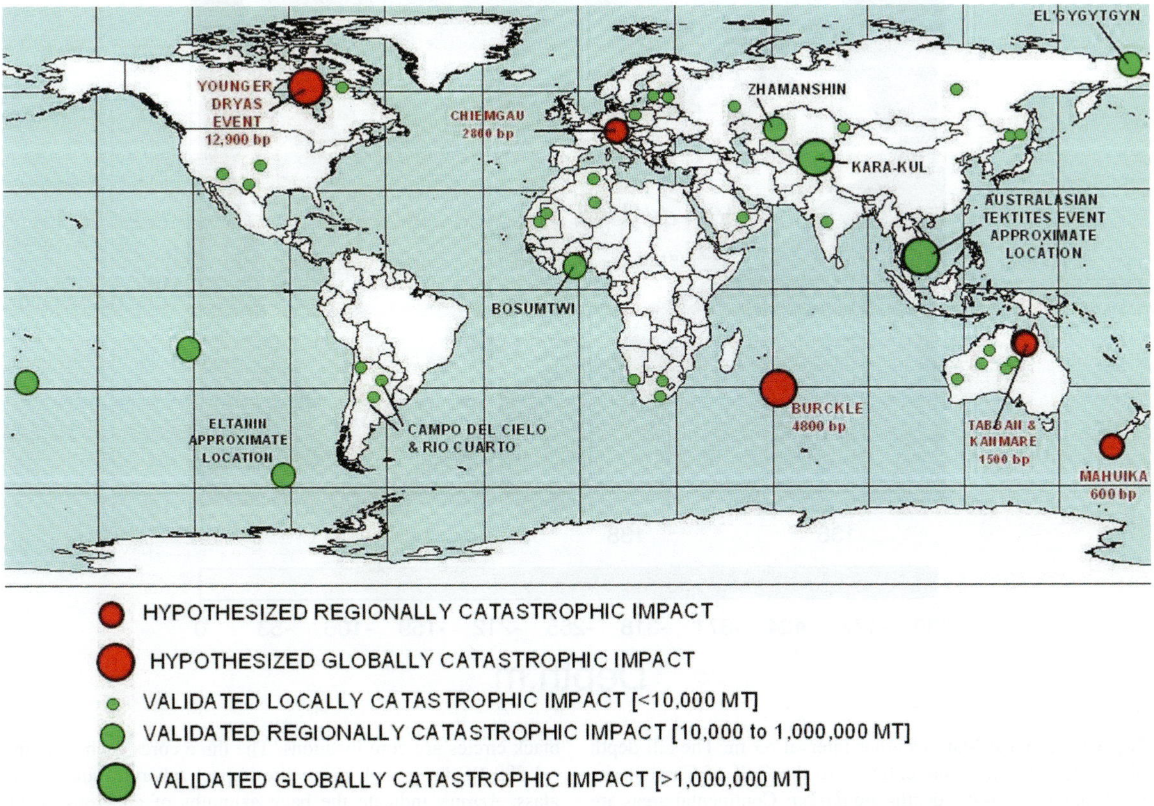

Fig. 9 Validated and selected hypothesized impacts by asteroids and comets during the past 5 million years

accumulate slowly, the background level of iridium from cosmic dust is high. Thus, carbonate poor deep-sea sediments are not suitable for verification of an impact layer by the detection of an iridium anomaly. The most common technique for verification of terrestrial impact craters is the confirmation of shocked quartz. Because deep sea basalts are not quartz normative, most contain no quartz at all. Thus, this third common technique is unsuitable for most oceanic impact cratering events.

Gulf of Carpentaria Craters

The Gulf of Carpentaria is a square marine basin on the north coast of Australia (Fig. 10). The Gulf of Carpentaria contains stable continental crust, with no evidence for tsunamigenic earthquakes (M>7.0), recent landslides, or volcanic eruptions (Drummond et al. 1985). The Gulf of Carpentaria is connected to the open ocean via the Arafura Strait with a sill depth of 53 m (Smart

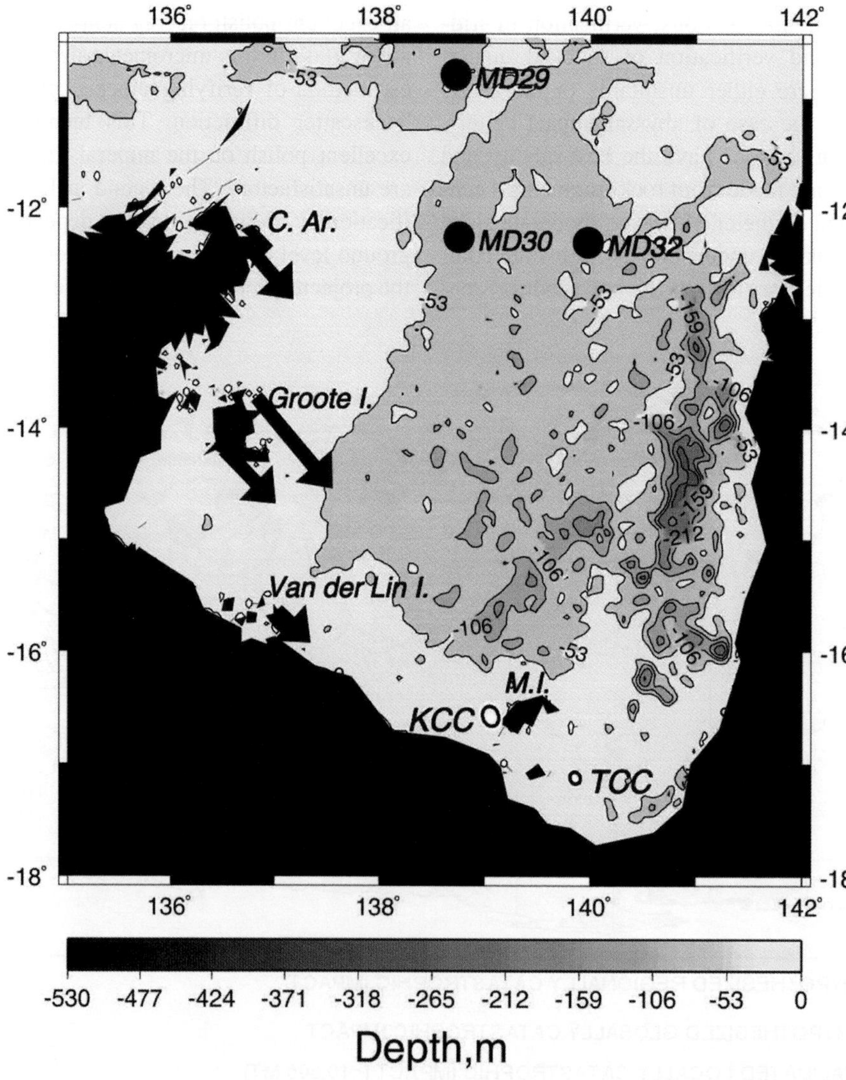

Fig. 10 Regional Map. Contour interval 53 m. The sill depth of the Arafura Strait, the entrance to the Gulf of Carpentaria, is 53 m. Greater water depths are *darker*. Continental areas are *black*. KCC: Kanmare Crater candidate. TCC: Tabban Crater candidate. M.I.: Mornington Island, C. Ar.: Cape Arnhem. Filled black circles are core locations. The three cores each contain a ~1,500 BP layer with magnetite impact spherules and impact glass. *Arrows* indicate the back azimuths of chevrons. These arrows are the inferred directions to the tsunami source of the chevrons (Kelletat and Scheffers, 2003)

1977; Torgerson et al. 1983). As a result, low sea level in the early Holocene turned the Gulf into a brackish lake (Chivas et al. 2001; Jones and Torgerson 1988; Reeves et al. 2007). The brackish water produced extensive deposition of siderite ($FeCO_3$), a mineral absent from the open ocean except in micrometer sized fragments. Around 10,800 years ago, sea level rose and the Gulf of Carpentaria became an open marine basin (Reeves et al. 2007). Since that time, the Gulf has had a high sedimentation rate due to rapid deposition of marine microfossils composed of $CaCO_3$.

We initially searched for impact craters in the Gulf of Carpentaria because Kelletat and Scheffers (2003) found chevron dunes with an orientation that implied a spatially restricted source area in the southeastern corner of the Gulf. They interpreted the chevron dunes as mega tsunami deposits. When we contoured bathymetry derived from satellite altimetry, we found two impact crater candidates that are 18 and 12 km in diameter (Fig. 11). Because of the unconsolidated nature of the sediment, we estimate that an impactor about 640 m in initial diameter could have fragmented to produce both craters.

Discussion of Impact Ejecta

Three cores from the Gulf of Carpentaria contain a thin layer with vitreous material, magnetite spherules, and occasional pure carbon and silicate spherules with an age of about 1500 B.P. (Martos et al. 2006; Abbott et al. 2007a, b). Although magnetite spherules can form as ablation products of meteorites, the magnetite spherules in the Gulf of Carpentaria are undoubtedly products of a terrestrial impact. The spherule in Fig. 12 has a quench texture and a perfect spherical form, implying that it is melting rather than eroding out of the sediment. The terrestrial origin of the sediment is clear in Fig. 12f. This figure shows that the sediment contains a replacement cast of a microfossil. The replacement cast is composed of iron oxide. The groundmass of the sediment is iron oxide with a Cl peak (Fig. 12c). The bright spots to the left of the spherule are barite (Fig. 12d). So far as we know, barite has never been found in a meteorite. Thus, the combination of fossil casts and barite points to a terrestrial source for the sediment. This implies that the magnetite spherules are the result of a terrestrial impact and are not ablation products of a meteorite.

Examples of Chevron Dunes

Figures 13 and 14 show examples of the chevron dunes first described in detail by Kelletat and Scheffers (2003). Figure 13 shows the chevron dunes on Groote Island, and Fig. 14 shows the chevron dunes on Van der Lin Island. The inferred maximum implied run-ups

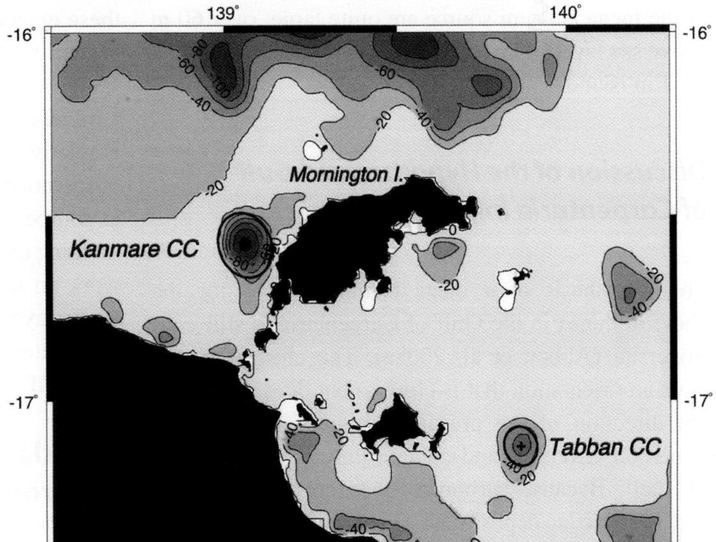

Fig. 11 Topographic map. Bathymetry inferred from satellite altimetry (Sandwell and Smith, 2005). Contour interval 20 m. Two crater candidates Kanmare and Tabban appear as elliptical topographic lows near Mornington Island. *Black ellipses* at the crater margins have average diameters of 18 and 12 km respectively

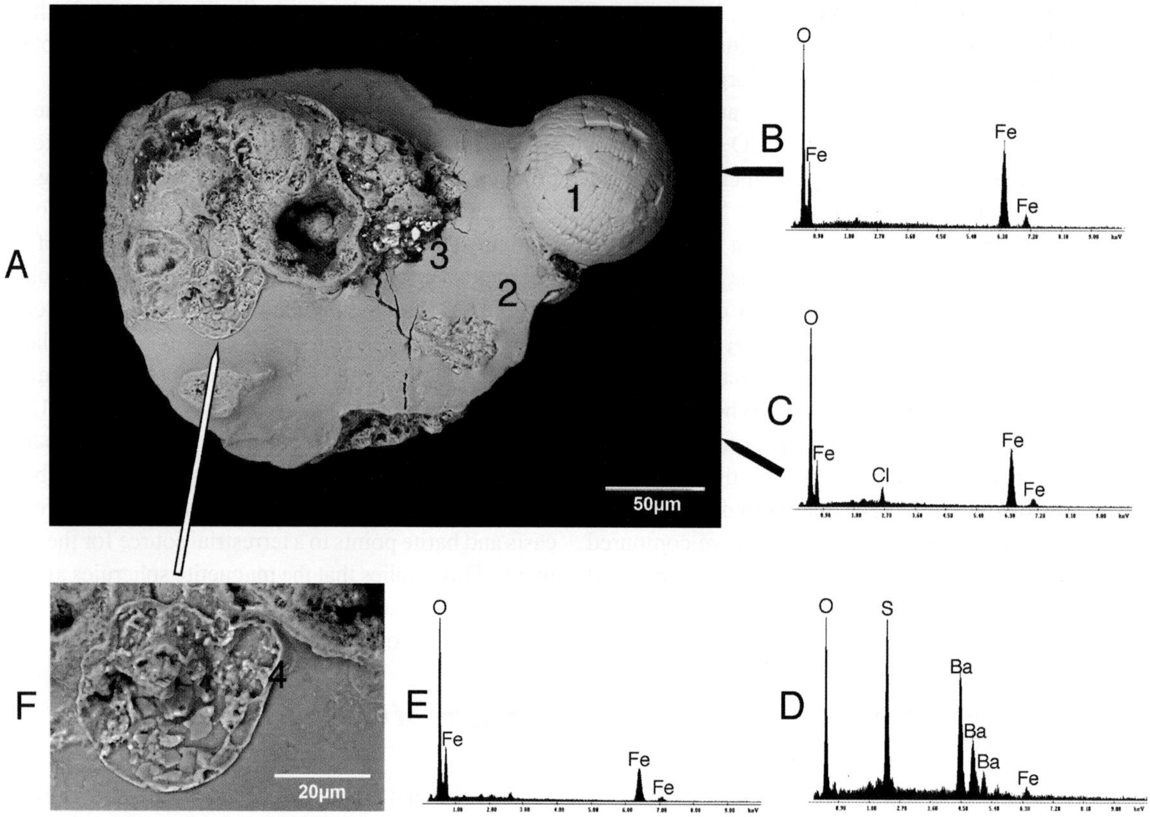

Fig. 12 Evidence for a terrestrial source rock for magnetite spherules. (**a**) Magnetite spherule with quench texture melting out of hematite rich marine sediment; (**b**) X-ray analysis of the composition of the quench textured spherule (1); (**c**) X-ray analysis of the composition of the groundmass (2); (**d**) X-ray analysis of the composition of brightest areas (3); (**e**) X-ray analysis of edge of fossil in F; (**f**) Enlarged image of fossil. (photo credit – D.Breger)

of the mega tsunami source are quite large, over 60 m above sea level on Groote Island and over 20 m on Van der Lin island.

Discussion of the Hypothesized Gulf of Carpentaria Impact Event

The hypothesis of a mega tsunami origin for the chevron dunes in the Gulf of Carpentaria is still controversial (Abbott et al. 2008a). The chevron dunes have an orientation that is close to but does not match the direction of the prevailing winds. The chevron dunes on Groote Island contain unusual lag deposits of shell. Because carbonate does not survive wind action, these lag deposits are a primary target for future investigation.

The circa 1,500 BP. spherules and glass we have found in three cores are undoubtedly impact ejecta from a terrestrial impact. The problem is to firmly link these ejecta to the two crater candidates in the southeastern part of the Gulf. In some cases, the magnetite spherules are melting out of rocks containing siderite, a mineral that is extremely rare in the open ocean and is relatively common in the Gulf of Carpentaria. This observation is consistent with the two crater candidates as source craters but is not definite proof. Geophysical imaging of the crater candidates and marine geological work on mapping changes in thickness of the impact ejecta layer are required to determine if the two crater candidates are the source of the impact ejecta layer in cores MD29, MD30 and MD 32.

Burckle Crater and the Madagascar Chevrons

Burckle Crater

We have identified an extremely large crater candidate (Burckle crater, 29 km in diameter) in the Indian

Fig. 13 Chevron dunes on Groote Eylandt in the Gulf of Carpentaria. The arrow depicts the direction of the wave source. Note the absence of large chevrons at the beach along the bay on the right. This part of the bay is protected from waves coming from the source area but not from the prevailing winds

Ocean at 31°S, 61°E (Fig. 15). We initially identified this crater in response to anthropological and archaeological work that pointed to the southern Indian Ocean as the source of a Holocene age impact (Masse 1998, 2007; Abbott et al. 2005). We found this crater using satellite altimetry, before there was any data on chevrons available on Google Earth. The latitude of this crater is identical to the latitude of Perth, Australia. Independently, Kelletat and Scheffers (2003) predicted that the tsunami source for chevrons in western Australia lay at the same latitude as Perth. Once we were aware of their work, we realized that there should be even larger chevrons in Madagascar if our inferences about Burckle crater were correct. When Google Earth imagery became available, we found chevrons that were over 45 km long and extending to over 200 m above sea level in southern Madagascar.

Burckle crater is a round hole 29 km in diameter on the wall of a fracture zone south of the Southwest Indian Ridge (Fig. 16). Its morphology is subtle but we are certain that it is not a submarine volcanic edifice or a submarine fault block basin. Volcanic

Fig. 14 Chevrons on Van der Lin Island, near south coast of the Gulf of Carpentaria

edifices have greater topographic relief. Volcanic edifices appear like flat-topped steep sided hills except when they are active. Submarine fault block basins are square rather than round. The low topographic relief of Burckle is expected for a submarine impact crater. Submarine impact craters form when an impactor with a diameter that is at least 1/10 of the water depth strikes the sea floor (Davison and Collins 2007; Gault and Sonnett 1982). The impact crater forms in the rock, sediment, and water. After the initial explosion, the impact water cavity and associated sediment collapse inward, filling the interior of the crater (Tsikalas et al. 1999). Thus submarine impact craters have much more modest relief than impact craters of comparable diameter on land. The crater rim and walls are also eroded by the resurge of water into the crater, producing gullies (Sturkell 1998; Tornberg 1997). Three such prospective gullies are evident for Burckle crater (Fig. 16). Unfortunately, there is no actual geophysical data from the location of the proposed Buckle crater impact structure itself. All of our inferences about Burckle crater are based on satellite gravity determination of sea floor bathymetry (Sandwell and Smith 2005; Smith and Sandwell 1997).

Burckle crater has several characteristics that would likely produce some impact ejecta comparable to a giant (>60 km) crater in the ocean floor. Normal oceanic crust is 6–7 km thick (White et al. 1992). However, fracture zone crust may be nonexistent or at most 1–2 km thick (Mutter and Detrick 1984). Fracture zone walls often expose mantle rocks (Bonatti 1990; Michael and Bonatti 1985). Thus, Burckle's location on a fracture zone wall implies that its impact ejecta will contain mantle rocks such as serpentinized peridotites. The process of serpentinization enriches mantle rocks in calcium, and deposits veins of calcite. It may also contain some oceanic crustal rocks with a high abundance of plutonic rocks (gabbros) and fractionated rocks (oxide gabbros and plagiogranites) (Dick and Fisher 1984). Because impact craters

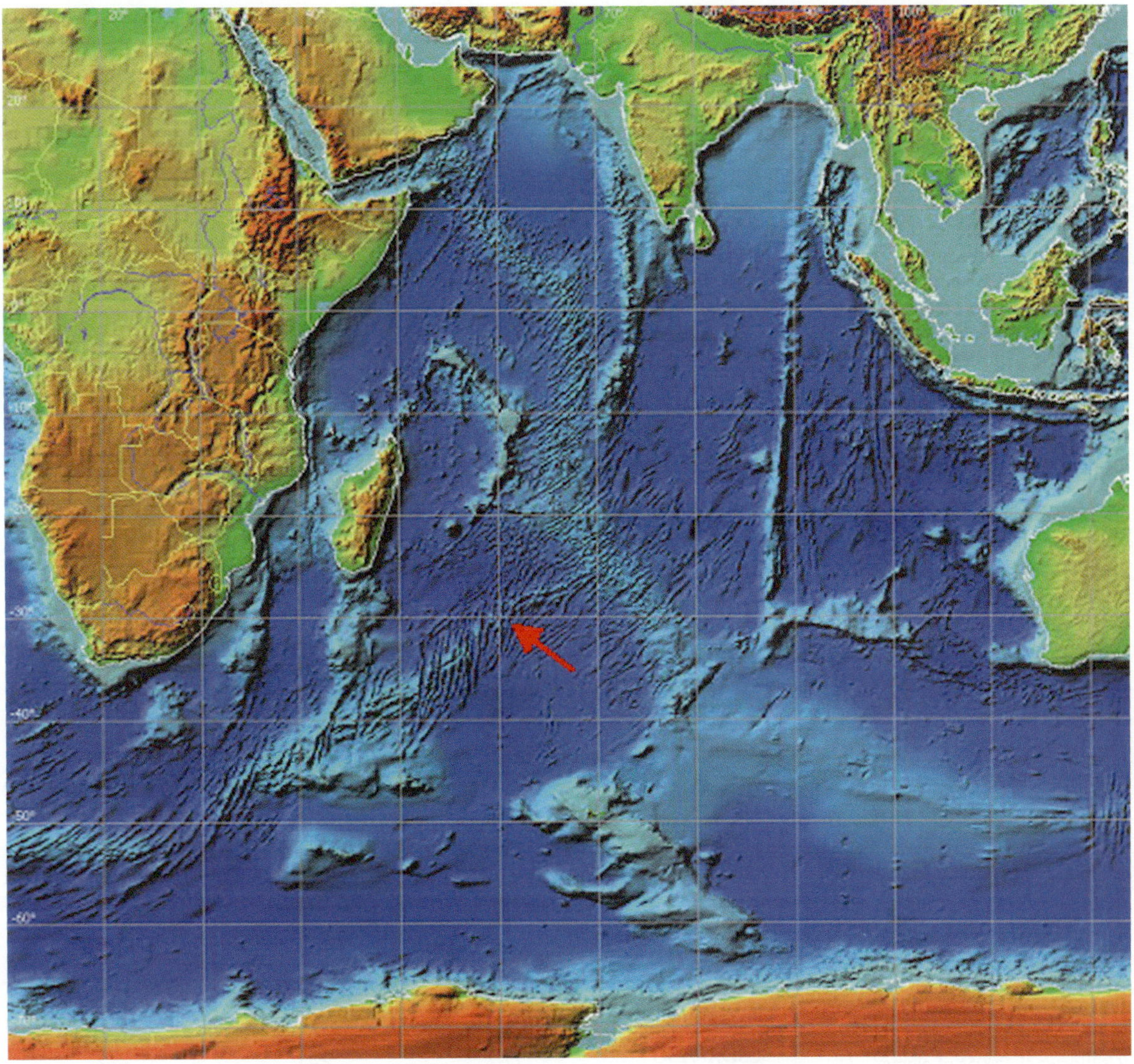

Fig. 15 Map of the approximate location of Burckle Crater candidate abyssal impact structure (*arrow*) along the southeast Indian Ridge. The map is adopted from the ETOPO5 topography coverage on the *Integrated Tsunami DataBase for the Pacific and Indian Oceans* CD-ROM (ITDB, 2005)

excavate rocks down to a depth that is about 1/10 of their diameter (Melosh 1989), an impact onto normal oceanic crust might eject small amounts of gabbro; however, its dominant ejecta should be oceanic sediment and oceanic basalt. Impact ejecta composed of oceanic mantle would be absent unless the crater was at least 60 km in diameter. As craters over 60 km in diameter have a repeat time of tens of millions of years (Collins et al. 2005), Burckle is most likely the only impact in the last 2 million years that produced ejecta containing mantle rocks.

Ejecta Mapped in Core Samples

We have made preliminary estimates of the thickness of impact ejecta in five cores near Burckle crater candidate (Table 4, Fig. 17). Our estimated thicknesses of the ejecta layers are in general agreement with Burckle as the source crater. However, there is still great uncertainty about the overall thickness of the layers. This is due to the small sample sizes that we have worked with and the unusual nature of the ejecta. We have

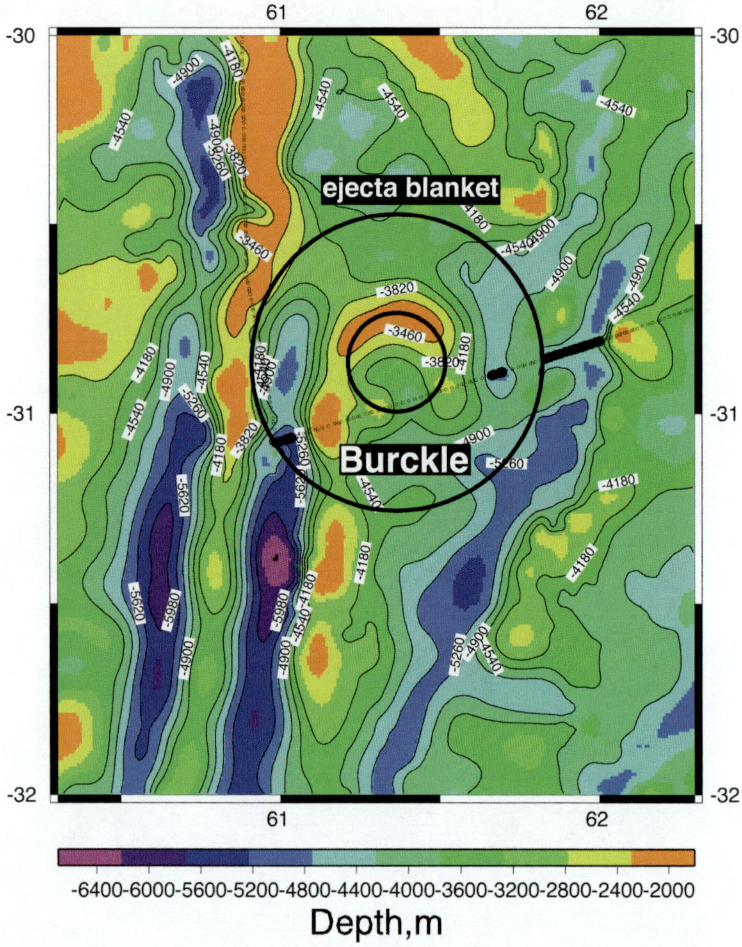

Fig. 16 (after Abbott et al., 2005) Topographic map. Contour interval: 360 m. *Large black circle*: calculated edge of continuous ejecta blanket for a crater on land. *Small black circle*: rim of Burckle crater. Note the three major lows in the crater rim that are possible research gullies Dotted line: portions of seismic line showing no sediment cover. *Solid black line*: portions of seismic line showing sediment cover. The sediment is ponded in topographic lows creating locally smooth topography. Area to N and W of Burckle crater shows smooth topography and may represent the continuous ejecta blanket of Burckle crater

Table 4 Observed thicknesses of impact ejecta and modeled thickness (Collins et al. 2005) of ejecta from Burckle crater

Core#	Distance to Burckle, km	Observed Thickness, cm	Modeled Thickness, cm
LSDA-123G	77.7	>54	279
LSDA-122G	118.7	67±5	78.3
LSDH-23G	154.5	32±5	32.8
LSDA-124G	228.8	> 6	10.9
DODO-132P	335.5	3±1	3.5

learned from our work on other oceanic impacts that the minimum weight of sample we need to confidently identify the presence (or absence) of impact ejecta is between 10 and 20 g. The cores we have been working with are old and heavily sampled. Thus, in many cases we could only obtain a 2–5 g sample. This size of sample is not large enough for us to confidently state that impact ejecta are absent from a given layer. Thus, our layer thicknesses are in many cases only minimum estimates.

Nature of the Burckle Ejecta

The ejecta from cores near Burckle crater are unusual. We find mineral fragments, rock fragments, glass fragments, nearly pure carbon impact spherules, and calcite rhombohedrons (Fig. 18). The latter are the most common type of impact ejecta. Because the water depth at all of the coring sites is deep, it is impossible for these calcite rhombohedrons to represent in

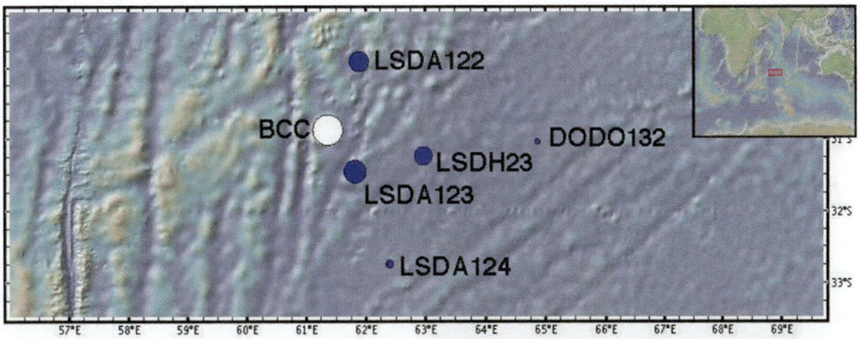

Fig. 17 Burckle Crater and Thickness of Ejecta Layer. Topographic map is from GeoMapApp. *White circle*: Location of Burckle Crater Candidate. *Blue circles*: Cores containing impact ejecta. Size of blue circle is directly proportional to the average inferred thickness of the impact ejecta layer. Ejecta layers are thicker closer to the crater candidate and thinner further away from the crater candidate

Table 5 Water depths and ejecta types of material in core samples from Indian Ocean

Core#	Depth, m	Ejecta Types
LSDA-123G	4,200	Carbonate rhombs, native Fe, pyroxene, high C silicate glass, C spherule
LSDA-122G	4,400	Carbonate rhombs, olivine, pyroxene, C spherule
LSDH-23G	4,910	Carbonate rhombs, C spherule, glass
LSDA-124G	4,780	Carbonate rhombs, C spherule
DODO-132P	4,805	Carbonate rhombs, native Ni, native Fe

situ precipitates (Table 5). Seawater is not super saturated in calcite at any water depth this far south in the Indian Ocean. Carbonate fossils found with the calcite appear quite corroded, as would be expected for a location with bottom water that is heavily undersaturated in calcite. Thus, the preservation of these calcite rhombohedrons requires two circumstances: a mechanism to produce them locally and a mechanism to bury them quickly. Because Burckle crater candidate is situated on a fracture zone, it is possible that the calcite rhombohedrons represent cleavage fragments of the fracture zone crust and mantle. The crust and mantle of fracture zones are heavily altered by hydrothermal circulation. During this process, calcite veins are precipitated. These calcite veins would be broken apart during an impact event and could be the source of the calcite rhombohedrons that we find.

The Madagascar Chevrons

Mega tsunami chevrons are ubiquitous along the coast of southern Madagascar, more so than on any other coastline in the world (Fig. 19). They extend east in a nearly unbroken chain 375 km long from Itampolo Bay to a point midway between Mandrare River and the city of Taolagnaro (Ft. Dauphin). In August-September 2006, D. Abbott, E. Bryant, and V. Gusiakov conducted two weeks of preliminary reconnaissance from Faux Cap to Ampalaza Bay. Participants made several traverses of the four largest chevron formations in this area-near Faux Cap, Cap St. Marie and along the coast of Fenambosy and Ampalaza Bays (Figs. 19, 20 and 21).

All investigated chevrons appear to consist of marine sand transported by water. In some cases, the chevrons overtopped the front edge of the Karimbola Plateau escarpment whose top is situated more than 125–150 m above the neighboring coastal plain (Fig. 21). Run-ups greater than 200 m have been documented throughout the Ambovombe area, although these features were erroneously considered wind-generated "paleo-dunes" by previous research (Clark et al. 1998). The Ambovombe chevrons have been largely masked due to intensive farming and wind activation of the original chevron dunes, but their general structure is clearly apparent in false color applications of satellite imagery (Clark et al. 1998: Fig. 10).

The two best delineated and preserved chevron dunes are those associated with Ampalaza and Fenambosy Bays (Figs. 20 and 21). The Ampalaza chevron

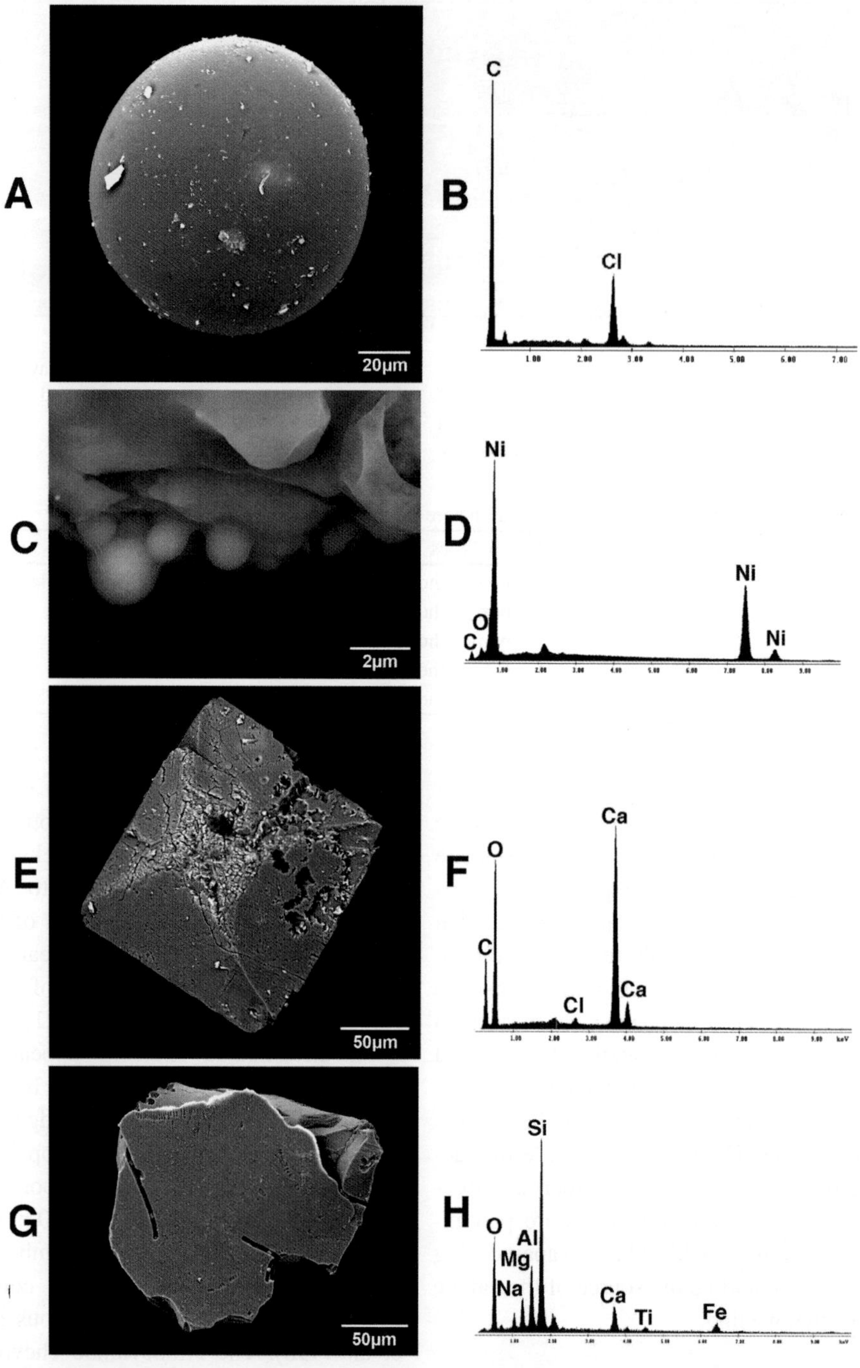

Fig. 18 Typical impact ejecta from cores near Burckle crater candidate. (**a**) Nearly pure c impact spherule from core LSDA122: 80–82 cm depth; (**b**) EDS (elemental X-ray) analysis of the c spherule in a; (**c**) Native Ni metal being melted (drops are NiO) from core DODO132: 14–16 cm depth; (**d**) X-ray analysis of the native Ni in c; (**e**) Calcite rhombohedron from core LSDH23: 21–23 cm depth. The rhombohedron surface appears etched and contains tiny craters on upper tip; (**f**) EDS analysis of the calcite rhombohedron in e; (**g**) Glass with holes like worm burrows from core LSDA123: 22–24 cm depth. The hole shapes are not typical of volcanic glass; (**h**) EDS analysis of composition of the glass in g. (photo credit – D.Breger)

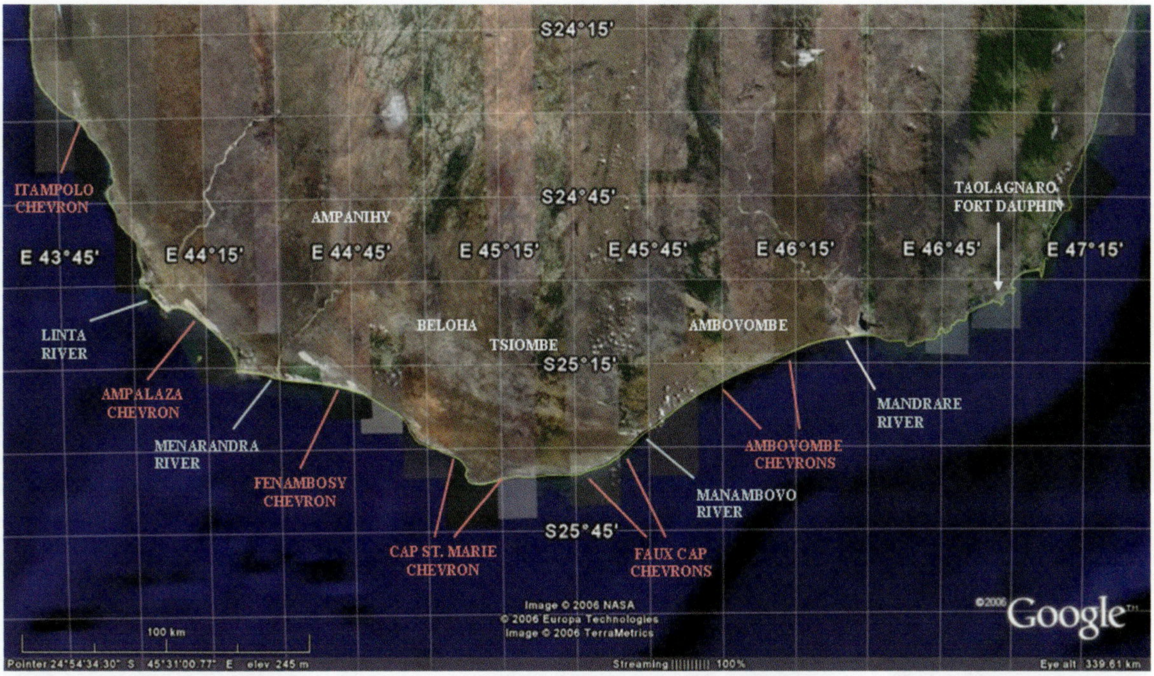

Fig. 19 Four systems of chevron dunes (Ampalaza, Cap St. Marie, Faux Cap, Fenambosy) on the southern coast of Madagascar studied during the 2006 expedition

is a remarkable 45 km in length from its base immediately to the east of Ampalaza Bay to its apex just short of confluence with the Linta River (Fig. 20). The tip of the apex (north of the village of Antanamilihitsa) reaches a height of approximately 80 m above present sea level.

There are a number of intriguing features associated with the chevron and its presumed formation by a mega tsunami wave(s). The first is that of the multiple sets of well-defined smaller internal chevrons which apparently resulted from the backwash of the largest mega tsunami wave. Also apparent is the fact that the Menarandra River delta appears to have lost much of its originally deposited sediment. We hypothesize that the mega tsunami wave scoured the delta, forming the lagoonal system west of the delta and ending up as part of the bulk of the Ampalaza chevron dune itself. That the chevron was created by surface water flow, is also seemingly evidenced by the orientation of river bottom sediments (probable silts) smeared across floodplain terraces northwest of the Linta River, and by the orientation of smaller chevrons near the bay. Together these features indicate both the complexity and magnitude of these mega tsunami deposits. [See Bourgeois and Weiss 2009 for a non-mega tsunami interpretation of the Madagascar chevrons]

As remarkable as is the Ampalaza chevron, in several respects the Fenambosy chevron (Fig. 21) is even more striking. Although the full length of the chevron is nearly 10 km less than that for Ampalaza, only approximately 35 km from base to apex, the height of run-up reaches a staggering 180 m above mean sea level at a point some 8 km inland from the coast. In reaching this elevation the mega tsunami wave fronted and eventually overtopped the Karimbola Plateau escarpment, the top of which averages about 140–150 m above mean sea level. Where the mega tsunami wave overtopped the escarpment rim, several sections show evidence of erosional channel scarring.

Critics of the mega tsunami genesis of the southern Madagascar chevrons point to what appears at first glance to be a match between the orientation of the chevrons and that of the prevailing winds. However, careful measurement of the orientation of each chevron instead reveals consistence with wave refraction patterns. Fig. 22 shows the systematic change in orientation of long axes of Madagascar chevrons consistent with the predicted azimuths of a tsunami wave com-

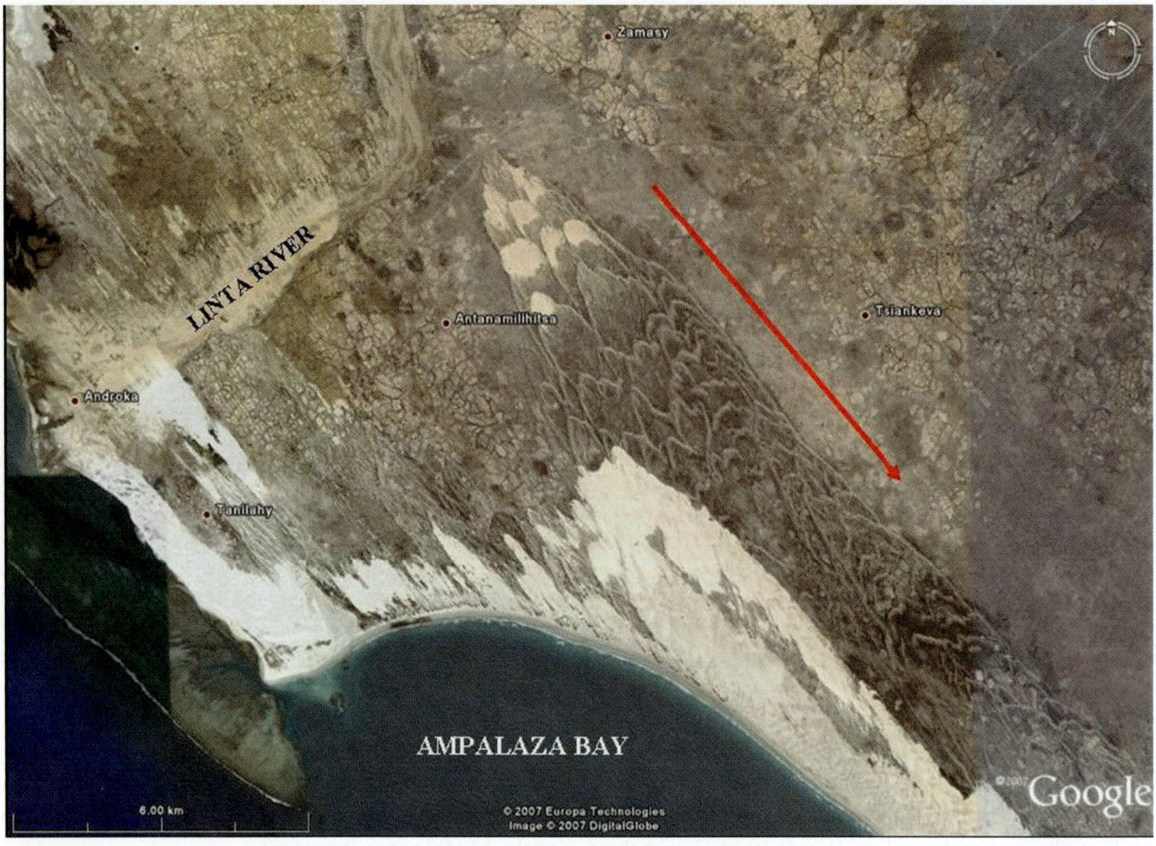

Fig. 20 Vertical Google Earth satellite image of the Ampalaza chevron, Madagascar. The *red arrow* points toward the tsunami source. The chevron apex (north of Antanamilihitsa) exhibits recent barchan dune activation

ing from a point source located in the vicinity of the Burckle crater candidate and refracted by sub-bottom topography near the southern coast of Madagascar.

In contrast to wind-blown dunes, which consist of a well sorted, unimodal size distribution, the Madagascar chevrons are unsorted with a broad range of particle sizes, from small boulders down to clay particles. They also include marine shells and microfossils (Fig. 23), with tentative species indentification being indicative of a source originally along the Madagascar outer shelf (Simon Haslett, personal communication 2008). We assume that a mega tsunami wave of the size indicated by the Madagascar chevrons should be able to entrain outer shelf sediments along with material closer to the coast. Dump deposits, containing rock fragments typical of mega tsunami processes, were found eastward from the Fenambosy chevron to Cap St. Marie. Many of the rock fragments were not locally derived. In addition to wave run-ups of 80 m and 180 m found respectively at Ampalaza and Fenambosy, we documented maximum run-ups of ca. 200 m above present day sea level at Faux Cap and 190 m at Cap St Marie. Each chevron represents lateral transport of sediment onto the coast over many kilometers: 20 km at Faux Cap, 30 km at Fenambosy, and 45 km at Ampalaza.

Analysis of sediment samples collected from the southern Madagascar chevron reveals several features possibly indicative of cosmic impact. These include marine microfossils with surface splash particles of molten native metals. In addition, a number of the marine microfossils give the appearance of being "melted" as if from high temperatures (Fig. 23), although this could simply represent normal chemical weathering.

The age of the Madagascar chevrons is unknown. We estimate mid-Holocene (ca. 7,000–4,000 years old)

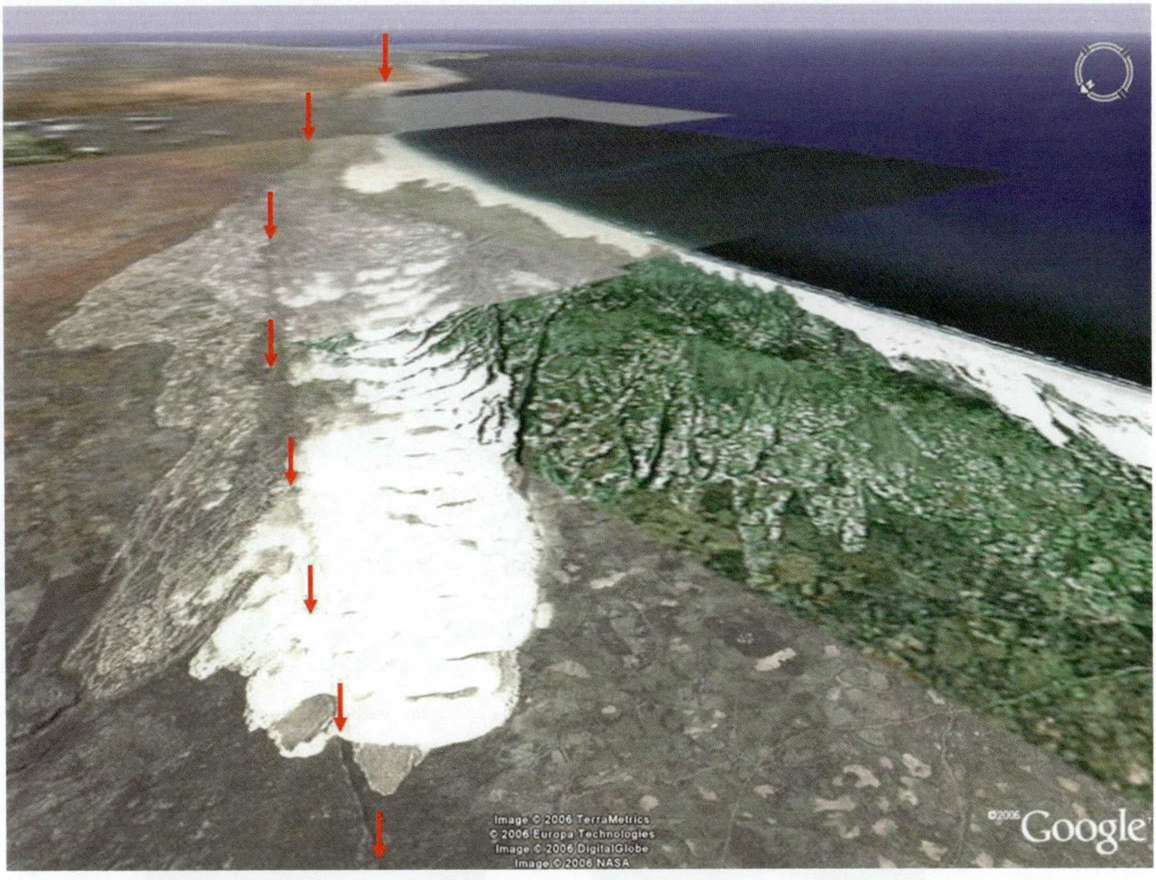

Fig. 21 Oblique view looking southeast at Fenambosy chevron. *Arrows* point to the top front edge of the 150-m-high Karimbola Plateau escarpment. *White areas* in the apex and the central portion of the chevron represent an extensive area of recently activated wind-blown dunes that are common also along coastal beaches. The *green rectangle* is an area of low resolution imagery

based on general appearance and vegetation cover. They are no younger than AD 500 based on the radiocarbon dating of archaeological sites situated on top of the dunes, and they are almost certainly not older than the beginning of the Holocene Period. Chevrons are relatively fragile geomorphic features, and they would not have survived the dramatic changes that occurred during the Younger Dryas climatic event between 12,000 and 13,000 years ago and subsequent sea level rise. Portions of the chevrons have been disturbed by historic human activity during the estimated 2,500 years that people have lived on Madagascar (Blench 2006; Burney et al. 2004), and are being even more severely damaged by wind activation and the formation of incipient barchan dunes. Because the Madagascar chevron dunes are older than the arrival of humans on the island, unlike the Gulf of Carpentaria impact event, there seemingly is no oral history specific to the formation of these features.

Rapid Climate Change

There is no question that the larger recorded impacts on Earth, such as the Chicxulub Cretaceous-Tertiary (KT) boundary event, signal major rapid climatic shifts, and in the case of the KT impactor, associated significant species extinction. However, smaller impact events, in the range of 10^4–10^6 Mt, are much less certain in terms of environmental and climatic effects, but they are of considerable concern due to the interconnected nature of modern society (MacCraken 2007). Some studies

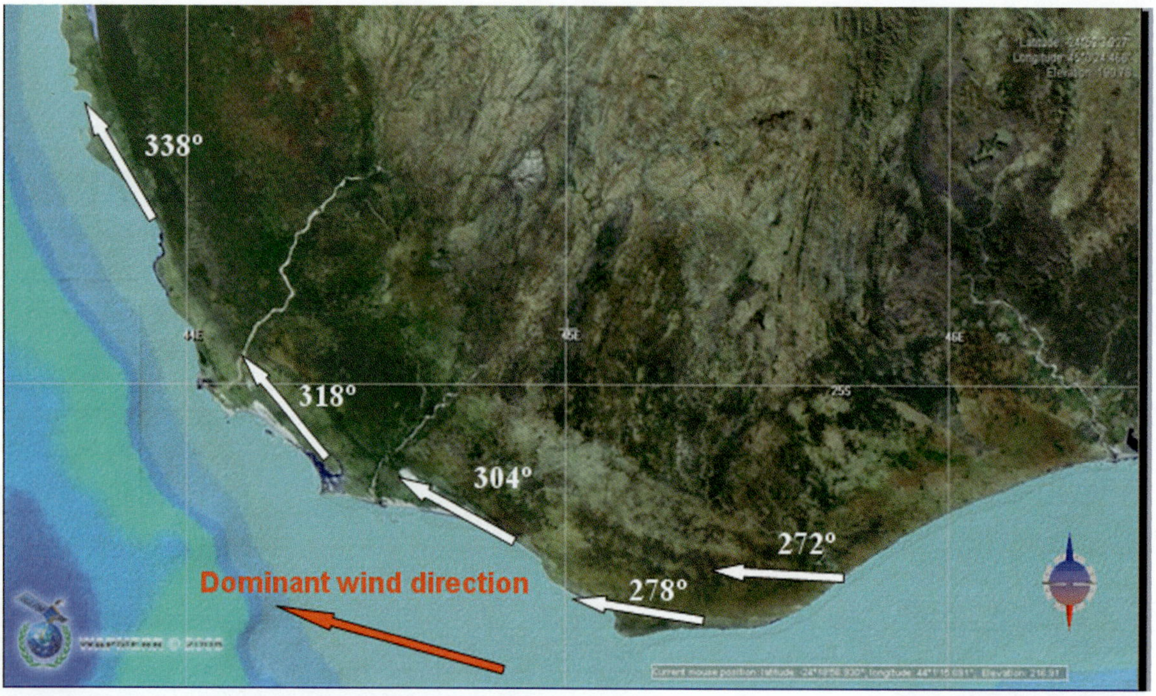

Fig. 22 Systematic change in orientation of Madagascar chevrons that is consistent with refraction pattern of incoming tsunami waves. Azimuth to the Burckle crater is 285°

Fig. 23 Examples of well preserved fragment of coral (**a**) and "melted" (**b**) and "unmelted" (**d**) benthonic foraminifera, found in sediment samples collected from the Madagascar chevrons. The other specimen (**c**) has not yet been identified (photo credit – D. Breger)

have even indicated that impactors between 150 and 1,000 m in diameter are capable of causing significant ozone depletion (Birks et al. 2007). Most studies suggest that significant environmental perturbations and short-term climatic effects would not be expected for impactors under about 1 km in diameter and yielding energies less than around 10^5 Mt (e.g. Chapman 2007), with long-term climatic effects being the consequence of "globally catastrophic" impactors between 2 and 3 km in diameter and an energy release between 10^6 and 10^7 Mt (Toon et al. 1997; Chapman 2007).

Of considerable interest for our research, then, is the possibility that the hypothesized Gulf of Carpentaria twinned impact and the Burckle crater-Madagascar chevron impact are temporally associated with environmental perturbations and the rapid onset of climate change.

The Burckle crater-Madagascar chevron impact is proposed to date to around 4,800 BP (Masse 2007), which would then make it roughly coincident with the climatic boundary shift between the middle to late Holocene, variously dated at between 5,000 and 4,800 BP. This climatic boundary shift is both poorly dated and its genesis uncertain; however, it generally represents a permanent change from warmer-dryer conditions to a cooler-wetter climatic regime at least for mush of the northern hemisphere, and also seems to signal Hong et al. 2005 a shift in the periodicity and intensity of the El Niño-Southern Oscillation pattern (e.g., Cane 2005; Hong et al. 2007). These effects are suggested as being consonant with the atmospheric injection of considerable water vapor and aerosols from an abyssal impact of around 10^6 Mt indicated by the apparent size of Burckle crater candidate impact structure, and from mega tsunami effects noted in southern Madagascar and on the western coast of Australia. Documentary and archaeological evidence (Masse 2007) suggests that the impact resulted in an extended period of virtually worldwide atmospheric rainout and associated cyclonic storms that stopped after 7–10 days once water vapor in the atmosphere approached pre-impact levels. Storm surges during the cyclonic activity would have devastated many oceanic coasts that normally are not subject to cyclone activity.

Our present dating for the Gulf of Carpentaria chevrons and the presence of likely impact microejecta in oceanic sediment cores places the hypothesized impact in the sixth century AD. This dating then suggests a potential relationship with the 536–545 "years without a summer" climatic perturbation (Abbott et al. 2008b). The Kanmare and Tabban hypothesized impact structures are indicative of an impact between approximately 10^4–10^5 Mt, perhaps suitable for the sun-obscuration and cooling temperatures historically documented for this period (Rigby et al. 2004). Unlike the Burckle crater event, the Gulf of Carpentaria impacts would have lofted less water and more sediment particles into the atmosphere, thus creating aerosol dispersion probably similar to that associated with massive ultra-Plinian volcanic eruptions. The environmental and climatic effects lasted several years, resulting in a period of crop failure and starvation, but were not permanent as was the apparent case with the hypothesized larger Burckle crater impact.

The Burckle crater and Gulf of Carpentaria impacts take on added significance in light of ongoing studies associated with the proposed Laurentide ice sheet impact at 12,900 BP that allegedly triggered the dramatic beginning of the Younger Dryas climatic cooling event (Firestone et al. 2007). The Younger Dryas perturbation lasted more than 1300 years, resulting in massive northern hemisphere vegetation change, megafaunal extirpations and extinctions, and apparent major effects on human populations.

Thus these three climatic events, that of 12,900, 4,800, and 1,500 BP, have potential major ramifications if caused by cosmic impact, rather than by more pedestrian terrestrial causes.

Conclusions

1. Available historical data show that the largest possible seismogenic and volcanic tsunamis can hardly exceed 45–50 m in their maximum run-up at the nearest coast with 15–20 m run-up in the far-field. Landslide-generated tsunamis can be highly destructive locally, but never flood any extended part of the coast. Meanwhile, the world ocean's coastline contains prominent erosional and depositional features of catastrophic water currents and waves of much higher magnitude.
2. The measured run-up heights (100–200 m) and inland penetration (tens of kilometers) of these features over the extended part of the coast are far beyond the range produced by the largest histor-

ically known tectonic tsunamis (seismic and volcanic). Such great run-ups can be produced only by large-volume submarine landslides or oceanic impacts. In the latter case, such impacts may also be responsible for major environmental downturns in the Holocene that are contemporaneous with those implied by tree-ring records, lake-bottom sediment analyses, and ice-core data.

3. The Gulf of Carpentaria crater candidates have several lines of evidence in favor of a bolide impact origin. The first is their overall morphology derived from satellite altimetry. The second is the occurrence of clear terrestrial impact ejecta in the form of impact spherules and vitreous material. The impact spherules are mostly iron oxide but a few silicate spherules and pure carbon spherules are also present. We have found iron oxide spherules with quench textures melting out of rocks that contain fossils, barite, and siderite. This allows us to rule out an origin for the iron spherules as ablation spherules derived from meteorites. The impact ejecta occur as a well defined, ~1 cm thick layer in three piston cores and have an age of about 1,500 BP. The third line of evidence is that the local chevrons all have azimuths that point back towards the locations of the crater candidates. The chevron orientations do not precisely match the direction of the prevailing wind. Finally, some chevrons contain 10 cm thick lag deposits of shell, and are locally absent where the coast is shielded by flat, off shore islands. Although these three lines of evidence are consistent with an impact into the Gulf of Carpentaria around 1,500 BP, they are not definitive proof. Geophysical surveys of the craters and marine geological mapping of the thickness variations and composition of the proximal ejecta blanket are needed to confirm the proposed impact origin of the Carpentaria crater candidates.
4. Our work on the Burckle crater candidate in the central Indian Ocean shows evidence of an impact ejecta layer that is thicker towards the crater candidate. The ejecta layer contains calcite rhombs, pyroxene, plagioclase, and olivine fragments, pure carbon spherules, and impact glass. Our findings are still preliminary due to the small sizes of available samples. These cores are heavily sampled and have been degraded during decades of storage. The bathymetry around Burckle crater is mostly derived from single line bathymetric tracks and satellite altimetry. It is located just at the edge of the ridge crest area that has been surveyed using modern multibeam bathymetric mapping. The hypothesis of an impact into this area could be tested by taking and sampling new kasten cores and by multibeam bathymetric mapping and a gravity/magnetic survey of the crater candidate. A modern marine geological and marine geophysical survey is needed to assess the nature and origin of the Burckle crater candidate.
5. The Madagascar chevrons, the largest on the planet, are situated on land that is closest to the Burckle crater candidate. Their systematic change in orientation is consistent with the refraction around the southern coast of Madagascar of a large tsunami whose source area approximates the Burckle crater candidate. The marine microfossils and coarse debris found in the Madagascar chevrons are derived from sites that are as much as 170 m above sea level and over 7 km in a direct line from the coast. The Ampalaza chevron is being more reworked by wind then surrounding soils and it is not suitable for farming. We therefore conclude that human beings did not bring in the marine microfossils and coarse debris we have found. In addition, the fossils are often concentrated in lag deposits and occur over 40 km along the strike of the chevrons. The fossils appear well preserved in these lag deposits. We find it highly unlikely that such well-preserved carbonate fossils could have been transported over a distance up to 40 km overland to these sites by the wind. Thus, our overall conclusion is that the Madagascar chevrons were most likely produced by a mega tsunami generated in a source area close to, or at the site of, the Burckle crater candidate.
6. In terms of rapid climate change, if even one of the three environmental/climatic events of ca. 12,900, 4,800, and 1,500 BP was caused by a cosmic impact, the concept of a cosmic impact during the terminal Pleistocene and late Holocene would have significant ramifications for how we currently understand and model past climate change. If all three were demonstrated to be caused by cosmic impact, there would be a crucial need to rethink everything from our current use of paleoclimate proxies to model global warming, to that of our understanding of human biological and cultural evolution. The risks and potential devastat-

ing consequences of oceanic impacts by comets and asteroids would need serious reconsideration with regard to future human coastal population and infrastructure.

With the data currently available, we cannot prove conclusively that the craters identified in the Gulf of Carpentaria or at the site of the Burckle candidate crater are impacts. But our study demonstrates that there is enough geomorphic and mineralogical evidence to suggest that such a hypothesis cannot be dismissed outright. Present disaster management is based on the type and frequency recurrence of the disasters that have been observed on the time scale of hundreds of years (e.g. Gad-el-Hak 2008). Disasters that occur, instead, on the scale of thousands of years, such as volcanic mega eruptions or significant cosmic impacts, are almost completely ignored despite being potentially massive in scope and effect. Moreover, a considerable part of the disaster community does not believe that catastrophic impacts by comets and asteroids have actually occurred during the course of human history. It is true that in terms of the evaluation of present day human loss statistics this type of global disaster has zero input. But cosmic impacts have so great a potential for large scale damage and fatalities, that their inclusion in the overall risk evaluation of natural hazards cannot be ignored.

Before 26 December 2004, a similar false premise existed in terms of the risk from tsunami. In the twentieth century, tsunamis were responsible for less than 0.5% of human fatalities resulting from natural disasters. The large death toll of over 230,000 victims that resulted from the 2004 Indian Ocean tsunami was neither predictable in terms of the total number of deaths or the location of the disaster. However in terms of twenty-first century risk from natural disasters, this statistic will remain a benchmark for a long time ... until the next great earthquake occurs in a large metropolitan area. Based on our field studies and laboratory analyses, the hypothesis of recent Holocene oceanic impacts is valid and requires serious consideration in the assessment of risks due to natural hazards.

Acknowledgements The authors wish to thank Mrs. T. Kalashnikova for assistance in preparing the figures and tables, and undertaking the final formatting of the manuscript. This work was partly supported by the RFBR grants 08-07-00105, 09-05-00294, and 07-05-13583, and NSF Grant OCE06-49024. Authors also appreciate the financial support provided by the WAPMERR (Geneva, Switzerland) for the 2006 Madagascar field trip, along with the field support provided by University of Antananarivo graduate students H. Razafindrakoto and A. Raveloson. D. Breger conducted the scanning electron microscopy and assisted with the X-ray analyses. We thank the centralized research facilities of Drexel University for the use of their SEM/EDS system.

References

Abbott DH, Biscaye P, Cole-Dai J, Breger D (2008b) Magnetite and silicate spherules from the GISP2 core at the 536 A.D. horizon. EOS Transactions, American Geophysical Union, Fall Meeting Supplement, Abstract PP41B-1454:89

Abbott DH, Bryant EA, Gusiakov V, Masse WB, Breger D (2008a) Impacts, mega-tsunami, and other extraordinary claims. Comment, GSA Today, 18(6):12

Abbott DH, Masse WB, Burckle L, Breger D, Gerard-Little P (2005) Burckle abyssal impact crater: did this impact produce a global deluge? Proceedings of Atlantis 2005 Conference: Milos, Greece

Abbott DH, Tester EW, Meyers CA (2007a) Impact ejecta and megatsunami deposits from a historical impact into the Gulf of Carpentaria. Geological Society of America, Abstracts with Programs, vol 39, p 312

Abbott DH, Tester EW, Meyers CA, Breger D, Chivas AM (2007b) Sediment transport, mixing, and erosion by an impact generated tsunami: Gulf of Carpentaria, Australia. EOS Transactions, American Geophysical Union, Abstract OS31B-07:88

Baillie MG (2007) Tree-rings indicate global environmental downturns could have been caused by comet debris. In: Bobrowsky, PT, Rickman H (eds) Comet/Asteroid Impacts and Human Society: An Interdisciplinary Approach. Springer, Berlin, pp 105–122

Baker VR (ed) (1981) Catastrophic flooding: the origin of the channeled scabland. Dowden Hutchinson & Ross, Stroudsburg, PA

Barrientos G, Masse WB (2009) Mid-Holocene cosmic impacts in central and northeastern Argentina: exploring probable effects on human population dynamics. Submitted to American Antiquity

Birks JW, Crutzen PJ, Roble RG (2007) Frequent ozone depletion resulting from impacts of asteroids and comets. In: Bobrowsky, PT, Rickman H (eds) Comet/Asteroid Impacts and Human Society: An Interdisciplinary Approach. Springer, Berlin, pp 225–245

Blench R (2006) New palaezoogeographical evidence for the settlement of Madagascar. Conference on The Maritime Heritage and Cultures of the Western Indian Ocean in Comparative Perspective, Zanzibar, Stone Town

Bobrowsky P, Rickman H (eds) (2007) Comet/asteroid impacts and human society: an interdisciplinary approach. Springer, Berlin

Bonatti E (1990) Subcontinental mantle exposed in the Atlantic Ocean on St. Peter-Paul islets. Nature, 345:800–802

Bourgeois J, Weiss R (2009) "Chevrons" are not mega-tsunami deposite—A sedimentologic assessment. Geology, 37:403–406.

Bryant E (2001) Tsunami: the underrated hazard, Cambridge University Press, Cambridge. Praxis, Chichester

Bryant E (2008) Tsunami: the underrated hazard, 2nd edition. Praxis, Chichester
Bryant E, Walsh G, Abbott D (2007) Cosmogenic mega-tsunami in the Australia region: are they supported by Aboriginal and Maori Legends? In: Piccardi, L, Masse, WB (eds) Myth and Geology. Geological Society of London Special Publication 273, London, pp 203–214
Bryant EA, Young RW (1996) Bedrock-sculpturing by tsunami, South Coast New South Wales, Australia. Journal of Geology, 104:565–582
Burney DA, Burney LP, Godfrey LR, Jungers WL, Goodman SM, Wright HT, Jull AJ (2004) A chronology for late prehistoric Madagascar. Journal of Human Evolution, 47:25–63
Cane MA (2005) The evolution of El Nino, past and future. Earth and Planetary Science Letters, 230:227–240
Chapman CR (2007) The asteroid impact hazard and interdisciplinary issues. In: Bobrowsky, PT, Rickman H (eds) Comet/Asteroid Impacts and Human Society: An Interdisciplinary Approach. Springer, Berlin, pp 145–162
Chivas A, Garcia A, van der Kaars, S, Couapel MJJ, Holt S, Reeves JM, Wheeler DJ, Switzer AD, Murray Wallace CV et al. (2001) Sea-level and environmental changes since the last interglacial in the Gulf of Carpentaria, Australia: an overview. Quaternary International, 83–85:19–46
Clark CD, Garrod SM, Parker-Pearson M (1998) Landscape archaeology and remote sensing in southern Madagascar. International Journal of Remote Sensing, 19:1461–1477
Collins GS, Melosh HJ, Marcus R (2005) Earth impact effects program: a web-based computer program for calculating the regional environmental consequences of a meteoroid impact on earth. Meteoritics and Planetary Science, 40:817–840
Davison T, Collins G (2007) Investigating the effect of water depth on marine impact crater morphology. Workshop on Impact Cratering II, 8041
Dick HJB, Fisher RL (1984) Mineralogic studies of the residues of mantle melting: abyssal and Alpine-type peridotites. In: Komprobst, J (ed) Kimberlites II: The Mantle and Crust-Mantle Relationships. Amsterdam, Elsevier, pp 292–308
Drummond BJ, Denham D, Michael-Leiba M (1985) Rheology of the lithosphere and Australian earthquakes. Geology and Geophysics, Bureau of Mineral Resources, p 60
Firestone RB, West A, Kennett JP, Becker L, Bunch TE, Revay ZS, Schulz PH, Belgya T, Kennett DJ, Erlandson JM, Dickensen OJ, Goodyear AC, Harris RS, Howard GA, Kloosterman JB, Lechler P, Mayewski PA, Montgomery J, Poreda R, Darrah T, Hee SSQ, Smith AR, Stich A, Topping W, Wittke JH, Wolbach WS (2007) Evidence for an extraterrestrial impact 12,900 years ago that contributed to the megafaunal extinctions and the Younger Dryas cooling. Proceedings of the National Academy of Sciences, 104:16016–16021
Gad-el-Hak M (ed) (2008) Large-scale disasters: predictions, control, and mitigation. Cambridge University Press, Cambridge
Gault DE, Sonnett CP (1982) Laboratory simulation of pelagic asteroidal impact: atmospheric injection, benthic topography, and the surface wave radiation field. In: Silver, LT, and Schultz, PH (eds) Geological Implications of Impacts of Large Comets and Asteroids on the Earth. Geological Society of America Special Paper 190, pp 69–92
Gisler GR, Weaver RP, Mader CL, Gittings MR (2003) Two and three dimensional simulations of asteroid ocean impacts. Science of Tsunami Hazards, 21:119
Gusiakov VK (2003) NGDC/HTDB meeting on the historical tsunami database proposal. Tsunami Newsletter, 35(4):9–10
Gusiakov VK (2009) Tsunami history. In: Robinson A, Bernard E (eds). The Sea, Tsunamis. Harvard University Press, Cambridge, vol 15, pp 23–53
Hearty PJ, Neumann AC, Kaufman DS (1998) Chevron ridges and runup deposits in the Bahamas from storms late in Oxygen–Isotope substage 5e. Quaternary Research, 50: 309–322
Hills JG, Mader CL (1997) Tsunami produced by the impacts of small asteroids. Annals of the New York Academy of Sciences, 822:381–394
Hong YT, Hong B, Lin QH, Shibata Y, Hirota M, Zhu YX, Leng XT, Wang Y, Yi L (2005) Inverse phase oscillations between the East Asian and Indian Ocean summer monsoons during the last 12,000 years and paleo-El Nino. Earth and Planetary Science Letters, 231:337–346
ITDB (2005) Integrated tsunami database for the Pacific and Indian Oceans, version 6.1 of 31 July 2005. Intergovernmental Oceanographic Commission – Tsunami Laboratory of the Institute of Computation Mathematics and Mathematical Geophysics, SD RAS, Novosibirsk, CD-ROM.
Iwasaki SI (1997) The wave forms and directivity of a tsunami generated by an earthquake and a landslide. Science of Tsunami Hazards, 15:23–40
Jones MR, Torgerson T (1988) Late quaternary evolution of Lake Carpentaria on the Australia-New Guinea continental shelf. Australian Journal of Earth Sciences, 35:313–324
Kawana T, Nakata T (1994) Timing of Late Holocene tsunamis originated around the Southern Ryukyu Islands, Japan, deduced from coralline tsunami deposits. Japanese Journal of Geography, 103:352–376
Kelletat D (2008) Comments to Dawson, A.G. and Stewart, I. (2007) Tsunami deposits in the geological record. Sedimentary Geology, 211(3–4):87–91
Kelletat D, Scheffers A (2003) Chevron-shaped accumulations along the coastlines of Australia as potential tsunami evidences. Science of Tsunami Hazards, 21:174–188
Kindler P, Strasser A (2000) Paleoclimatic significance of co-occuring wind and water induced sedimentary structures in the last interglacial coastal deposits from Bermuda and the Bahamas. Sedimentary Geology, 131:1–7
Kor PSG, Shaw J, Sharpe DR (1991) Erosion of bedrock by subglacial meltwater, Georgian Bay, Ontario: a regional view. Canadian Journal of Earth Science, 28:623–642
Lander JF (1996) Tsunamis affecting Alaska, 1737–1996. United States National Geophysical Data Center. Key Geophys. Research Document 31
MacCraken MC (2007) The climatic effects of asteroid and comet impacts: Consequences for an increasingly interconnected society. In: Bobrowsky P, Rickman, H (eds) Comet/Asteroid Impacts and Human Society: An Interdisciplinary Approach: Berlin. Springer, Berlin, pp 277–289
Martos SN, Abbott DH, Elkinton HD, Chivas AR, Breger D (2006) Impact spherules from the craters Kanmare and Taban in the Gulf of Carpentaria. Geological Society of America, Abstracts with Programs, 38:299–300
Masse WB (1998) Earth, air, fire, and water: the archaeology of Bronze Age cosmic catastrophes. In: Peiser BJ, Palmer T, Bailey ME (eds) Natural catastrophes during Bronze Age civilizations: archaeological, geological, astronomical,

and cultural perspectives. BAR International Series 728, Archaeopress, Oxford, pp 53–92

Masse WB (2007) The Archaeology and Anthropology of Quaternary Period Cosmic Impact. In: Bobrowsky P, Rickman H (eds) Comet/Asteroid Impacts and Human Society: An Interdisciplinary Approach. Springer, Berlin, pp 25–70

Masse WB, Masse MJ (2007) Myth and catastrophic reality: using myth to identify cosmic impacts and massive Plinian eruptions in Holocene South America. In: Piccardi, L, Masse, WB (eds) Myth and Geology. Geological Society of London Special Publication 273, London, pp 177–202

Masse WB, Weaver RP, Abbott DH, Gusiakov VK, Bryant EA (2007) Missing in action? Evaluating the putative absence of impacts by large asteroids and comets during the quaternary period. Proceedings of the Advanced Maui Optical and Space Surveillance Technologies Conference, Wailea, Hawaii, pp 701–710

Maxwell TA, Haynes CV (1989) Large-scale, low amplitude bedforms (chevrons) in the Selima Sand Sheet. Science, 243:1179–1182

Melosh HJ (1989) Impact cratering: a geologic process. Oxford University Press, New York

Michael PJ, Bonatti E (1985) Peridotite composition from the North Atlantic: regional and tectonic variations and implications for partial melting. Earth and Planetary Science Letters, 73:91–104

Morrison D, Harris AW, Sommer G, Chapman CR, Carusi A (2002) Dealing with the impact hazard. In: Bottke W, Cellino A.,Paolicchi P, Binzel RP (eds) Asteroids III, University of Arizona Press, Tucson, pp 739–754

Mutter JC, Detrick RS (1984) Multichannel seismic evidence for anomalously thin crust at Blake Spur Fracture Zone. Geology, 12:534–537

Pinegina TK, Bourgeois J (2001) Historical and paleo-tsunami deposits on Kamchatka, Russia: long-term chronologies and long-distance correlations. Natural Hazards and Earth System Sciences, 1:177–185

Reeves JM, Chivas AR, Garcia A, Deckker PD (2007) Paleoenvironmental change in the Gulf of Carpentaria (Australia) since the last interglacial based on Ostracoda. Paleogeography, Paleoclimatology, Paleoecology, 246:163–187

Rigby E, Symonds M, Ward-Thompson D (2004) A comet impact in AD 536? Astronomy and Geophysics, 45: 1.23–1.26.

Sandwell DT, Smith WHF (2005) Retracking ERS-1 altimeter waveforms for optimal gravity field recovery. Geophysical Journal International, 163:79–89

Scheffers A (2004) Tsunami imprints on the Leeward Netherlands Antilles (Aruba, Curaçao, Bonaire) and their relation to other coastal problems. Quaternary International, 120: 163–172

Scheffers AM, Kelletat DH, Scheffers SR, Abbott DH, Bryant EA (2008) Chevrons-enigmatic sedimentary coastal features. Zeitschrift für Geomorphologie, 52:375–402

Schultz PH, Lianza RE (1992) Recent grazing impacts on the Earth recorded in the Rio Cuarto crater field, Argentina. Nature, 355:232–237

Smart J (1977) Late quaternary sea-level changes, Gulf of Carpentaria, Australia. Geology, 5:755–759

Smith WHF, Sandwell DT (1997) Global seafloor topography from satellite altimetry and ship depth soundings. Science, 277:1956–1962

Sturkell E (1998) The marine Lockne impact structure, Jamtland, Sweden: a review. Geological Rundschau, 87:253–267

Toon OB, Zahnle K, Morrison D, Turco RP, Covey C (1997) Environmental perturbations caused by the impacts of asteroids and comets. Review of Geophysics, 35:41–78

Torgerson T, Hutchinson MF, Searle DE, Nix HA (1983) General bathymetry of the Gulf of Carpentaria and the quaternary physiography of Lake Carpentaria. Palaeogeography, Palaeoclimatology, Palaeoecology, 41:207–225

Tornberg R (1997) Impact-related resurge sediments as exemplified by the Lockne, Tvaeren and Kaerdla structures. Geological Society of America, Abstracts with Programs, 29(6):80

Tsikalas F, Gudlaugsson ST, Faleide JI, Eldholm O (1999) Mjolnir Structure, Barents Sea; a marine impact crater laboratory. In: Dressler, BO, Sharpton, LV, (eds) Large meteorite impacts and planetary evolution. Geological Society of America, Special Paper 339, pp 193–204

White RS, McKenzie D, O'Nions RK (1992) Oceanic crustal thickness from seismic measurements and rare earth element inversions. Journal of Geophysical Research, 97: 19683–19715

Young R, Bryant E, Price DM (1996) Catastrophic wave (tsunami?) transport of boulders in southern New South Wales, Australia. Zeitschrift für Geomorphologie, 40: 191–207

Understanding Slow Deformation Before Dynamic Failure

G. Ventura, S. Vinciguerra, S. Moretti, P.H. Meredith, M.J. Heap, P. Baud, S.A. Shapiro, C. Dinske, and J. Kummerow

Abstract Slow deformation and fracturing have been shown to be leading mechanisms towards failure, marking earthquake ruptures, flank eruption onsets and landslide episodes. The common link among these processes is that populations of microcracks interact, grow and coalesce into major fractures. We present (a) two examples of multidisciplinary field monitoring of characteristic "large scale" signs of impending deformation from different tectonic setting, i.e. the Ruinon landslide (Italy) and Stromboli volcano (Italy) (b) the kinematic features of slow stress perturbations induced by fluid overpressures and relative modelling; (c) experimental rock deformation laboratory experiments and theoretical modelling investigating slow deformation mechanisms, such stress corrosion crack growth. We propose an interdisciplinary unitary and integrated approach aimed to:

(1) transfer of knowledge between specific fields, which up to now aimed at solve a particular problem; (2) quantify critical damage thresholds triggering instability onset; (3) set up early warning models for forecasting the time of rupture with application to volcanology, seismology and landslide risk prevention.

Keywords Stromboli volcano · Landslides

Introduction

Slow deformation and fracturing are the leading mechanisms towards failure episodes, such as earthquake ruptures, flank eruptions and landslides (e.g., Scholtz 1968; Cruden 1974; Varnes 1989; Kilburn and Voight 1998; Brehm and Braile 1999; Di Giovambattista and Tyupkin 2001; Kilburn 2003). Field observations routinely monitor the slow deformation and accelerating seismic event rates that precede macroscopic failure of the crust. Laboratory experiments allow to relate applied stress and the evolution of crack damage by describing the microcrack interaction, growth and coalescence into major fractures (Heap et al. 2009). Theoretical modelling of field (Shapiro et al. 2007) and laboratory data (Main 2000) has been proposed to interpret the observed patterns. Hereafter we present data from two selected sites (Ruinon landslide, Alps, and Stromboli volcano, Italy) where multidisciplinary monitoring networks run. We first show how remotely acquired data by SAR interferometry and seismic arrays can be used to monitor deformation processes in different volcanic settings. We then report data from fluid-induced microseismicity and laboratory investigations that allow us to better understand the physical processes (e.g. crack growth and interaction) that generally proceed larger scale failure events like earthquakes, landslides, and volcanic eruptions and sector collapses.

G. Ventura (✉)
Department of Seismology and Tectonophysics, Istituto Nazionale di Geofisica e Vulcanologia, 00143 Roma, Italy
e-mail: ventura@ingv.it

T. Beer (ed.), *Geophysical Hazards*, International Year of Planet Earth,
DOI 10.1007/978-90-481-3236-2_14, © Springer Science+Business Media B.V. 2010

Monitoring Sites

Ruinon Landslide

A landslide is a mass of rock, debris, or earth, which moves down a slope due to gravity (Cruden 1991). Despite this simple definition, a landslide is a complex phenomenon (Cruden and Varnes 1996). It is characterized by five fundamental mechanisms of movement (fall, topple, slide, spread, and flow) and their combinations. Landslides can occur on material that can range in size and consistency from hundreds of millions of cubic meters of solid rock to single particles of earth or debris. The rate of movement ranges from imperceptible creeping (velocity $<10^{-10}$ m/s) to catastrophic failures (velocity >10 m/s), while the material can move as a whole like a solid block or can flow like a fluid, depending on the water content and other factors. In addition, landslide activity can vary in space and through time and even between different parts of the same displaced mass.

Here, a rather new specific application to a landslide site is described, the synthetic aperture radar (SAR). SAR data are collected by a ground-based radar system forming the synthetic aperture by the sliding of the antennas on a linear rail. Coherent SAR processing converts the raw data into a complex image. The phase of each image pixel contains information on the target sensor distance and can be exploited as a ranging tool. The interferometric technique (InSAR), based on the comparison between paired and coherent SAR images taken at different times, gives extremely precise quantitative information of displacement, thus allowing the monitoring of the morphological changes. One of the main applications of synthetic aperture radar (SAR) interferometry (InSAR) is the monitoring of natural hazards, in particular of those phenomena producing ground displacements such as landslides.

In this specific case we considered a large rockslide in the Italian Alps, where the presence of an independent monitoring system allowed us to validate the results through "ground truth" measurements.

The Ruinon landslide is one of the most hazardous slope movements in the Italian Alps (Fig. 1). The word ruinon literally means "huge ruin" this name gives the idea that the area has been unstable since the past. It is located in Valfurva (Middle Valtellina), near the village of Bormio, in the Rhaetian Alps on the hydrographic right of the Frodolfo stream. The landslide has a continuous movements affecting an estimated volume of rock of 30 million m^3, representing a serious threat to human lives and socioeconomic activities in the area. Its rapid collapse would destroy the road connecting Bormio and Santa Caterina Valfurva and, furthermore, would block the Frodolfo stream, with the consequent formation of a highly unstable landslide dam.

The landslide is located in the lower portion of a southwest-facing slope, with an average inclination of 36°. The slope consists of pre-Permian metapelites (phyllites), belonging to the Upper Austroalpine basement of the Campo-Ortles Nappe, and glacial deposits and debris produced by rock falls locally cover it. The landslide, which has a total length of 770 m and a

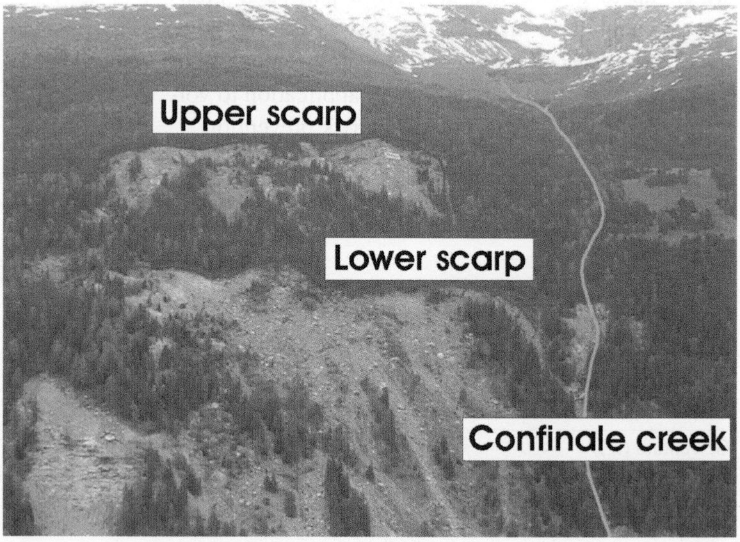

Fig. 1 Aerial picture of the Ruinon rock slide

width of 410 m, is characterized by two main scarps northwest southeast oriented, parallel to the main fracture system (Fig. 2). The "upper scarp" is located at an elevation of 2,100 m above sea level (asl), the "lower scarp" at 1,900 m asl. The Confinale creek, a right hand tributary of the Frodolfo, the course of which is controlled by a northeast-southwest master joint, abruptly cuts the southeastern border of the landslide. Figure 3 shows the geological section of the area and gives a possible interpretation of the landslide geometry and movement mechanism. Four deep (120 m) boreholes, equipped with inclinometers, were done to determine the section. The data from boreholes show the presence of weak and cataclastic zones at a depth of more than 90 m. The system employed is a portable SAR device known as Linear SAR (LISA),

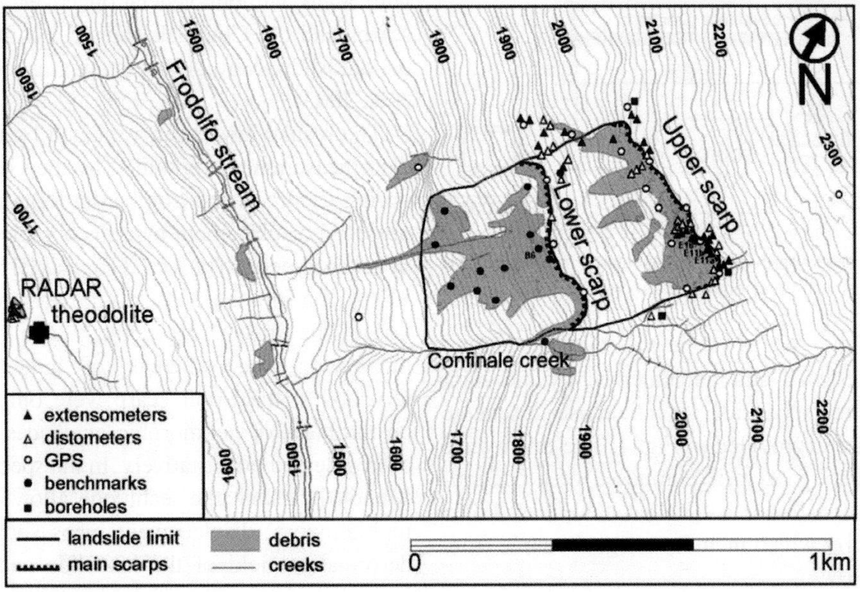

Fig. 2 General plan showing the main geomorphic elements and the location of monitoring instrumentation. The position of the radar is also shown. Labels E11a, E11b, E16, and B6 indicate the sensor used for data validation

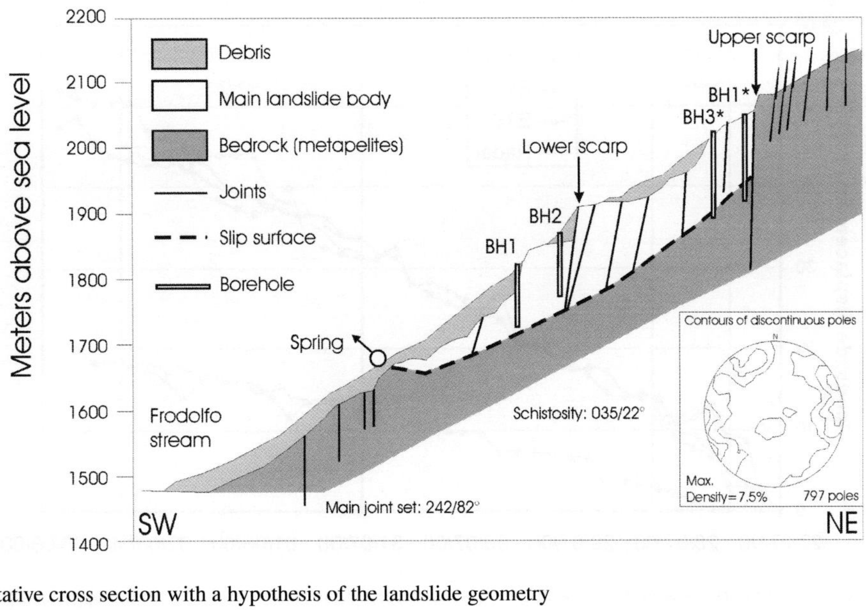

Fig. 3 Interpretative cross section with a hypothesis of the landslide geometry

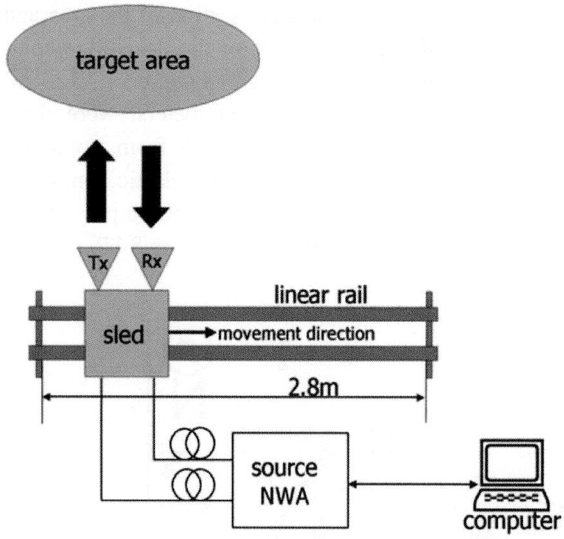

Fig. 4 Scheme of the hardware components of the linear SAR (LISA). Tx, transmitting antenna; Rx, receiving antenna; and NWA, network analyzer

specifically designed and implemented by the Joint Research Centre, Ispra, Italy, for infield use. A scheme of the hardware is provided in Fig. 4. The microwave component of the system is composed of a continuous wave, stepped-frequency scatterometer based on a network analyzer which includes a signal source from 30 kHz to 6 GHz.

An additional module from coherent up-and-down frequency conversion allows measurements in the frequency band from 14 to 18 GHz. The antenna synthesis is obtained by moving a motorized sled, hosting the antennas and other microwave components, along a 2.8 m long linear rail system (Rudolf et al. 1999; Rudolf and Tarchi 1999).

SAR power images of the landslide and a coherence map was produced and a direct comparison with the optical image was done showing a close correlation between areas of higher radar backscattering with areas without vegetation cover. The two main scarps and the Confinale stream, on the left flank of the landslide, were clearly distinguishable too (Tarchi et al. 2003).

The results have been validated by using independent measurements provided by the Geological Monitoring Centre of the Lombardia Region (Fig. 5). The observed displacement patterns obtained by radar and by extensometers (E11a, E11b, and E16) on the upper scarp, confirm the activity of the southeastern portion of the upper scarp, already documented by past monitoring data, whereas the large displacements recorded at the lower scarp were only hypothesized on the basis of geomorphic evidences but had never been assessed quantitatively. In the specific case of the Ruinon landslide the technique allowed us to derive multitemporal displacement maps showing the LOS deformation field of the landslide and providing an immediate indication of the state and distribution of activity.

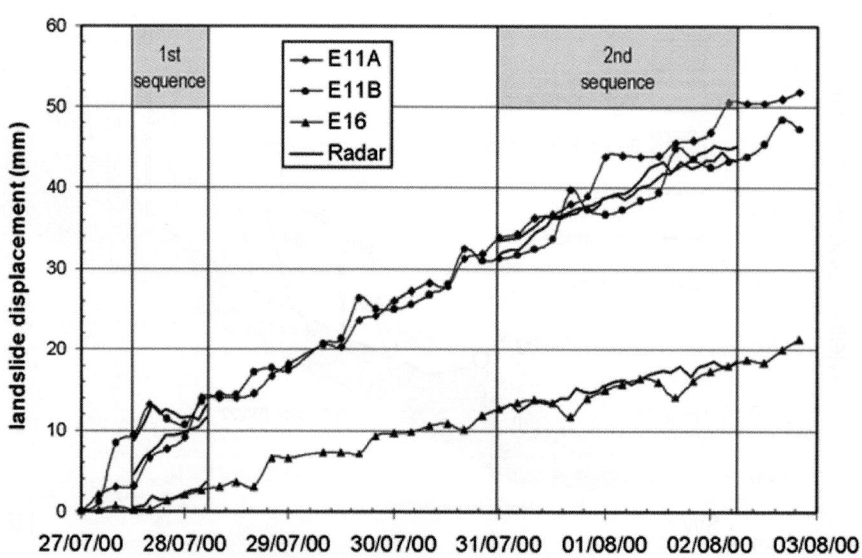

Fig. 5 Cumulated displacements assessed by radar and by extensometers E11a, E11b, and E16 on the upper scarp

Stromboli Volcano

Stromboli Island (924 m asl) is the northernmost volcano of the Aeolian Islands (Southern Italy) and is characterized by a persistent explosive activity over the last 1,400–1,800 years (Rosi et al. 2000). Each episode consists of "strombolian" events every 10–20 min and is associated with streaming of gas from the summit craters (output ~6–12 × 10^2 t/day (Allard et al. 1994)). Following Chouet et al. (2003), the gas slug responsible for the explosions form at 220–260 m beneath the summit craters. The persistent activity is periodically interrupted by more energetic explosions (paroxysms), which consist of shortlived blasts fall and ballistic ejecta that reach several hundreds of meters from the craters. One or two events of this type occur per year and the erupted material sometimes affects the two villages on the coast (Barberi et al. 1993). Lava emissions occur about every 10–20 years.

From a structural point of view, magma emissions develops from a NE-SW elongated conduit feeding the three summit craters, which are also NE-SW aligned (Fig. 6). The craters are located at the top of the Sciara del Fuoco (SDF) depression, a morphological expression of sector collapses. This depression, which dip towards NW, is filled with loose deposits (Tibaldi 2001).

Since the summer of 2002, Strombolian activity from the summit craters have shown increased intensity. On 28 December 2002, the magma overflowed, accompanied by an eruptive fissure propagating laterally towards NE. This fissure marked the onset of the effusive eruption, and the end of strombolian activity. On 30 December 2002, a sector of Sciara del Fuoco collapsed, reaching the coastline (Pompilio 2003). The landslide involved a volume of 3 10^7 m^3 and consisted of a submarine and subaerial episodes. The subaerial landslide generated a tsunami, with waves reaching 5–8 meters (Bonaccorso et al. 2003). A vent was

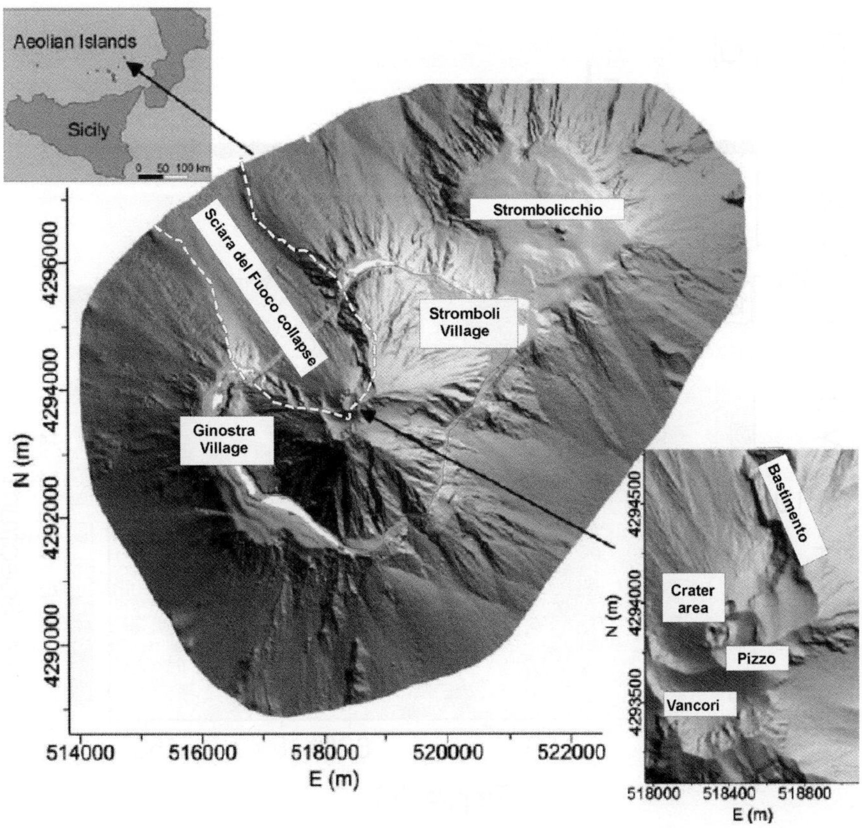

Fig. 6 Digital terrain model of Stromboli Island and offshore with location of the crater area and Sciara del Fuoco collapse (modified from Baldi et al. 2008)

formed between 30 December 2002 and 1 January 2003 within the collapsed area, at 470 m asl. The magma supply decreased in early February, reaching a temporary minimum around mid-February (Ripepe et al. 2005). On 15 February 2003, the NW-SE fissure ended its activity followed by ash emissions from the summit craters occurred. On 3 April 2003, an explosion occurred from the summit crater. This was followed, on 5 April 2003 , by a paroxysm producing pyroclastic density currents and fall deposits (Fig. 7). Strombolian activity at the summit craters resumed after the end of the effusive activity (21 July 2003) (Acocella et al. 2006).

Thus, strombolian activity poses a severe hazard to the population. It is therefore particularly important to understand how magma emplacement and landslides occur at Stromboli and its possible impact. The present-day Istituto Nazionale di Geofisica e Vulcanologia monitoring network at Stromboli include thermal, infrared, and conventional cameras, broadband and short period seismic and acoustic emission

Fig. 7 Picture of the April 5 paroxysm (modified from Rosi et al. 2006)

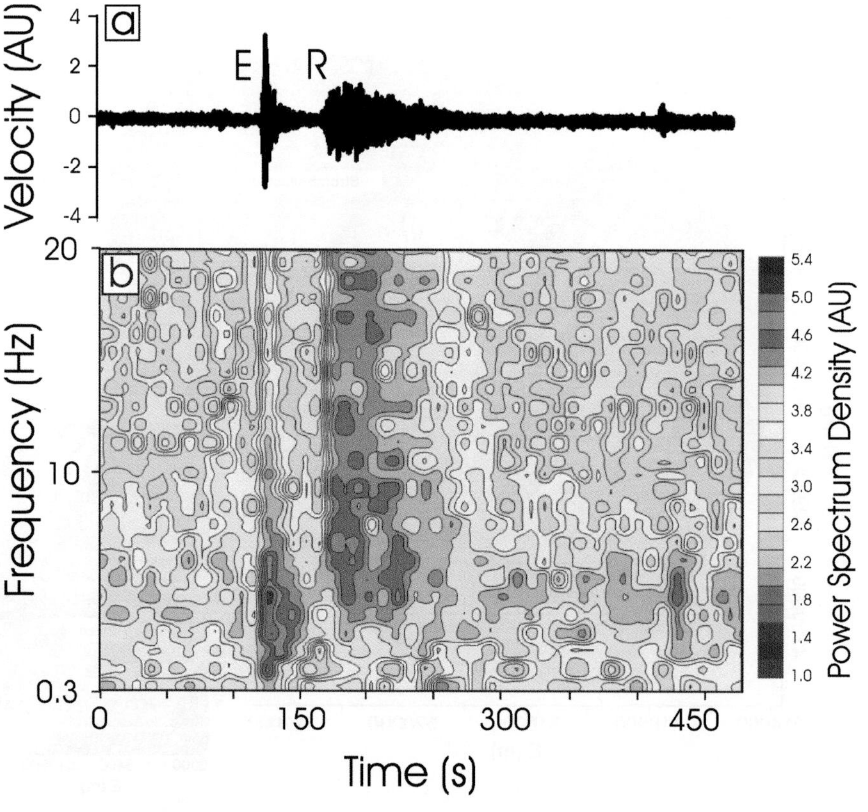

Fig. 8 (**a**) Seismogram and (**b**) relative spectrogram of a time series recorded at the vertical component of STR8 on 27 August 2004. An explosion quake (E) shortly precedes a rockfall episode (R) (modified from Falsaperla et al. 2006)

sensors, spectrometers for the detection of SO_2 flux, gravimetric, magnetic, GPS and clinometric networks. The CO_2 flux, the water and gas chemistry are also monitored (http://www.ingv.it and related pages).

According to Falsaperla et al. (2006), the 30 December 2002 landslides represent the only major collapse episodes instrumentally recorded at Stromboli. The seismic monitoring with permanent stations close (<2 km) to the summit craters and Sciara del Fuoco has highlighted minor episodes of flank instability throughout and well after the 2002–2003 lava effusion. Landslides, rockfalls, and debris flows have a typical seismic signature, which allows distinguishing these signals from earthquakes and explosion quakes. An example of time series and relative spectrogram of an explosion shortly preceding a rockfall episode is shown in Fig. 8. Besides the diverse waveform, the relatively brief (30 s) signature of the explosion quake differs strongly from the long (>100 s) duration of the rockfall. The spectrogram analysis highlights the different frequency content of each transient event also with respect to volcanic tremor, i.e., the continuous background signal recorded at Stromboli. The long-term seismic and video monitoring of Stromboli (Falsaperla et al. 2006) suggests that (a) the beginning of a remodeling process of the unstable sectors of the collapsed area after 2002, and (b) the enlargement of the head scarp between 2002 and 2004 involved a negligible rock volume. Nevertheless, if the sliding reach the rim of the crater, this might change the eruptive behavior owing to the asymmetric depressurization of the northwestern portion of the cone. As a result, the AE monitoring of the Sciara del Fuoco collapse is of primary importance for the detection of creep-like signals possibly related to the early phases of sliding episodes.

Kinematic Features and Magnitudes of Microseismicity Slow Stress Perturbations Fluid Pressure Induced: Case Studies and Their Interpretation

Sometimes fluid injections are characterized by a risk to induce a seismic event of a significant magnitude. Here we address magnitude distribution of seismicity induced by borehole fluid injections. Firstly we give a short introductory review of microseismicity interpretation in geothermic reservoirs by hydraulic fracturing. Then, we introduce a simple theoretical model, which predicts the earthquake magnitude distributions for fluid injection experiments. The temporal distribution of microearthquake magnitudes depends on the injection pressure, the size of the borehole injection section, the hydraulic diffusivity of rocks, and is also inherited from the statistics of pre-existing crack/fracture systems controlling the local seismicity. We consider different case studies and show how our model can be used to optimise the design of fluid injection experiments and reduce their seismic risk.

In the following, we give a short review of recent research toward establishing physical fundamentals for microseismic investigations of borehole fluid injections. Experiments with borehole fluid injections are typical for exploration and development of hydrocarbon and geothermal reservoirs. The fact that fluid injection causes seismicity has been well-established for several decades. Current on going research is aimed at quantifying and control of this process (Shapiro et al. 2006a, b, 2007).

The fluid induced seismicity covers a wide range of processes between two in the following described asymptotic situations. In liquid-saturated rocks with low to moderate permeability the phenomenon of microseismicity triggering by borehole fluid injections is often related to the pore pressure diffusion. Fluid induced seismicity typically shows then several diffusion specific features, which are directly related to the rate of spatial grow-, to the geometry-, and to the spatial density of microseismic clouds.

Another extreme is the hydraulic fracturing of rocks. Propagation of a hydraulic fracture is accompanied by creation of a new fracture volume, fracturing fluid loss and its infiltration into reservoir rocks as well as diffusion of the injection pressure into the pore space of surrounding rocks and inside the hydraulic fracture. Some of these processes can be seen from features of spatio-temporal distributions of the induced microseismicity. Especially, the initial stage of fracture volume opening as well as the back front of the induced seismicity starting to propagate after termination of the fluid injection can be well identified.

Here, we describe main quantitative features of the both types of induced microseismicity, where triggering is controlled by the pore pressure diffusion and by the process of new volume opening in the rocks. We also address also magnitude distribution of seismicity induced by borehole fluid injections. Evidently, this is an important question closely related to seismic risk of injection site.

Pore Pressure Diffusion Controlled Seismicity

If the injection pressure (i.e., the bottom hole pressure) is less than the minimum principal stress, then, at least in the first approximation, the behaviour of the seismicity triggering in space and in time is controlled by the process of linear relaxation of stress and pore pressure perturbations initially created at the injection source.

The spatio-temporal features of the pressure-diffusion induced seismicity can be found in a very natural way from the triggering front concept (Shapiro et al. 2002). At a given time t it is probable that events will occur at distances, which are smaller or equal to the size of the relaxation zone (i.e., a spatial domain of significant changes) of the pore pressure. The events are characterised by a significantly lower occurrence probability for larger distances. The surface separating these two spatial domains is the "triggering front." In a homogeneous and isotropic medium the triggering front has the following form:

$$r = \sqrt{4\pi Dt} \tag{1}$$

where t is the time elapsed from the injection start, D is the hydraulic diffusivity and r is the radius of the triggering front (which is a sphere in a homogeneous isotropic medium). Because a seismic event is much more probable in the relaxation zone than at larger distances, Eq. (1) corresponds to the upper bound of the cloud of events on a plot of r versus t (see Fig. 9).

If the injection stops at time t_0 then the earthquakes gradually stop to occur. For times larger then t_0 a surface can be defined which describes propagation of a maximal pore pressure perturbation in the space. This surface (also a sphere in homogeneous isotropic rocks) separates the spatial domain which is still seismically active from the spatial domain (around the injection point) which is already seismically quiet. This surface has been firstly described in (Parotidis et al. 2004) and termed "back front" of induced seismicity:

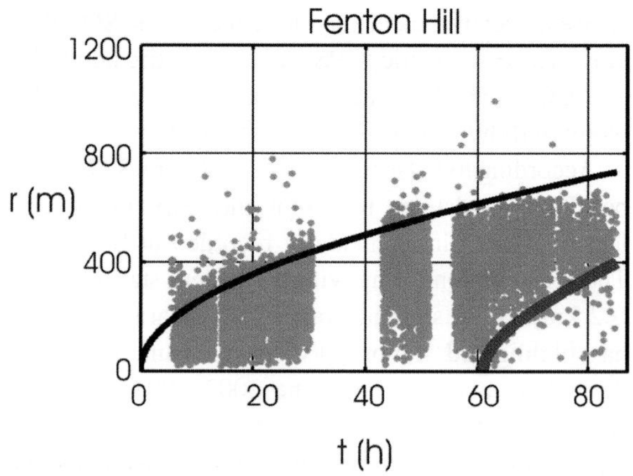

Fig. 9 Fluid-injection induced microseismicity at Fenton Hill (data courtesy of Michael Fehler). *Grey points* – r-t plot of induced microseismic events (the *black line* is a triggering front; the *grey line* is a back front)

$$r_{bf} = \sqrt{2dDt\left(\frac{t}{t_0} - 1\right)\ln\frac{t}{t - t_0}}. \tag{2}$$

Here d is the dimension of the space where the pressure diffusion occurs. The back front is often observed on real data and provides estimates of hydraulic diffusivity consistent with those obtained from the triggering front (see Fig. 9).

Hydraulic Fracturing Controlled Seismicity

During the hydraulic fracturing a fluid is injected through a perforated domain of a borehole into a reservoir rock under the bottom pressure larger than the minimum principal stress. In order to understand the main features of the induced seismicity by such an operation we apply a very simple and rough approximation of the process of the fracture growth resulting from a volume balance for a straight planar (usually vertical – this is the case for the real-data example given here) fracture confined in the reservoir layer. This is the so-called PKN model known from the theory of hydraulic fracturing (Economides and Nolte 2003, pp. 5-1–5-14). Basically, the half-length r of the fracture (which is assumed to be symmetric in respect to the borehole) is approximately given as a function of the injection time t by the following expression:

$$r = \frac{Q_I t}{4\,h_f C_L \sqrt{2t} + 2\,h_f w}. \quad (3)$$

where Q_I is an injection rate of the treatment fluid, C_L is a fluid-loss coefficient, h_f is a fracture height and w is a fracture width. The first term in the denominator describes the fluid loss from the fracture into surrounding rocks. It is proportional to $\sqrt{t}$ and has a diffusion character. The second term, $2\ h_f w$, represents the contribution of the effective fracture volume and depends mainly on the geometry of the fracture vertical cross-section. In the case of hydraulic fracturing of a formation with a very low permeability (e.g., tight gas sandstones) the fracture body represents the main permeable channel in the formation. The propagating fracture changes in its vicinity the effective stress and activates mainly slip events in the critical fracture systems existing in surrounding rocks (Rutledge and Phillips 2003).

During the initial phase of the hydraulic fracture growth the process of the fracture opening is dominant. This can often lead to a linear expansion with time of the triggering front. If the injection pressure drops the fracture will close. A new injection of the treatment fluid leads to reopening of the fracture, and thus, to a repeated linear propagation of the triggering front. A long term fluid injection leads to domination of diffusion processes. The growth of the fracture slows down and becomes approximately proportional to $\sqrt{t}$. After termination of the fluid injection the seismicity is mainly triggered by the process of the pressure relaxation in the fractured domain. Correspondingly, the back front of the induced microseismicity can be observed, which is described by the Eq. (2) with $d=1$ (i.e., approximately, a 1-D diffusion along the hydraulic fracture). Figure 10 shows an example of data demonstrating all the mentioned features of the induced seismicity during hydraulic fracturing.

Magnitudes of Induced Seismicity

Sometimes fluid injections are characterized by a risk to induce a seismic event of a significant magnitude. The magnitudes M of the stimulated seismicity are usually in the range $-3 < M < 2$. Nevertheless, especially

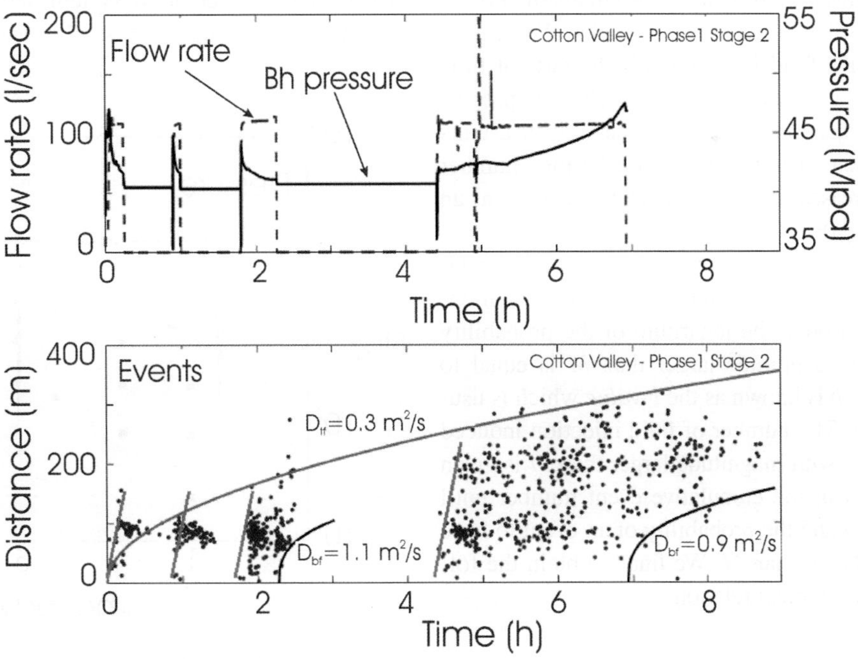

Fig. 10 Hydraulic fracturing induced microseismicity at the Carthage Cotton Vally gas field (data courtesy of James Rutledge). *Top*: Borehole pressure (measured at the injection domain) and fluid flow rate. *Bottom*: r-t plot of induced microseismic events (the parabolic grey line – a diffusion type approximation of the triggering; two other parabolic (*black*) lines – back fronts; straight (*grey*) lines – fracture opening and reopening and correspondingly, linear with time triggering fronts propagation)

for long-term injections with durations of m onths or even years, earthquakes with larger magnitudes ($M = 4$ or even larger) have been observed (Majer et al. 2007). So far, little effort has been undertaken to estimate the probability for these events to occur. Here we present a model which allows to calculate the expected number of events with a magnitude larger than a given magnitude value M. It also enables us to identify the main factors which affect the magnitude probabilities.

A basic assumption made here is that the seismicity is induced by pore pressure relaxation in a homogeneous medium, where the hydraulic diffusivity is independent of position and time. For simplicity we also assume a point source for the injection and a constant injection pressure. Wether or not an earthquake occurs on a pre-existing crack depends on the pore pressure and the criticality of the crack (Rothert and Shapiro (2007)). We define the critical value C for a crack as the pore pressure necessary to induce slip along the crack according to the Coulomb failure criterion. The criticality C usually spans several order of magnitudes. Typical ranges are 0.001–1 MPa (see Rothert and Shapiro (2007) and Brodsky et al. (2000)). Note, that a higher value of C means that the crack is more stable. We define n as the density of statistically homogeneously distributed pre-existing cracks. We furthermore consider a fluid injection, which starts at time $t = 0$ and has a constant strength q (which is proportional to the injection pressure and has physical units of power). Then it can be shown that the total number of events increases linearly with injection time at an event rate of $(qn)/C_{\max}$.

We postulate that the frequency magnitude relation is consistent with the Gutenberg-Richter relationship. In other words, the logarithm of the probability of events with magnitude larger than M is equal to $a - bM$. Here, b is known as the *b value* which is usually close to *1*. The number of fluid injection induced events N (M, t) with magnitude larger than M is given by the product of the cumulative event number until injection time t and the probability of an event to have a magnitude larger than M. We finally obtain the following bi-logarithmical relation

$$\log N(M,t) = \log\left(\frac{qt}{F}\right) - bM + a \qquad (4)$$

We denote the parameter $F = C_{max}/n$ as tectonic potential of the injection site. This quantity has physical units of energy and characterizes how easy or difficult seismicity can be induced at a particular location. A detailed analysis of Eq. (4) indicates that the number of earthquakes of magnitudes greater than a given one increases with the duration of the injection, the injection pressure, the flow rate and the borehole radius. It also depends on the hydraulic diffusivity and the crack concentration divided by the maximumcritical pore pressure.

We compare the number of events as a function of time as predicted by our formulation (Eq. 4) with observations from data sets at injection sites in Japan (Ogachi geothermic site). During an experiment at Ogachi geothermic site in 1991, a volume of more than 10,000 cubic meters of water was injected at a depth of 1,000 m into hard rock (granodiorite). The pressure remained relatively stable throughout the experiment (Fig. 11). A microseismic event cloud of about 500 m thickness and 1,000 m length with nearly 1,000 detected events was stimulated (Kaieda et al. 1993). The magnitudes were determined by measuring velocity amplitudes and alternatively seismogram oscillation durations (Kaieda and Sasaki 1998).

Magnitude statistics were biased by the performance of the observation system and processing in the magnitude ranges $M < -2.5$ and $M > -1.5$.

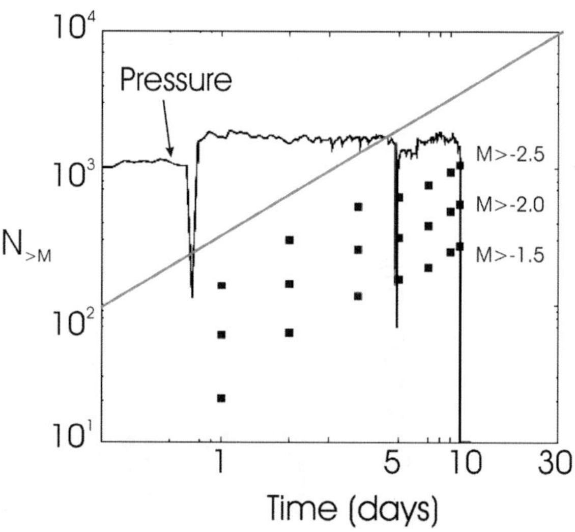

Fig. 11 Data of the Ogachi 1991 borehole injection experiment. The squares are observed cumulative numbers of earthquakes with a magnitude large than the indicated one as a function of injection time (bi-logarithmic plot). The straight line shows the theoretically predicted (see Eq. 4) proportionality coefficient 1. Shown is also the injection pressure as a function of time

When the injection pressure is close to constant, the $N(M, t)$ functions are nearly linear in the bilogarithmic plot. The steps between lines corresponding to different magnitudes M are regularly distributed and time-independent. These two prominent observations are as predicted by our model (Eq. 4). The steps between the lines can be used to estimate the b value. In 1993 an injection into two open borehole sections at different depth levels was performed at Ogachi injection site. The magnitude distributions exhibit a similar behaviour as for the experiment in 1991 (not shown here).

The magnitude distributions observed in Fig. 11 agree quite well with the predictions of Eq. 4, which postulates a linear relation between log $N(M, t)$ and log t.

Spatio-temporal dynamics of microseismic clouds contributes to characterization of hydraulic properties of reservoirs and to monitoring and description of hydraulic fractures. For example, r-t-plots show signatures of fracture volume growth, of fracturing fluid loss, as well as of diffusion of the injection pressure into rocks and inside the fracture. Diffusion controlled triggering is often observed at geothermic reservoirs. New volume creation controlled triggering is usually observed at hydraulic fracturing of tight gas reservoirs.

We have furthermore considered a poroelastic medium with randomly distributed sub-critical cracks obeying a Gutenberg-Richter statistics. Based on that we have derived a simple theoretical model, which predicts the earthquake magnitude distributions for fluid injection experiments. The temporal distribution of microearthquake magnitudes depends on the injection pressure, the size of the borehole injection section, the hydraulic diffusivity of rocks, and is also inherited from the statistics of pre-existing crack/fracture systems controlling the local seismicity. Our parametrization can be used to optimize the design of fluid injection experiments and reduce their seismic risk.

Experimental Study of Brittle Creep in Crustal Rocks

The understanding of slow, time-dependent crack growth is crucial to unraveling the complexities of the evolution and dynamics of the brittle crust. The presence of cracks allows the crust to store and transport fluids, and even modest changes in crack size, density or linkage can produce major changes in fluid transport properties. Time-dependent rock deformation therefore has both a scientific and a socio-economic impact since it exerts a key influence on the precursory phase of important geo-hazards such as earthquake rupture and volcanic eruption.

The majority of rocks forming the Earth's crust, even those at depth, contain microporosity comprising some or all of open pores between grains, triple-junction voids between crystalline phases, grain boundary voids and open microcracks. Water and other aqueous solutions are ubiquitous in the upper crust, and below a few hundred metres these void spaces in most rocks are saturated. The presence of a fluid phase not only affects the mechanical behaviour of rock, but also allows chemical rock-fluid interactions to occur. In a purely mechanical sense, a pressurized pore fluid acts to reduce all the applied normal stresses and thus allows rocks to fail at lower differential stress than would otherwise be the case (Terzaghi 1943; Jaeger et al. 2007). Chemically, aqueous solutions affect the deformation of rock in two main ways: (1) they act to weaken the rock via the reduction of surface free energy as the result of the absorption of pore fluid onto the internal pore surfaces (Orowan 1944; Rehbinder 1948; Andrade and Randall 1949), and (2) they also weaken rocks by promoting subcritical crack growth, of which stress corrosion is considered the most important mechanism under upper crustal conditions (Anderson and Grew 1977; Atkinson 1984; Atkinson and Meredith 1987; Costin 1987). Under these conditions, the (intensified) stress concentration (K) at crack tips that is responsible for crack growth is known to be a function of both applied stress (σ_a) and crack length (c) (e.g. Lawn 1993; Paterson and Wong 2005). Hence, since the relationship between the crack growth velocity (V) and K is very non-linear, the crack growth rate will accelerate as the cracks extends leading eventually to dynamic failure (Fig. 12). However, the majority of experimental data on stress corrosion cracking has been derived from experiments on single cracks at ambient pressure, and few data exist on the bulk behaviour of rock containing a population of cracks. Nevertheless it has been hypothesized that, for bulk rock deforming in a brittle manner under triaxial stress conditions, stress corrosion will lead to highly non-linear time-dependent deformation (Main et al. 1993;

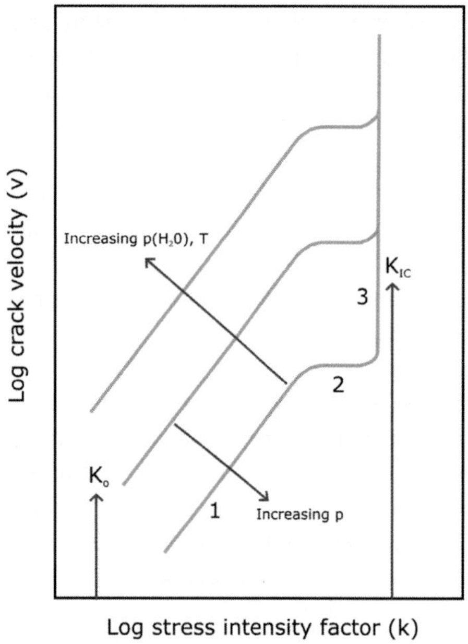

Fig. 12 Schematic stress intensity factor (K)/crack velocity (V) diagram for subcritical tensile crack growth due to stress corrosion. K_o – the stress corrosion crack growth limit; K_{IC} – the critical stress intensity factor or fracture toughness; p – effective confining pressure; p (H_2O) – pore fluid pressure; T – temperature

Main 2000). This allows rocks to deform even under a constant applied differential stress over extended periods of time; a phenomenon known as *brittle creep* (or *static fatigue* in the engineering literature). This style of deformation has conventionally been described as exhibiting an apparent trimodal behaviour when axial strain is plotted against time (commonly known as a *creep curve*). The three stages of the creep curve have conventionally been described as; (1) primary or decelerating creep, (2) secondary or steady-state creep, and (3) tertiary or accelerating creep (as illustrated in Fig. 13).

Here we present illustrative results from an experimental study of brittle creep in sandstone under triaxial stress conditions. The experiments were performed at room temperature in the servo-controlled 400 MPa triaxial rock deformation apparatus in the Rock & Ice Physics Laboratory (RIPL) at University College London (Fig. 14a). The internal sample assembly is shown in Fig. 14b. During all experiments, axial strain was measured continuously using LVDT displacement transducers, and pore volume change was measured continuously using a servo-controlled pore fluid pressure intensifier and volumometer (Benson et al. 2007). Microseismicity in the form of acoustic emission (AE) was recorded continuously via ten piezo-electric transducer crystals embedded within the rubber. The measurement of axial strain, pore volume change and the output of acoustic emission (AE) energy during experiments is important because all three parameters are considered as proxies for the accumulation of crack damage during deformation, as previously demonstrated by and Baud and Meredith (1997).

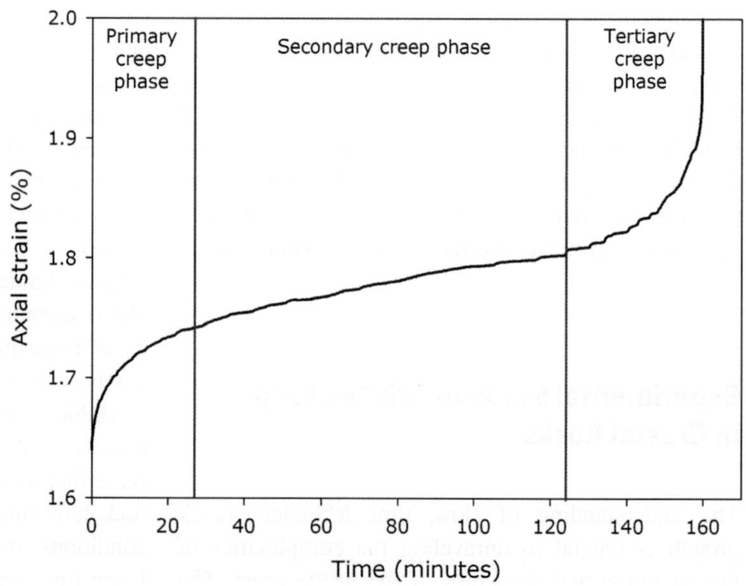

Fig. 13 The classic trimodal creep curve for brittle material at a constant applied differential stress. The curve shows the three stages of brittle creep (1) primary or decelerating (2) secondary or steady-state and (3) tertiary or accelerating creep. The secondary creep phase is where creep strain rates are calculated

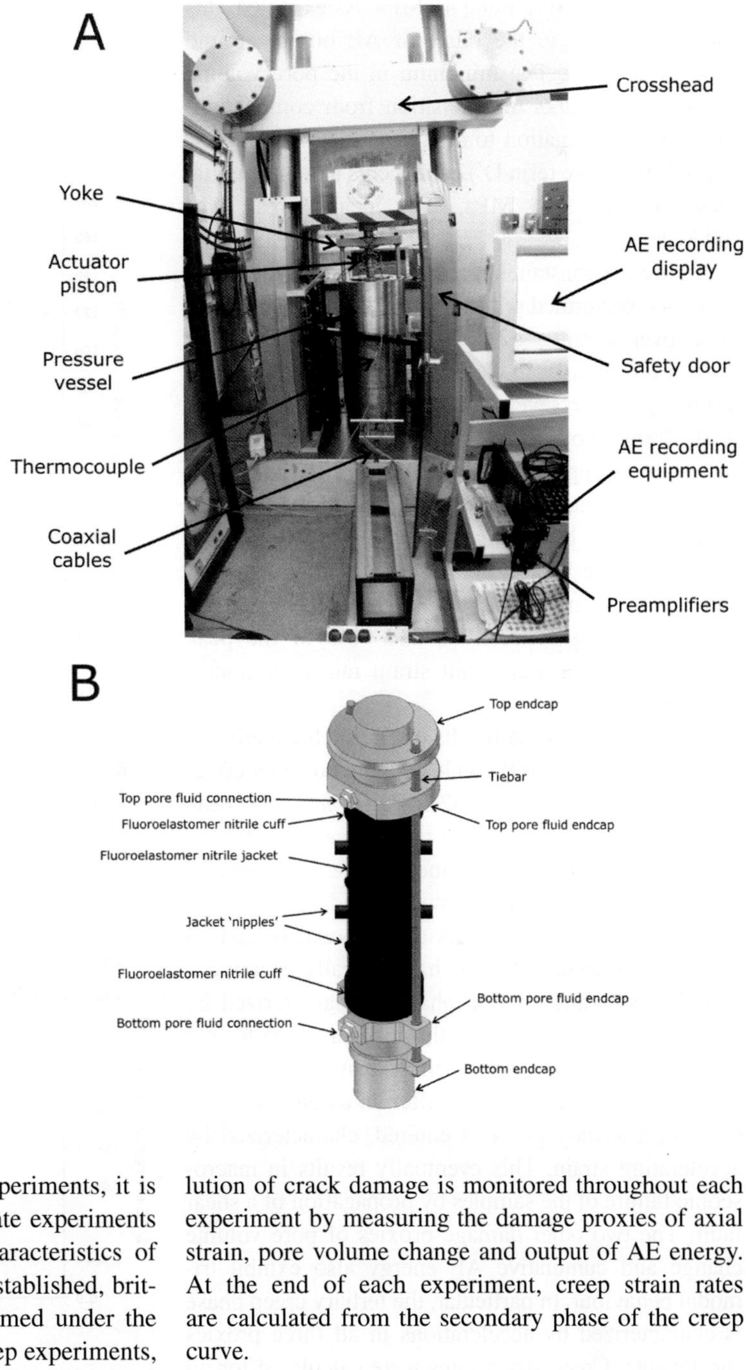

Fig. 14 (**a**) Photograph of the servo-controlled 400 MPa triaxial rock deformation apparatus at the Rock & Ice Physics Laboratory, UCL (**b**) three-dimensional Autodesk Inventor picture of the jacketed sample setup

Prior to performing brittle creep experiments, it is necessary to perform constant strain rate experiments to establish the short-term failure characteristics of the test material. Once this has been established, brittle creep experiments can then performed under the same effective stress conditions. In creep experiments, samples are first loaded to a pre-determined percentage of the peak stress (or strength) established during the constant strain rate experiments. Generally, following Baud & Meredith (1997), this is in the range 80–90% of the short-term strength. After this preloading, samples are allowed to deform under constant effective differential stress until failure. The evolution of crack damage is monitored throughout each experiment by measuring the damage proxies of axial strain, pore volume change and output of AE energy. At the end of each experiment, creep strain rates are calculated from the secondary phase of the creep curve.

Figure 15 shows the stress-strain curve and the damage proxies of pore volume change and cumulative AE energy for a constant strain rate experiment on a sample of Darley Dale sandstone conducted at an effective confining pressure of 30 MPa and a strain rate of 1.0×10^{-5} s^{-1}. Following an initial period of compaction, the onset of dilatancy occurs at a differential stress of around 40 MPa. As expected, this also corresponds to the onset of AE output around 0.5% axial strain. The minimum in the pore volume change curve marks the transition from compaction-dominated deformation to dilatancy-dominated deformation (which we term D′), and occurs at a differential stress around 130 MPa. The peak stress was 160 MPa.

A series of constant stress brittle creep experiments were then performed on samples of Darley Dale sandstone over a range of constant applied differential stress levels in order to yield times-to-failure and creep strain rates over several orders of magnitude (Heap et al. 2009). Following Baud and Meredith (1997), the levels of applied differential stress were selected to be greater than that corresponding to the onset of dilatancy (C′), but lower than the stress level that would generate very rapid failure within a few seconds or a few minutes. Such stress levels generally corresponded to between 80 and 90% of the peak stress in short-term constant strain rate experiments (Fig. 15).

Figure 16 shows the results of one such experiment (from Heap et al. 2009) conducted at a constant effective differential stress of 132 MPa (85% of the peak stress). The three independent proxy measures of damage (axial strain, pore volume change and cumulative AE energy) are also shown plotted against time. The axial strain curve shows the trimodal behaviour via which creep deformation has generally been interpreted. The primary creep phase is characterized by an initially high strain rate that decreases with time to reach a linear secondary phase that is often interpreted as steady-state creep. After an extended period of time, a tertiary phase is entered, characterized by accelerating strain. This eventually results in macroscopic failure of the samples by propagation of a shear fault. The two other damage proxies of pore volume change and cumulative AE energy also exhibit trimodal behaviour. In particular, the tertiary creep phase is characterized by accelerations in all three proxies for damage. Creep strain rates were calculated for all experiment in the series from the linear portions of the strain-time curves. These are shown in Fig. 17, plotted as a function of differential stress. The data demonstrate that the creep strain rate depends strongly and non-linearly on the level of applied differential stress. Even modest changes in applied stress results in order of magnitude changes in creep strain rate. For example, at 141 MPa (90% of peak stress), the creep strain rate was 3.6×10^{-6} s^{-1} and the time-to-failure was approximately 10 min. At 132 MPa (85% of peak stress; Fig.

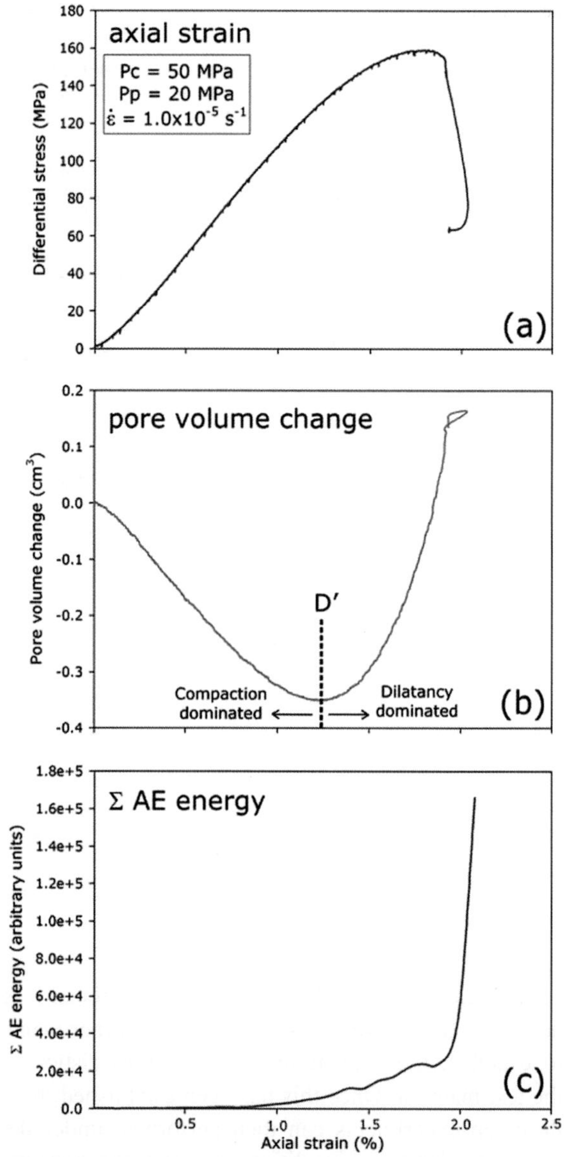

Fig. 15 Conventional constant strain rate experiment on a water-saturated sample of Darley Dale sandstone showing variability in the three proxies for damage within the rock: (**a**) stress-strain curves, (**b**) the pore volume change curves and (**c**) the AE energy output curves. Experimental conditions are indicated on the figure. D′, the stress at which a dilatant-dominated regime dominates, is indicated in (**b**). Pc – confining pressure; Pp – pore fluid pressure

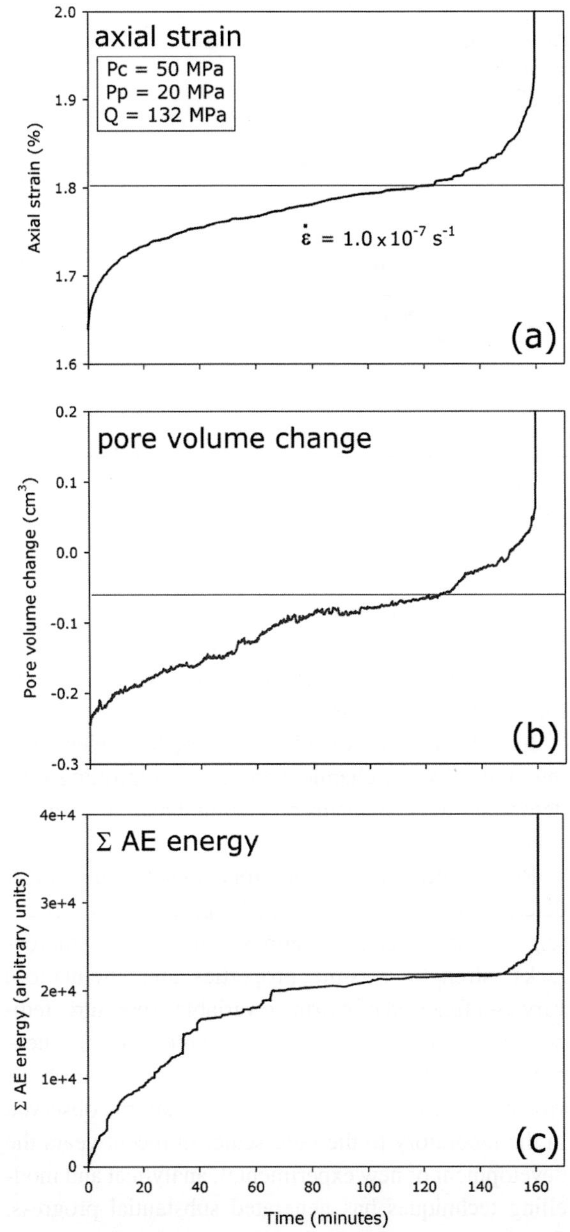

Fig. 16 Graphs of the three damage proxies for a conventional brittle creep experiment on water-saturated Darley Dale sandstone that yielded a creep strain rate of 1.0×10^{-7}. (**a**) time-strain (creep) curve (**b**) pore volume change curve and (**c**) output of AE energy. The position of the onset of accelerating tertiary creep is indicated in each figure by a horizontal line. Experimental conditions are indicated on the figures. Pc – confining pressure; Pp – pore fluid pressure; Q – applied differential stress

16), the creep strain rate decreased to 1.0×10^{-7} s^{-1} and the time-to-failure increased to 160 min (2.6 h). At the lowest stress of 125 MPa (80% of peak stress), the creep strain rate was 1.3×10^{-8} s^{-1} and the time-to-failure was 3,600 min (60 h). In summary, a 10% reduction in applied differential stress resulted in a decrease in creep strain rate and an increase in time-to-failure of approximately 2.5 orders of magnitude.

Furthermore, we also observe that all three measured proxies for crack damage (axial strain, pore volume change and output of AE energy) have approximately equal values at the onset of tertiary creep in all experiments, even though the creep strain rates vary over about three orders of magnitude. This implies that there is a critical level of damage required before the onset of acceleration to failure. The observation of a critical damage threshold has previously been proposed in a number of experimental studies (Griggs 1939, 1940; Wawersik and Brown 1973; Cruden 1974; Kranz and Scholz 1977; Baud and Meredith 1997) and has also been predicted by modelling (e.g., Amitrano and Helmstetter 2006). Kranz & Scholz (1977) suggested that the tertiary creep phase will not commence until the rock has sustained a critical amount of inelastic volumetric strain. More recently, in creep experiments on sandstone, Baud and Meredith (1997) showed that not only was the onset of tertiary creep marked by a critical level of strain, but that it was also marked by critical levels of AE energy output and change in damage volume.

More recently, mean-field theories of damage mechanics have been developed (e.g., Lockner 1998; Main 2000) that invoke a two-stage, rather than a three-stage creep process: (1) a phase of strain hardening involving distributed crack damage, followed by (2) a phase of strain softening involving crack interaction and coalescence. It is proposed that phase (1) dominates in the early stage of deformation and phase (2) dominates in the later stage of deformation. Such as models also therefore postulate a critical damage threshold, where crack interaction leads to a rapid acceleration to failure on a localized fault plane. As the stress intensity at crack tips is influenced by the lengths of the cracks, given suitable conditions, the non-linear relation between strain rate and stress given by stress corrosion theory means that the rate at which a crack grows can accelerate even under a constant boundary stress. For a population of cracks growing in a rock sample held at constant stress, this process is embodied by accelerating strain rate and AE emission with time.

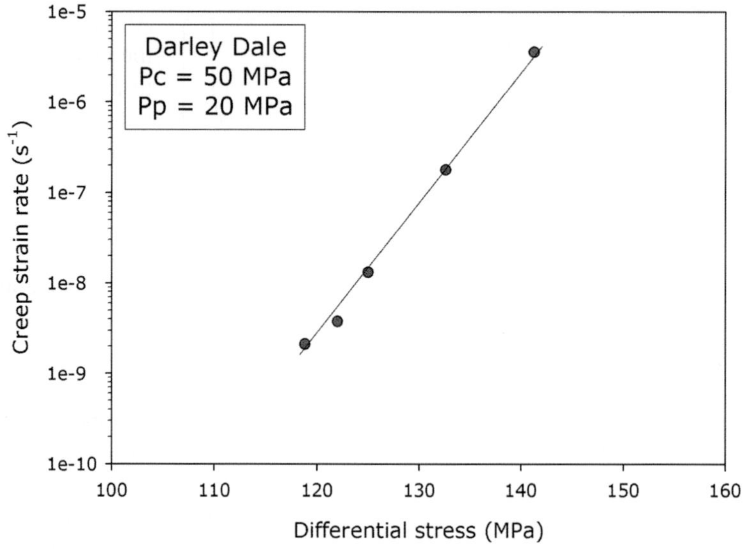

Fig. 17 Creep strain rate data from multiple conventional brittle creep experiments on water-saturated Darley Dale sandstone plotted on a log scale against applied differential stress. Experimental conditions are displayed on the figure. Pc – confining pressure; Pp – pore fluid pressure

Proposal Objectives and Conclusions

Natural hazards such earthquakes, volcanic eruptions and landslides are a severe and permanent threat to human life and sustainable economic development. Dynamic ruptures are preceded by slow deformation mechanisms, in which fractures extend by the growth and coalescence of smaller cracks. Understanding the slow deformation mechanisms preceding dynamic failure is thus a crucial goal to design innovative strategies for forecasting seismic, volcanic and landslide behaviour. Advances towards this goal strongly depend on developing a better comprehension of the flow and fracture behaviour of rocks on various scales in space and time. In order to do so a large range of skills needs to be developed and applied, from a deep fundamental understanding of the physical processes involved up to their monitoring and crisis management.

At the field scale a multidisciplinary monitoring is now required, aimed to apply new technologies and integrated methodologies to real time monitoring of ground deformation (GPS, SAR, InSAR, clinometry, cameras), seismic and microseismic ruptures (broadband, short period seismic and acoustic emission sensors), along with gravimetric and magnetic networks. In volcanic areas real time monitoring of temperature, gas and fluid chemistry (CO_2 and SO_2) is also required (Infrared cameras, spectrometers for the detection of gas and fluid flux). Thanks to the fast technological developments taking place, new instrumentation and the relative techniques can be potentially utilized for interpreting deformation mechanisms in space and time and to design an alarm and early warning system for civil protection purposes.

In order to achieve this goal a better understanding of the physico-chemical processes controlling slow deformation mechanisms preceding dynamic failure is needed.

Rock deformation laboratory experiments allow determination of how intrinsic properties of rocks (e.g., microstructure, composition, fabric, microcracks, strength, seismic properties and attenuation) vary as a function of extrinsic variables (pressure, temperature, pore fluid pressure, differential stress). Theoretical modelling can validate the experimental observations and can be used to scale the patterns observed in the laboratory to the field scale. In recent years the development of new experimental, analytical and modelling techniques has generated substantial progress. Despite the smaller length and time scale of laboratory experiments compared to nature, field observations can be quantitatively reproduced and modelled, thus significantly contributing to the understanding of the dynamics of tectonic and transport processes. Thanks to laboratory experiments and the work at different length scales (from micro to field scale) is now possible to estimate key parameters such as the strength of the lithosphere.

Environmental monitoring, laboratory experiments, theoretical modelling and hazard assessment has now reached a stage where further development of any specific field requires a much tighter and active collaboration among various groups of experimentalists, field-scale scientists and theoreticians both from the geological and geophysical communities to upscale from the micro to the field scale the results obtained in the laboratory.

One of the main issues is that current monitoring systems are mostly calibrated for the detection of ruptures ranging at "large scale" from hundred of meters to kilometres, thus they cannot detect slow deformation nucleating at "small scale" (centimetres to tens of meters), which have been therefore defined so far as "silent" process (e.g., Ozawa et al. 2003). The developing and deploying of specific sensors devoted to the "slow deformation" the a tight connection with the know-how from laboratory experiments is needed. Indeed through triaxial rock-deformation experiments is possible to analyse how data comparable to field observations such seismic event rates, seismic wave velocities, and ratios of seismic P- to S-wave velocities evolve with respect carefully simultaneously measured swith mechanical parameters of the rock (e.g., stress, strain and crack damage).

We then propose a new departure based on an interdisciplinary unitary and integrated approach aimed to

(1) transfer of knowledge between specific fields, which up to now aimed at solve a particular problem;
(2) quantify critical damage thresholds triggering dynamic failure;
(3) set up early warning models for forecasting the time of rupture with application to volcanology, seismology and landslide risk prevention.

In order to do so we propose to combine the outcome of two neat examples of multidisciplinary (e.g. remote sensing by SAR interferometry and high frequency sensors for microseismicity) field monitoring of characteristic "large scale" signs of impending deformation from different tectonic setting, i.e. the Ruinon landslide (Italy) and Stromboli volcano (Italy) with the kinematic features of slow stress perturbations induced by fluid overpressures and relative modelling and the results of experimental rock deformation laboratory experiments and theoretical modelling investigating slow deformation mechanisms, such stress corrosion crack growth.

The expected benefits of this approach will allow us to develop new technologies aimed to monitor the "slow deformation" preceding the dynamic failure signals in unstable sectors of the Earth, and will provide invaluable data for theoretical modeling of critical damage thresholds triggering instability onset. This "topic" answers two of the main questions of the Hazards Theme of the International Year of the Planet Earth (http://www.yearofplanetearth.org/), and, in particular, the questions 2 "What technologies and methodologies are required to assess the vulnerability of people and places to hazards – and how might these be used at a variety of spatial scales?" and 3 "How does our current ability to monitor, predict and mitigate vary from one geohazard to another? What methodologies and new technologies can improve such capabilities, and so help civil protection locally and globally?" (Beer 2008, and http://www.yearofplanetearth.org/content/downloads/Hazards.pdf).

Acknowledgments This work has been supported by the sponsors of the PHASE university consortium project. The NEST Pathfinder Program Triggering Instabilities in Materials and Geosystems (contract NEST-2005-PATH-COM-043386) is acknowledged. The microseismic data from the Ogachi site are courtesy of Dr. H. Kaieda (Central Research Institute of Electric Power Industry, Japan). Help and assistance of Dr. T. Ito (Institute of Fluid Science, Tohoku) is greatly appreciated. This paper is related to the project "Creep" accredited by the Hazards Theme of IYPE.

References

Acocella V, Neri M, Scarlato P (2006) Understanding shallow magma emplacement at volcanoes: orthogonal feeder dikes during the 2002–2003 Stromboli (Italy) eruption. Geophys Res Lett. DOI:10.1029/2006GL026862

Allard P, Carbonelle J, Metrich N, Loyer H, Zattwoog P (1994) Sulfur output and magma degassing budget of Stromboli volcano. Nature 368:326–330

Amitrano D, Helmstetter A (2006) Brittle creep damage and time to failure in rocks. J Geophys Res. DOI:10.1029/2005JB004252

Anderson OL, Grew PC (1977) Stress corrosion theory of crack propagation with applications to geophysics. Rev Geophys 15:77–104

Andrade EN, Randall RFY (1949) The Rehbinder effect. Nature 164:1127

Atkinson BK (1984) Subcritical crack growth in geological materials. J Geophys Res 89:4077–4114

Atkinson BK, Meredith PG (1987) The theory of subcritical crack growth with applications to minerals and rocks. In:

Atkinson BK (ed) Fracture Mechanics of Rock. Academic Press, London

Baldi P, Coltelli M, Fabris M, Marsella M, Tommasi P (2008) High precision photogrammetry for monitoring the evolution of the NW flank of Stromboli volcano during and after the 2002–2003 eruption. Bull Volcanol 70:703–715

Barberi F, Rosi M, Sodi A (1993) Volcanic hazard assessment at Stromboli based on review of historical data. Acta Volcanol 3:173–187

Baud P, Meredith PG (1997) Damage accumulation during triaxial creep of Darley Dale sandstone from pore volumometry and acoustic emission. Int J Rock Mech Min Sci 34:3–4

Beer T (2008) Minimising risk maximising awareness: the hazards theme of the international year of the planet Earth. 33IGC Oslo 2008 PEH01204L 105 Abstract volume

Benson PM, Thompson AB, Meredith PG, Vinciguerra S, Young RP (2007) Imaging slow failure in triaxially deformed Etna basalt using 3D acoustic-emission location and X-ray computed tomography. Geophys Res Lett. DOI:10.1029/2006GL028721

Bonaccorso A, Calvari S, Garfi' G, Lodato L, Patane D (2003) Dynamics of the December 2002 flank failure and tsunami at Stromboli volcano inferred by volcanological and geophysical observations. Geophys Res Lett. DOI:10.1029/2003GL017702

Brehm DJ, Braile LW (1999) Intermediate-term earthquake prediction using the modified time-to-failure method in South California. Bull Seis Soc Am 89:275–293

Brodsky EE, Karakostas V, Kanamori H (2000) A new observation of dynamically triggered regional seismicity: earthquakes in Greece following the August 1999 Izmit Turkey earthquake. Geophys Res Lett 27:2741–2744

Chouet B, Dawson P, Ohminato T, Martini M, Saccorotti G, Giudicepietro F, De Luca G, Milana G, Scarpa R (2003) Source mechanisms of explosions at Stromboli Volcano Italy determined from moment-tensor inversions of very-long-period data. J Geophys Res. DOI:10.1029/2002JB001919

Costin LS (1987) Time-dependent deformation and failure. In: Atkinson BK (ed) Fracture Mechanics of Rock. Academic Press, London

Cruden DM (1974) Static fatigue of brittle rock under uniaxial compression. Int J Rock Mech Min Sci 11:67–73

Cruden DM (1991) A simple definition of a landslide. Bull Int Assoc Eng Geol 43:27–29

Cruden DM, Varnes DJ (1996) Landslides types and processes. In: Land-slides: Investigation and Mitigation. Transportation Research Board, National Research Council, National Academies Press, Washington DC

Di Giovambattista R, Tyupkin YS (2001) An analysis of the process of acceleration of seismic energy emission in laboratory experiments on destruction of rocks and before strong earthquakes on Kamchatka and in Italy. Tectonophysics 338: 339–351

Economides MJ, Nolte KG (2003) Reservoir Stimulation. Wiley, Chichester

Falsaperla S, Neri M, Pecora E, Spampinato S (2006) Multidisciplinary study of flank instability phenomena at Stromboli volcano Italy. Geophys Res Lett. DOI:10.1029/2006GL025940

Griggs D (1939) Creep of Rocks. J Geol 47:225–251

Griggs D (1940) Experimental flow of rocks under conditions favouring recrystallization. Bull Seis Soc Am 51:1001–1022

Heap M, Baud JP, Meredith PG, Bell AF, Main IG (2009) Time-dependent brittle creep in Darley Dale sandstone. J Geophys Res 114, B07203, doi:10.1029/2008JB006212.

Jaeger JN, Cook GW, Zimmerman R (2007) Fundamentals in Rock Mechanics (4th Edition). Blackwell Publishing, London

Kaieda H, Kiho K, Motojima I (1993) Multiple fracture creation for hot dry rock development. Trends Geophy Res 2:127–139

Kaieda H, Sasaki S (1998) Development of fracture evaluation methods for Hot Dry Rock geothermal power – Ogachi reservoir evaluation by the AE method. CRIEPI report U97107 (in Japanese with English abstract)

Kilburn CRJ (2003) Multiscale fracturing as a key to forecasting volcanic eruptions. J Volcanol Geotherm Res 125:271–289

Kilburn CRJ, Voight B (1998) Slow rock fracture as eruption precursor at Soufriere Hills volcano: Montserrat. Geophys Res Lett 25:3665–3668

Kranz R, Scholz CH (1977) Critical dilatant volume of rocks at the onset of tertiary creep. J Geophys Res 82:4893–4898

Lawn B (1993) Fracture of Brittle Solids. Cambridge University Press, Cambridge

Lockner D (1998) A generalized law for brittle deformation of Westerly granite. J Geophys Res 103:5107–5123

Main IG (2000) A damage mechanics model for power-law creep and earthquake aftershock and foreshock sequences. Geophys J Int 142:151–161

Main IG, Sammonds PR, Meredith PG (1993) Application of a modified Griffith criterion to the evolution of fractal damage during compressional rock failure. Geophys J Int 115: 367–380

Majer EL, Baria R, Stark M, Oates S, Bommer J, Smith B, Asanuma H (2007) Induced seismicity associated with enhanced geothermal systems. Geothermics 36:185–222

Orowan E (1944) The fatigue of glass under stress. Nature 154:341–343

Ozawa S, Miyazaki S, Hatanaka Y, Imakiire T, Kaidzu M, Murakami M (2003) Characteristic silent earthquakes in the eastern part of the Boso peninsula Central Japan. Geophys Res Lett. DOI:10.1029/2002GL016665

Parotidis M, Shapiro SA, Rothert E (2004) Back front of seismicity induced after termination of borehole fluid injection. Geophys Res Lett. DOI:10.1029/2003GL018987

Paterson MS, Wong TF (2005) Experimental Rock Deformation – The Brittle Field. Springer, New York

Pompilio M (2003) Eruzione Stromboli 2002–2003: Cronologia dell'eruzione localizzazione e migrazione delle bocche eruttive. Internal report Ist Naz di Geofis e Vulcanol Catania, Italy

Rehbinder PA (1948) Hardness Reducers in Drilling (Translated from Russian). CSIR, Melbourne

Ripepe M, Marchetti E, Ulivieri G, Harris AJL, Dehn J, Burton M, Caltabiano T, Salerno G (2005) Effusive to explosive transition during the 2003 eruption of Stromboli volcano. Geology 33:341–344

Rosi M, Bertagnini A, Harris AJL, Pioli L, Pistolesi M, Ripepe M (2006) A case history of paroxysmal explosion at Stromboli: timing and dynamics of the April 5 2003 event. Earth Planet Sci Lett 243:594–606

Rosi M, Bertagnini A, Landi P (2000) Onset of the persistent activity at Stromboli volcano (Italy). Bull Volcanol 62: 294–300

Rothert E, Shapiro SA (2007) Statistics of fracture strength and fluid – induced microseismicity. J Geophys Res. DOI:10.1029/2005JB003959

Rudolf H, Leva D, Tarchi D, Sieber AJ (1999) A mobile and versatile SAR system. Int Geosc Rem Sens Symp, Hamburg

Rudolf H, Tarchi D (1999) LISA: the linear SAR instrument. Tech Rep I 99 126 Eur Comm Joint Res Cent, Ispra

Rutledge JT, Phillips WS (2003) Hydraulic stimulation of natural fractures as revealed by induced microearthquakes Carthage Cotton Valley gas field east Texas. Geophysics 68:441–452

Shapiro SA, Dinske C, Kummerow J (2007) Probability of a given magnitude earthquake induced by a fluid injection. Geophys Res Lett. DOI:10.1029/2007GL031615

Shapiro SA, Dinske C, Rothert E (2006a) Hydraulic-fracturing controlled dynamics of microseismic clouds. Geophys Res Lett. DOI:10.1029/2006GL026365

Shapiro SA, Kummerow J, Dinske C, Asch G, Rothert E, Erzinger J, Kümpel HJ, Kind R (2006b) Fluid induced seismicity guided by a continental fault: injection experiment of 2004/2005 at the German deep drilling site (KTB). Geophys Res Lett. DOI:10.1029/2005GL024659

Shapiro SA, Rothert E, Rath V, Rindschwentner J (2002) Characterization of fluid transport properties of reservoirs using induced microseismicity. Geophysics 67: 212–220

Scholz CH (1968) The frequency–magnitude relation of microfracturing in rock and its relation to earthquakes. Bull Seismol Soc Am 58:399–415

Tarchi D, Casagli N, Moretti S, Leva D, Sieber AJ (2003) Monitoring landslide displacements using groung-based synthetic aperture radar interferometry: application to the Ruinon landslide in the Italian Alps. J Geophys Res. DOI:10.1029/2002JB002204

Terzaghi K (1943) Theoretical Soil Mechanics. John Wiley and Sons, New York

Tibaldi A (2001) Multiple sector collapses at Stromboli volcano Italy: how they work. Bull Volcanol 63:112–125

Varnes DJ (1989) Predicting earthquakes by analyzing accelerating precursory seismic activity. Pageoph 130: 661–686

Wawersik WR, Brown WS (1973) Creep Fracture in Rock. Utah University, Department of Mechanical Engineering, Salt Lake City

Landslides in Mountain Regions: Hazards, Resources and Information

Raisa Gracheva and Alexandra Golyeva

Abstract The role of landslides in mountain regions is complicated. Landslides are considered as a mechanism of loss and the accumulation of dispersed mineral matter. They also play a role in the formation of new mountain relief and ecosystems, and the conservation of archeological and environmental information. The loss by landslides of fine earth, an irreversible resource, reduces the life-supporting resources of mountain regions. To investigate this a 15–100 year chronosequence of landslides was studied in the West Caucasus (Georgia), using case studies of stabilised slumps within newly settled agricultural areas with accumulated fine earth. Past landslides that buried settlements can be considered as "keepers" of scientific information recorded in cultural layers and fossil soils. The archeological site Gruzinka (North Caucasus, Russia) is such an example.

Keywords Landslides · The Caucasus · Fine earth · Use of landslides · Buried soils and cultural layers

Introduction

Landslides are counted among the most disastrous natural processes; they damage and often completely destroy human settlements and affect economic activities, in mountains in particular. Recognizing that more than 10% of the earth's population resides in mountains or piedmont regions (Gerrard 1990; Klupt 2008), one may easily visualize degree of this hazard. Most giant landslides that occurred in the past or recently are related to earthquakes and volcanic eruptions, though heavy rainfall and human activity may also induce widespread mass movements that remove soils and sediments from mountain slopes and exert an adverse influence over vast areas in mountains (Bolt 1975; Varnes 1978; Barsh and Caine 1984; Butler et al. 2003; Cruden and Varnes 1996; Allen 1997; Alexander 1998).

The last few decades are marked by an increase in the frequency of natural catastrophes, including landslides, mudflows, rockfalls, etc. (Osipov 2001). That tendency becomes a growing concern to both specialists and decision makers. There are a great number of publications that deal with landslides as natural phenomena, as well as the damage to the society. We refer to some of publications for examples of the most disastrous landslides resulting either from natural factors or human activities; those authors present estimates of socio-economic losses inflicted by landslides (Hewitt 1992; Jones 1992; Sassa 1999; Report on Landslides-Damages 2001; Schuster and Highland 2001; Evans and Alcántara-Ayala 2007).

When considering economical losses and casualties due to landslides, we have to emphasize another very important consequence of mass wasting processes, namely *irreversible losses* of great volumes of loose mineral material, including soils and subsoils. As part of a general tendency toward degradation of the Earth's surface by gravitational mass wasting and erosion, the processes are of particular importance for mountain landscapes and their inhabitants (Varnes 1978; Eisbacher and Clague 1984; Dunning and Cole 2002).

R. Gracheva (✉)
Institute of Geography of RAS, Moscow 119017, Russia
e-mail: gracheva04@list.ru

T. Beer (ed.), *Geophysical Hazards*, International Year of Planet Earth,
DOI 10.1007/978-90-481-3236-2_15, © Springer Science+Business Media B.V. 2010

In mountains, loose sediments and their *fine components* in particular are a life-supporting resource for mountain biota as well as for human habitatation and economic activities (agriculture, forestry, etc) (Dobrovolsky and Nikitin 1986; Ilychev and Gracheva 1998; Frossard et al. 2006). It is well known that fine-grained soils and mantles of loose material on slopes, once eroded, cannot be renewed within a single or even several generations. According to very rough estimates, the proportion of *inherited* fine earth in the existing pedosphere is about 80–90% of the total pedosphere resource. The *real input* of fine earth produced by weathering and pedogenesis during the Holocene is about 10–20% or even less (Targulian 2008). The mountains once devoid of loose cover soon loose their population. Examples of such situations are historically abundant (Messerli and Ives 1999).

There is, however, an opposite tendency in the population behavior: in many mountain regions inhabitants that left their settlements destroyed by a disastrous landslide soon return in the damaged district. For one example, Karmadon gorge in the Northern Caucasus suffered from destructive landslides, slumps and mudflows, which tend to occur in clusters every 30–60 years. And after each of these events, even those with the loss of human life, new settlements and farm lands reappear shortly thereafter. It is possible, however, that the Kolka glacier collapse and catastrophic debris flow that modified the local topography (Kotlyakov et al. 2004) put an end to the tradition. There is similar tendency in the Southern Caucasus, in Mountain Adzharia (Georgia): of the 15 000 residents of landslide-damaged areas in the 1980s, more than a half have returned (Gracheva 2004a). The same processes are characteristic in some regions of Mexico (Alcantara-Ayala et al. 2003) and elsewhere.

Why do people continue to return to the landslide-hazardous areas in mountains?

Leaving aside economic, ethnic and other reasons for such a behavior, we can return to the role of landslides in the transport of loose material.

The role played by landslides in the life of people living in mountains is much more complicated than is evident in the disastrous consequences for landscapes and humans. The sliding rock mass exposes fresh surfaces for weathering and accumulates loose material, thus starting new cycles in ecosystems development, forming new habitats for biota and new lands for agricultural ecosystems. Soils, sediments and archaeological evidence buried under landslides may provide a relatively complete record of past events and environments.

In this paper, we consider landslides as a potential hazard to loose sedimentary cover in mountains, as a resource of loose mineral material, and as a factor for scientific information preservation, with special reference to mountain regions of the Caucasus within the boundaries of Russia and Georgia.

Landslides as Hazard: Loss of Loose Sediment

It is beyond the scope of this study to provide a detailed review of sediment transfer processes in mountains. We consider but a few examples of landslide importance in mass wasting and sediment loss in mountains.

When discussing natural processes modeling in mountain environments and their deposits, specialists emphasize difficulties in distinguishing between various factors responsible for loose material losses (Dedkov and Moszherin 1992). By way of example, J. Warburton (2007) cites experiments performed by Page et al. (1994, 1999) with the aim of finding the contributions from each individual landslide event to the loose sediment movements.

A short-term sediment budget was constructed by Page et al. (1994) to assess the response of a small catchment (3,208 hectares) to an intense rain storm event. A total of 1.35 million cubic metres of sediment moved during the storm (420 m^3 per hectare). Of this total, 21% was stored on hillslopes, 22% deposited on valley floors, 51% was deposited in lakes and the remaining 6% was discharged at the catchment outlet. Approximately 89% of the sediment generated during the storm was from landslide erosion on the slopes.

Page et al. (1999) developed a method for assessing sediment production from landsliding and applied this to the Cyclone Bola event (New Zealand) with the aim of determining the contribution of landslides to suspended sediment output from the event. Shallow landslides are responsible for approximately 64% of the load exiting the catchment. In a few years immediately following an event of this magnitude suspended sediment concentrations are 100% greater than in the years preceding the event owing to continued erosion of landslide scars and stored sediments

(Warburton 2007). Therefore, erosion would continue to remove fine components from a stabilized landslide accumulations.

To assess the extent of landslide hazard, we used relation of landslide-damaged area (including deposits and landforms) to total area under consideration ($K = S_{landslide}/S_{total}$) (Emelyanova 1964; Sheko 1982).

In Russia the most hazardous region with respect to landslides is the Caucasian coast of the Black Sea where K is estimated as $\sim$0.7 (Milkov and Gvozdetski 1969; Dubrovin and Klimenko 1973; Sheko 1976, 1982; Razumov 2008). There, mountains come close to the coastline, leaving only a narrow strip of sandy or gravel shingled beach. The deeply dissected mountains are composed of limestones, flysch and schists; loose surficial mantle varies in thickness from 0.5 to 2–3 m. The climate of the region is humid, annual precipitation amounts to 1,400–1,600 mm. Within 10 km from the coastline between Adler and Anapa cities there are more than 1,000 landslides recorded, including block landslides (rockfalls and rockslides) in solid rock, slumps in Quaternary deposits, debris flows, shallow landslides; besides, creep of soil and subsoil which occurs all over the area. The rate of soil wasting by small landslides and creep exceeds that of surface wash by two (and more) orders of magnitude and amounts to 600–800 kg/ha per year (Azhigirov 1987). Each of the larger landslides or slumps accounts for sediment loss between 2×10^3 and 5×10^3 m^3.

Marine terraces, slopes and valleys are well populated and have a dense network of roads. Agricultural activities by the local population are adapted to the natural hazards and aimed at their alleviation. During the 1990s the influx of migrants from various regions of the former USSR changed the composition of the population. Many of the migrants were unaware of the measures applied in environmental protection. For example, people from arid and semiarid regions that settled down in the rural mountain and coastal areas continued to apply traditional techniques of farming, such as slope terracing and excessive irrigation; and when constructing their dwellings they did not take into consideration the slope state and engineering properties of rocks. In the humid landslide-hazardous region, such activities triggered both shallow and deep-seated landslides. At present, the recreation development (together with related infrastructure) is in process; it enhances environmental stress, the more so as the landslide hazard is rarely taken into account. Against the background of current climatic changes, the socioeconomic processes act as a catalyst promoting various natural hazards in the region, their contribution to landslide activation increasing dramatically (Gracheva 2004b).

It is not always the case, however, that landslide deposits are completely eroded and removed by water flow. In the case of large volumes of sliding mass, a part of it can be stopped and stabilized, forming a new element of the mountain topography. These are steps, terrace-like or fan-like, on slopes, at the base of scarps or in river valleys; and comprise enormous volume of loose material, both coarse and fine (Easterbrook 1993). Such landforms resulting from sliding, slumping and other processes of mass movement are found in many mountain regions, and in the coastal regions in particular including submarine landforms (Carlson and Karl 1988; Hampton et al. 1996; Brundsen and Moore 1999). Thus, they are quite typical of the Russian coast of the Black Sea (Dubrovin and Klimenko 1973; Sheko 1976). An example of an old landslide is the Utrish headland on the Abrau peninsula. Large landslides dated to the Pleistocene and Holocene are known from the coasts of the Sea of Japan, coasts of the Amur and Ussuri bays, and particularly on the eastern Sikhote-Alin macro-slope (Korotki et al. 2005). Some examples of the largest landslides in the Western Hemisphere, both prehistoric and historic, as well as their topographic effects, are cited by Schuster and Highland (2001).

There are many cases of mountain inhabitants actively using not only ancient landslide deposits and landforms, but quite recent ones; this is illustrated by examples near Mountain Adzharia, in the western part of the Republic of Georgia.

Landslides as Hazard and Resource

Mountain Adzharia belongs to the western part of the Adzharian-Trialet mountain system. At present it is a densely populated region with a long history of farming (Dzhaoshvili 1968; Putkaradze 1996). The area under consideration is bounded by the Meskheti Ridge in the north, by the Shavsheti Ridge in the south, and Arsian Ridge in the east; on the whole, it reflects an amphitheater of stepped configuration descending westwards and to the Adzharistskali river valley. The ridge altitudes vary from 1,000 to 2,000 m a.s.l., the highest points (mountain tops and crests) are more than 2,000 m high, relative heights above valley floor are about 400–600 m. The mountain slopes are steep, often in excess of 20°. Talus and rockfall slopes are widely distributed in the region, within the uppermost belt in particular. Those are areas of neotectonic uplifts intensifying erosion and mass-wasting processes. The region is for the greater part composed of Paleogene rocks with widely distributed volcanic formations, such as lava, tuffolava, tuffs and rocks of mostly intermediate composition (andesites and andesite porphyrites). Intensive postmagmatic processes altered the initial rock composition and caused either argillization or silicification of the rocks (Razumova 1977).

The climate of the area is controlled by air masses coming from the Black Sea; which is noted for high humidity and changes from subtropical to temperate with altitude. The mean annual precipitation amounts to 1,000–1,600 mm. Typical for the entire area are summer storms, their intensity and duration increasing with altitude, and thick snow cover (up to 3–4 m) (Dzhavahishvili 1964).

Prolonged weathering of the rocks under conditions of wet mild climate resulted in weathering crusts 3–10 m thick, clayey, loamy and silty in composition, often with small-size debris and rock fragments. The latter are supplied to the slope mantle by isolated massifs of compact quartz rich rock highly resistant to weathering.

According to data obtained by the Adzharian Service of Regular Observations and Prevention of Hazardous Geological Processes, the proportion of landslide-damaged area (*K*) varies from 0.1 to 0.7 and increases eastwards (Fig. 1). This region is noticeable for the presence of every factor controlling sliding processes, namely steep slopes; water saturation of slope material during snow melting and frequent rainfalls; weak slope cover consolidation due to fractured rocks and poorly cemented deeply weathered sediments (Emelyanova 1964; Tsereteli and Tsereteli 1985).

There are various types of landslides recorded in the region, including tectonic and shallow landslides, debris flow, earthflow, mudflow, creep, slumps, often forming a natural landslide complex. The most active removal of loose material takes place on steep slopes of gorges with a stream channel at their base. In typical cases of undisturbed blocks of regolith, together with forest slumping into the river, loss of the material is estimated at 100–1,200 m^3.

All the forested slopes display microrelief of shallow slumps; it is most clearly pronounced on south-facing slopes where sites of so-called "drunken forests" are occasionally found. The shallow landslides are distinctly seen in the soil profile. They are mostly responsible for buried humus horizons, or for "inverted" layers, with coarser material overlying fine soil.

Due to heavy dissection of the land surface, only 10–11% of the total Mountain Adzharia area is suitable for settlement and farming (not including mountain grazing land). The shortage of cultivable lands is one of the most crucial problems of this region. Many settlements are located on large bodies of regolith slowly moving downslope. The inhabitants have been adapting to the landslide hazard in the course of centuries. They construct light wooden pile-dwellings ("oda"). The rear wall of the dwellings typically rests against the mountain slope, so they do not require clearing a large building site, nor laying foundation; the piles act as vibration absorbers. Under conditions of land shortage, such a method of building was most useful and relatively safe.

In the twentieth century, and since the 1950s in particular, the human impact on the environment increased dramatically (Gigineishvili and Nahutsrishvili 1998). During the last 50 years the population in three administrative districts – Keda, Shuakhevi, Khulo – increased twofold. The population density in settled and cultivated areas expanded and now varies from 400–450 to 750–800 persons per square km.

Housing density increased and a network of motor roads was built, often using rock explosions. With an increase of the population income, some changes were noted in the building traditions; wooden houses being replaced by heavy constructions of brick or stone calling for a large site are to be cleared. Water pipes were laid to supply water to all the households and vegetable gardens; the water use, however, was not controlled and water often flowed onto the slopes unimpeded.

The 1980s in Mountain Adzharia were marked by an extremely rapid snow melting and by an increase in rain frequency and duration in the spring (Berdzenishvili 2000). This decade was also distinct for more frequent occurrences of sliding and slumping, as well as small mudflows. People did not pay attention to the hazard until fissures began to appear in the stone walls

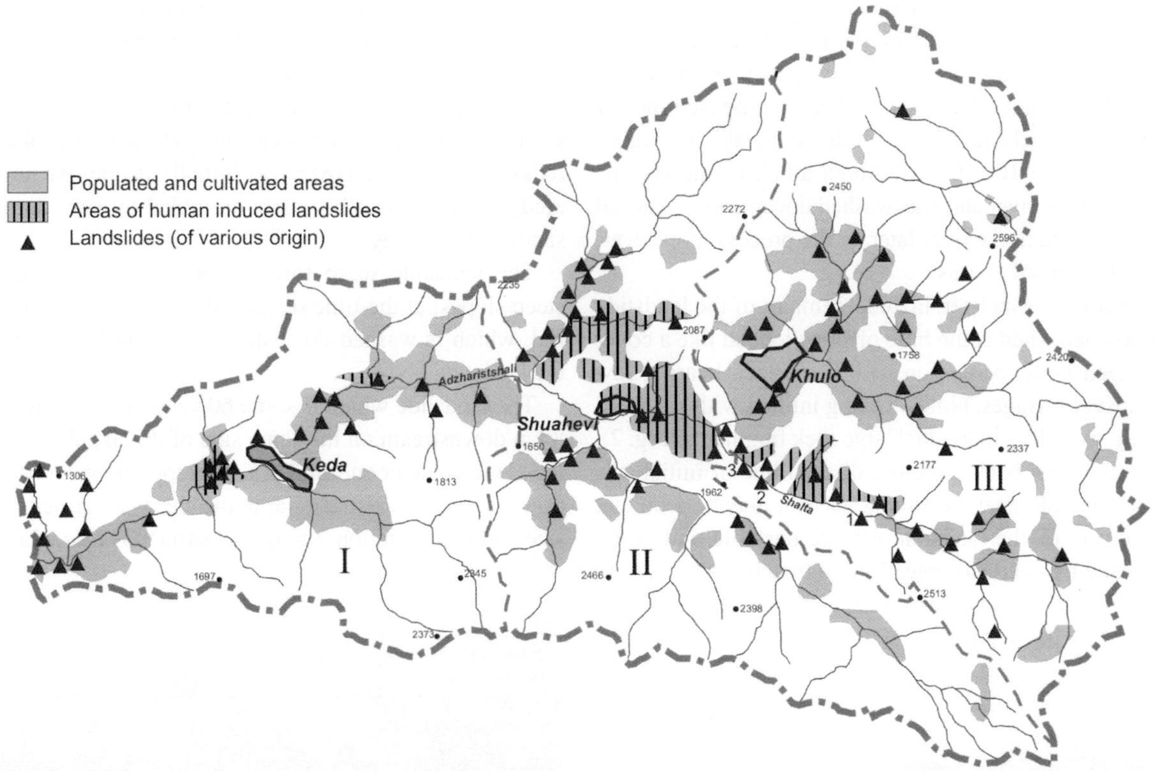

Fig. 1 Landslide occurrence in Mountain Adzharia (based on field observations and on the data kindly provided by the Department of Geology and Mining, Batumi, Georgia). Explanations: Coefficient of landslide hazard (K=$S_{landslide}/S_{total}$): I – K=0.2–0.3; II – K=0.3–0.5; III – K= 0.5–0.7. Landslides under study: 1 – landslide of 1989; 2 – landslide of ~1950; 3 – landslide of ~ 1910

of their houses, small slumps and slides damaged farmyards and vegetable gardens, as well as asphalt roads, and micro-mudflows blocked water sources. The processes of activation may be inferred from the statistics of the residents' appeals to the Service of Hazardous Geological Process Monitoring and Prevention and to other state agencies, including those responsible for property damage insurance. Thus, in the spring of 1988 officers of the state agencies visited 5–10 households daily to record damage due to landslides and small mudflows. It is clear that although landslides themselves are practically unavoidable, the damage to infrastructure could be far less if better practices were used in construction and other economic activities.

The land shortage accounts for people's intention to use every parcel of land. As is shown by a geomorphological survey in Mountain Adzharia (on the northern slopes of Shavsheti Ridge, Skhalta River canyon), many settlements in the lower portions of valley slopes are located on landslide toes. In 2003 a chronological sequence of landslides was analyzed over a 100 yr interval; the interval between landslide events was about 40 years. The sequence is as follows:

1. The landslide of 15 April 1988 (the date is precise).
2. The landslide occurring about 60 yrs ago (the age based on eye witnesses' account).
3. The landslide about a century old (dated by circumstantial evidence, including age of the forest on the scarp surface).

On 15 April 1988, a large mass of loose sediments slumped from the north-facing slope and impounded the river. A few minutes later, the second landfall covered the opposite river bank and buried a part of Tsablana village together with residents under masses of loam and debris.

The slump was detached from the mountain slope at an altitude of about 1,800 m, the height of fall was approximately 400 m, and the travel distance exceeded

3 km. The river was impounded with a dam 15–20 m in height. Luckily, the dam erosion proceeded at a slow rate, so no catastrophic flood happened downstream. Total volume of the landslide body (composed mostly of mixed rock fragments, rubble and clay) is roughly estimated at 10×10^4 m^3. About a half of the volume entered the river and was washed down; some material was moved to the river later, in the process of excavating the buried houses.

Fifteen years later, in 2003, remains of the landslide mass deposited at the base of slope looked like a cone-shaped body, partly forested, its surface bearing low transverse ridges, hollows being infilled with sand and clay, as well as individual large rock fragments (Fig. 2). On the river bank there is a distinct line of initial surface buried under the sliding mass; it may be seen that deposits of the landslide toe are unstable and being eroded by overland flow. Several trenches were dug to a depth of 1 m in order to study granulometric composition of the deposits. The uppermost 5 cm consisting of mixed small debris and loam with coarse organic matter may be regarded as the initial stage of soil formation. Below the deposits are more closely packed and even cemented into non-sorted mixture of rubble, sand, loam and clay with occasional fragments of solid rocks. As the slope cover was initially deeply weathered, the debris can be easily crumbled by hand into sandy loam or clay.

The landslide scarp forms poorly vegetated open steep slopes; at the time of rains they yield fine material, which is washed downslope, onto the surface of the landslide body.

The landslide which is some 60 years old is located 8 km downstream on the same side of the Skhalta valley as the younger one, and in similar topographic position (Fig. 3). It is also similar to the above-described in size and configuration, though its surface is more flat-

Fig. 2 Landslide of 1988 (the photo taken in 2003). Mountain Adzharia, Valley of Skhalta River, Rep. of Georgia

Fig. 3 Landslide of about 60 years age used for hay harvesting (after soil remediation). Mountain Adzharia, Valley of Skhalta River, Rep. of Georgia

tened and covered with coniferous forest. In the lower portion of the landslide the forest is clearcut and the area turned into hay field. Stones removed from the surface and the uppermost soil horizon are piled up in heaps (as seen in Fig. 3), the surface was repeatedly ploughed and sown with perennial grasses. According to the local residents, the landslide surface recultivation began about 20 years after event and took approximately 10 years. Stones have to be removed from the area continuously.

As has been shown by field studies, there exists a relatively well developed soil, with a structured humus horizon up to 10 cm thick and a loam horizon 30–40 cm thick; the latter contains debris in small proportion and is penetrated by grass roots. The soil granulometric composition has been altered by human activities aimed at creating favorable soil conditions for a hay field, and in general – at agricultural landscape development.

The tip of the landslide body is overhanging the river, so the area may be considered as landslide-hazardous. The open surface of landslide rupture is completely forested but for several steep scarps; on the slope above the hay field, under the forest canopy, there are accumulations of stones probably transported from the scar surface.

The landslide of estimated age 100 years is located still farther down stream, it is identical to the above in topographic position on the same river bank (Fig. 4). Its toe has reached the river and the landslide itself is almost imperceptible against the slope background. Its surface, except for its lowermost part, is covered completely with thick coniferous forest practically undistinguishable from the surrounding forests. The lower part presents a cultivated landscape typical of Mountain Adzharia (domestic buildings, vegetable gardens, orchards). The landslide cultivation started about 60–70 years ago.

At that part of the landslide soils consist of a thick (up to 20–30 cm) humus horizon underlain by a loamy horizon (to a depth of 60 cm or more) bearing all the signs of alteration by pedogenic processes. The soils are very fertile because of an abundance of nutritional minerals supplied by debris weathering.

Residents and local authorities explain that the described plots are very suitable from settlement and cultivation. The land is considered top-quality, and nobody is concerned by its origin considering the

Fig. 4 Built-up and cultivated landslide of approximately 100 years age. Mountain Adzharia, Valley of Skhalta River, Rep. of Georgia

shortage of cultivable lands. Before the event, there were forested mountain slopes quite unsuitable for habitation and other land use. The landslides created relatively gentle slopes and, which is more, they accumulated great masses of loose material – a valuable agricultural resource in the mountains. Some 20–30 years after the event the landslide surface solidifies, its sides spread laterally and become gentler, whereas mountain dwellers became quite accustomed to stone removal from the fields for many centuries. Approximately 40 years after the event the catastrophe is forgotten and the landslide surface is used for settlement and agriculture.

There are other regions of the Caucasus where landslides have been long used for agriculture. Landslide bodies were used for slope modeling into agricultural terraces in the central part of the North Caucasus, in North Jurassic intermountain depression, where the slopes are composed of schists and bear a loamy and crushed stony cover of 1–2 m thick. As a rule, people used small slumps moving almost undisturbed regolith and forming bench-like steps. They flattened the slump surface, strengthened the base and used it for growing cereals. The landslide bodies transformed in this way are usually resistant to surface wash and erosion, though repeated sliding and slumping cannot be excluded. Terrace steps still in existence are 5–25 m wide, 1–5 m high, and 20–100 m long. They are identifiable as landslides by their irregular distribution over the slope surface and by preserved scarps upslope. Another evidence of their genesis is original slope surface and buried soil locally preserved under the terrace body. In case of a constructed terrace, not resulting from slumping, the original slope surface is disturbed.

Studies of soils and deposits buried under landslides allow the event to be dated; besides, they give an insight into environments of the past and the history of the region settlement. Researches along these lines have been carried out in the Northern Caucasus, in the Abinsk intermountain depression (Krasnodar Territory).

Landslides as Storage of Scientific Information

The Abinsk intermountain depression is located in western part of the Northern Caucasus at altitudes of 500–600 to 1,000 m a.s.l. The area is composed of flysch, limestone and sandstone, flysch being absolutely dominant; the slopes of 15–25° steepness are covered with loose deluvial material up to 2 m thick notable for a considerable proportion of fine constituent. At present the main factors triggering slumping, debris flow and mudflow processes are forest felling and overgrazing. Degree of landslide hazard (proportion of landslide-damaged area) we estimate at 0.1–0.3.

The Abinsk depression abounds in archeological monuments (Krupnov 1960). Numerous dolmen and barrow (burial mound) complexes were studied in collaboration with the North Caucasian expedition of the Institute of Archaeology, Russian Academy of Sciences. Here we give some results obtained by explorations of multi-layered archeological sites at the foot of Gruzinka Mountain.

Excavations and a number of pits 150–200 cm deep revealed a series of fossil soils and cultural layers alternating and overlain with interbeds of loam and debris 20–40 cm thick. Therein have been found dolmen stone slabs covered with debris layers; the latter consist of non-sorted mixtures of rubble and loam, with individual stone fragments randomly oriented. Judging from the morphology of the layers, they may be considered as remains of micro-slumps or small mudflows with variable ratio between debris and fine earth constituents.

Humic acids from the soils and cultural layers under the dolmen constructions, as well as from those overlain by loam with debris have been dated by radiocarbon in the Institute of Geography of RAS. A series of ^{14}C dates have been obtained and calibrated within a 1σ confidence interval. The ^{14}C dates were calibrated using the University of Washington program (Stuiver and Reimer 1993); the latter permits one to take into account changes of ^{14}C content in the atmosphere through the Holocene. Pedology and paleopedology methods have been applied to the identification and study of buried soils and cultural layers (Yaalon 1971; Birkeland 1984).

The chronological sequence of buried soils, cultural layers and slope deposits have been found as follows (Table 1).

For the lower buried soil the age was determined at the lower boundary of a humus horizon as 5380 BP (calibrated data are shown in Table 1). Its' topsoil is a cultural layer and contains artifacts attributable to a Bronze Age settlement. As shown by the dates obtained for a series of samples from top to bottom,

Table 1 Sequence and radiocarbon age of buried soils and cultural layers. Archaeological site Gruzinka, North Caucasus (Krasnodar Territory, Russia)

Index IGAN	Sample	Depth beneath surface, cm	Radiocarbon age, BP	Calibrated age 1σ confidence interval: [start: end] relative area
–	Humic horizon of present soil	0–10	–	–
–	Cultural layer (middle ages, archaeological dating)	10–20(30)	–	–
–	Carbonate debris	30–45(50)	–	–
3050	Cultural layer* (early iron age)	60–77	2150±100	[2037 BP:2184 BP] 0,590026
3045	Buried soil	50–55	2420±100	[2349 BP: 2518 BP] 0,6432
–	Carbonate debris	55 (60)–95	–	–
3043	Buried soil (under the buried dolmen slab)	90–95	3260±100	[3381 BP:3588 BP] 0,973113
–	Brown carbonate-free loam	95 (100)–105	–	–
3049	Cultural layer (Bronze age)	105–115	3710±80	[3959 BP:4155 BP] 0,895528
3048	Cultural layer (Bronze age)	115–127	3940±110	[4229 BP:4527 BP] 0,944171
3054	Buried soil (humic horizon)	127–147	5380± 490	[5599 BP:6679 BP] 0,98301

*Sample taken from next excavation with the same layer consequence but different in depth.

the cultural layer developed within the interval of 3710 BP – 3940 BP. The range in age suggests the surface to be have been stable for a long time. The final stage of the site functioning has not been established at all the sites in the region; locally the upper part of the layer was destroyed by slope processes, as suggested by the morphology of the upper boundary and by the overlying carbonate-free loam.

Higher in the sequence are soils and cultural layers which are overlain and separated by at least two beds of carbonate debris about 20–40 cm thick each. The pattern is traced all over the area of dolmens. Some pits revealed the initial soil surface and dolmens buried under the lower debris layer. Radiocarbon age of the soil sampled directly under the dolmen slab was determined to be 3260 BP. Evidently, the dolmen and soil (cultural layer) were buried by a rapid flow of debris-laden material after that date.

The importance of finding of dolmen buried by landslide and radiocarbon data obtained is extremely high for archaeological sites of the North Caucasus. 3260 BP is a first date approximating the time of dolmen appearance in this region.

Complexes of soil and cultural layers of Early Iron Age are also overlain by carbonate debris. Radiocarbon dates of samples taken directly from under the debris layer provide evidence that they had not been buried by the debris until 2150–2400 yr BP.

The soil overlying the debris deposits includes numerous artifacts dated to the Middle Ages. This cultural layer is slightly disturbed by interbeds of the loam with carbonate debris and by processes of the modern soil formation.

The alternating fossil soils, cultural layers and slope deposits of various origin preserve a record of extensive information on natural and historical processes and events; so, the landslide processes contribute both to preservation and loss of information.

No fossil soils contain carbonates, though initial parent rocks are mostly calcareous. It follows that the soils developed under conditions of humid climate and leaching soil regime, which accounts for carbonate leaching from the soil matter. Furthermore, differences in this characteristic may be used as a basis for defining origin of the overlying layers: they could result either from slow movement of carbonate-free surficial loams, or from rapid sliding of calcareous debris from steep slopes stripped of soils by surface wash.

Dating the buried soils and cultural layers showed that the region has been populated since at least the mid-Holocene, and stages of active settling and cultivation of the area alternated with those of depopulation and decay. It seems quite probable that the people abandoned the cultivated lands because of periodical activation of rapid slope processes, as indicated by numerous traces of the latter. That activation could be due to high seismicity of the region, or to human activities. We can suppose the repeated circles: colonization of area – human impact on the slopes – frequent landslides – abandoning the area.

A high density of dolmens and settlements, as well as hundreds of dolmen slabs found, strongly suggest that human activities since the Bronze Age (including deforestation, stone quarrying and transportation of stone slabs) could seriously affect slope stability and promote landslide activation. The material brought downslope by slumping and sliding covers settlements and makes people leave their territory; at the same time it "seals" and preserves information on the past environments and historical events. The informative significance of landslides is confined to events of such a magnitude that buried surface could be preserved and found later on.

Conclusion

Transportation of loose material by landslides contributes to global process of the earth surface denudation; at the same time, the consequences of the process are of vital importance for particular mountain areas, for their environments and population. One of the consequences consists in destruction of cultivated lands, houses, infrastructure, and occasional losses of human life; it is of great importance and requires a vast amount of financial aid for damage compensation. There is another consequence – that is irrecoverable losses of loose material, soils and their most important constituent – fine earth. Eventually, that loss signifies a dramatic reduction of biodiversity, ecosystem productivity, depopulation of the territory and conversion into badlands. There is information on loose material losses for many landslide events in the world. It is possible to calculate amount of the lost fine earth washed away by rivers; such calculations have been performed, though in a few cases. But it is still uncertain how to estimate damage caused by loss of the loose cover and its fine constituents. Is it possible to forecast environmental evolution and social development in the mountains under conditions of catastrophic loss of loose material from vast areas? No such studies and calculations have yet been performed.

It should be noted, however, that the role of landslides in mountains goes beyond destructive activity and removal of regolith. By transportation and deposition of the loose material, landslides promote rejuvenation of mountain ecosystems and development of new soils as well as build up of new sites for settlement. On the one hand, new surfaces become exposed to weathering, on the other – new landforms are brought into being and stabilized landslide deposits provide a potential resource of fine earth. So, landslides destroy inhabited localities, but they also create new habitats for biota and humans, and the newly formed landforms are being adapted to the needs of the people (terrace cultivation, etc.).

In those cases sliding processes may be considered as a mechanism of long-term compensation for rapidly inflicted damage. As follows from pedological and geomorphological studies of landslide deposits in chronological sequence in the Adzharian-Trialet mountain system (Georgia), under conditions of cultivable land shortage people would colonize and cultivate landslide accumulations within a few decades. Then 50–60 years after the event, granulometric composition of the sediments appears to be essentially altered by remediation measures and the resulting soils are suitable for agriculture. Evidence of landslide processes may be found only in landforms and soils not affected by human activities.

Though landslides bury fertile soils and destroy settlements, they simultaneously serve to seal and to preserve information on past environments and colonization of the region in fossil (buried) soils and cultural layers. It is evident, however, that landslides may act in this way only in case of small-scale events when the underlying surface is only buried, and not destroyed. In such cases landslides may be considered as places of stored information on the past stages of evolution of environments and human society.

In considering landslides not only as a hazard, but also their significance as resource and information record, we gain a deeper insight into complicated natural processes in mountains, into history of development and cultivation of the mountain environment.

Acknowledgement The authors would like to thank sincerely Prof. G. Khomeriki, senior geologist of Department for Geology and Mining of Republic of Adzharia (Georgia), and O. Papidze, Chief of Land Department of Khulo district, for their assistance and providing information. Research in the Georgia could not have been undertaken without the goodwill and generosity of local administrations and the numerous landowners of Skhalta Sakrebulo.

References

Alcántara-Ayala I, Alcantara-Garsia D, Torns JB (2003) Erosion, deforestation and landslides: method of geomorphologic modelling. In: Society and environment interaction under global and regional changes. Institute of Geography of RAS, Moscow-Barnaul.

Alexander D (1998) Natural disasters. UCL Press, London.

Allen P (1997) Earth surface processes. Blackwell Science, Oxford.

Azhigirov AA (1987) Soil-destructive processes on the mountain slopes. Vestnik Moskovskogo universiteta. Geography 6: 53–56 (in Russian).

Barsh D, Caine N (1984) The nature of mountain geomorphology. Mountain Research and Development 4 (4):287–298.

Berdzenishvili D (ed) (2000) Information bulletin of ecological state of groundwater, studies of hazardous geological processes and their forecast. Georgian State Department of Geology, Tbilisi.

Birkeland PW (1984) Soils and geomorphology. Oxford University Press, Inc., New York.

Bolt BA (1975) Geological hazards: earthquakes, tsunamis, volcanoes, avalanches, landslides, floods. Springer-Verlag, New York.

Brundsen D, Moore R (1999) Engineering geomorphology on the coast: lessons from the west Dorset. Geomorphology 31:391–409.

Butler DR, Walsh SJ, Malanson GP (2003) Introduction to the special issue: mountain geomorphology – integrating earth system. Geomorphology 55:1–4.

Carlson PR, Karl HA (1988) Development of large submarine canyons in the Bering Sea, indicated by morphologic, seismic, and sedimentologic characteristics. Geological Society of America Bulletin 100:1594–1615.

Cruden DM, Varnes DJ (1996) Landslide types and processes. In: Turner AK, Schuster RL (eds) Landslides – investigation and mitigation, National Research Council, Washington, D.C., Transportation Research Board, Specification Report 247, pp. 36–75.

Dedkov AP, Moszherin VI (1992) Erosion and sediment yield in mountain regions of the world. In: Walling DE, Davies TR, Hasholt B (eds) Erosion, debris flows and environment in Mountain Regions. IAHS Publication 209, pp. 29–36.

Dobrovolsky GV, Nikitin ED (1986) Environmental functions of soils. MGU, Moscow.

Dubrovin NI, Klimenko VI (1973) Principal factors of landslide origin and development on the Black Sea coast of the Caucasus. Problemy Inzhenernoi Geologii Severnogo Kavkaza 5:37–68 (in Russian).

Dunning S, Cole P (2002) Rock avalanches in high mountains – A sedimentological investigation. In: Leroy S, Stewart IS (eds) Environmental catastrophes and recovery in the Holocene. Abstract Volume. Brunel University, West London, 28 August–2 September 2002.

Dzhaoshvili VSh (1968) Population of Georgia. Metsniereba, Tbilisi (in Russian).

Dzhavahishvili AN (ed) (1964) Atlas of the Georgian Soviet Socialist Republic. GUGK GGK of the USSR. Tbilisi-Moscow (in Russian).

Easterbrook DJ (1993) Surface processes and landforms. Macmillan Publishing Company, New York.

Eisbacher GH, Clague JJ (1984) Destructive mass movements in high mountains: hazard and management. Paper 84-16, Geological Survey of Canada, Vancouver, British Columbia.

Emelyanova EP (1964) The main regularities in landslide processes. Nedra, Moscow (in Russian)

Evans SG, Alcántara-Ayala I (2007) Disasters resulting from landslides, snow avalanches, and geotechnical failures in North America (Canada, United States, and Mexico) 1841–2006: a first assessment. In: Turner K, Schuster RL (eds) Landslides and Society. Association of Environmental & Engineering Geologists, Denver, CO.

Frossard E, Blum WEH, Warkentin BP (eds) (2006) Function of soils for human societies and the environment. Geological Society, London, Special publications 266.

Gerrard AJ (1990) Mountain environments. Belhaven Press, London.

Gigineishvili GN, Nahutsrishvili GSh (1998) Problems of sustainable development of the mountain regions of Georgia Izvestiya Akademii Nauk Series Geografia 6:95–101 (in Russian).

Gracheva RG (2004a) Land use transformation in Mountain Adzharia and its possible consequences (the last 15 years). Annals of Agrarian Science 4:7–15.

Gracheva RG (2004b) Unexpected transformation of nature use and soil resources in the eastern coast of the Black sea (the last 10 years) In: Barbanente A, Borri D, Camarda D et al. (eds) Local Resistance to Global Pressure: A Mediterranean Social/Environmental Planning Perspective. L'Harmattan, Paris.

Hampton MA, Lee HJ, Locat J (1996) Submarine landslides. Reviews in Geophysics, 34(1):33–59.

Hewitt KK (1992) Mountain hazards. GeoJournal 27:47–60.

Ilychev BA, Gracheva RG (1998) Conditions of loose mantle of mountain regions as a criteria of their sustainable development. Izvestiya Akademii Nauk Series Geografia 6:48–59 (in Russian).

Jones DKC (1992) Landslide hazard assessment in the context of development. In: McCall H, Laming DJC, Scott SC (eds) Geohazards: Natural and man-made. Chapman and Hall, London.

Klupt M (2008) Demography of regions of earth. Piter press, St. Petersburg (in Russian).

Korotki AM, Korobov VV, Shornikova et al. (2005) Activation of catastrophic landslides in the coastal zone. Vestnik DVO RAN 5:48–52 (in Russian).

Kotlyakov VM, Rototaeva O V, Osokin NI (2004) Surging glaciers and glacial catastrophe in the northern Caucasus. Vestnik Vladikavkazskogo nauchnogo tsentra 4(3):65–71 (in Russian).

Krupnov EI (1960) Ancient history of the North Caucasus. Nauka, Moscow (in Russian).

Messerli B, Ives JD (eds) (1999) Mountains of the world. A global priority. Publishing house "Noosphere", Moscow (in Russian).

Milkov FN, Gvozdetski NA (1969) Physical geography of USSR. Mysl, Moscow (in Russian).

Osipov VI (2001) Natural catastrophes at the turn of the century. Geoekologiya, Inzhenernaya Geologiya, Gidrogeologiya, Geokriologiya 4:293–301 (in Russian).

Page MJ, Reid LM, Linn IH (1999) Sediment production from Cyclone Bola landslides Waipaowa catchment. Journal of Hydrology (New Zealand) 38(2):289–308.

Page MJ, Trustrum NA, Dymond JR (1994) Sediment budget to assess the geomorphic effect of a cyclonic storm, New Zealand. Geomorphology 9(3):169–188.

Putkaradze MSh (1996) Economic-geographical problems of population in Mountain Adzharia. Adjaria Press, Batumi.

Razumova VN (1977) Ancient crusts of weathering and hydrothermal processes. Nauka, Moscow (in Russian).
Razumov VV (ed) (2008) Hazardouz natural processes in the south of European Russia. Publishing house “Design Cartography Information”, Moscow (in Russian).
Report on Landslides-Damages, Costs & Casualties, U.S. Geological Survey (2001) http://geology.cr.usgs.gov/pub/open-file-reports/ofr-01-0276 /
Sassa K (ed) (1999) Landslides of the world. Japan Landslide Society, Kyoto University Press, Kyoto.
Schuster RL, Highland LM (2001) Socioeconomic and environmental impacts of landslides in the western hemisphere. In: Castaneda Martinez JE, Olarte Montero J (eds) Proceedings of the Third Panamerican Symposium on Landslides, 29 July–3 August 2001, Cartagena Colombia.
Sheko AI (1976) Landslides. In: Recent geological processes on the Black Sea coasts of the USSR. Nedra, Moscow (in Russian).
Sheko AI (1982) Theory and methods of forecasting exogenous geological processes. In: Sheko AI (ed) Landslides and mudflows. Center of International Projects, GKNT, Moscow (in Russian).
Stuiver M, Reimer PJ (1993) Radiocarbon calibrating program Rev. 3.0.3 Radiocarbon 35:215–230.
Targulian V (2008) Soils and society: human impact and soil responses. In: Dazzi C, Costantini E (eds) The soils of tomorrow. Soils changing in a changing world. Advances in GeoEcology 39, pp. 13–26
Tsereteli ED, Tsereteli DD (1985) Geological setting of mudflows in Georgia. Metsniereba, Tbilisi (in Russian).
Varnes DJ (1978) Slope movement types and processes. In: Schuster RL, Krizek RJ (eds) Landslides – analysis and control, National Academic Sciences, Washington, D.C., Transportation Research Board Specifiction Report 176, pp. 11–33.
Warburton J (2007) Mountain environments. In: Perry C, Taylor K (eds) Environment sedimentology. Blackwell Publishing, Malden, MA .
Yaalon DH (ed) (1971) Paleopedology – origin, nature and dating of paleosols. ISSS and Israel University Press, Jerusalem.

Index

NOTE: The letter 't' and 'f' followed by the locators denotes 'table' and 'figure'

T. Beer (ed.), *Geophysical Hazards*, International Year of Planet Earth,
DOI 10.1007/978-90-481-3236-2, © Springer Science+Business Media B.V. 2010